BRITISH WILDLIFE PHOTOGRAPHY AWARDS

COLLECTION 3

Editor: Donna Wood
Designer: Tracey Butler
Image retouching and colour repro: Jacqueline Street
Production: Lorraine Taylor

Produced by AA Publishing

Published by AA Publishing (a trading name of AA Media Limited, whose registered office is Fanum House, Basing View, Basingstoke RG21 4EA; registered number 06112600).

A04933

ISBN: 978-0-7495-7364-5

A CIP catalogue record for this book is available from the British Library.

Printed and bound in Italy by Printer Trento SRL

theAA.com/shop

SUE DALY > HIGHLY COMMENDED

Soapy Starfish
(Soapy Starfish, *Echinaster sepositus)*
Pavlaison, Sark, British Channel Islands
See page 166

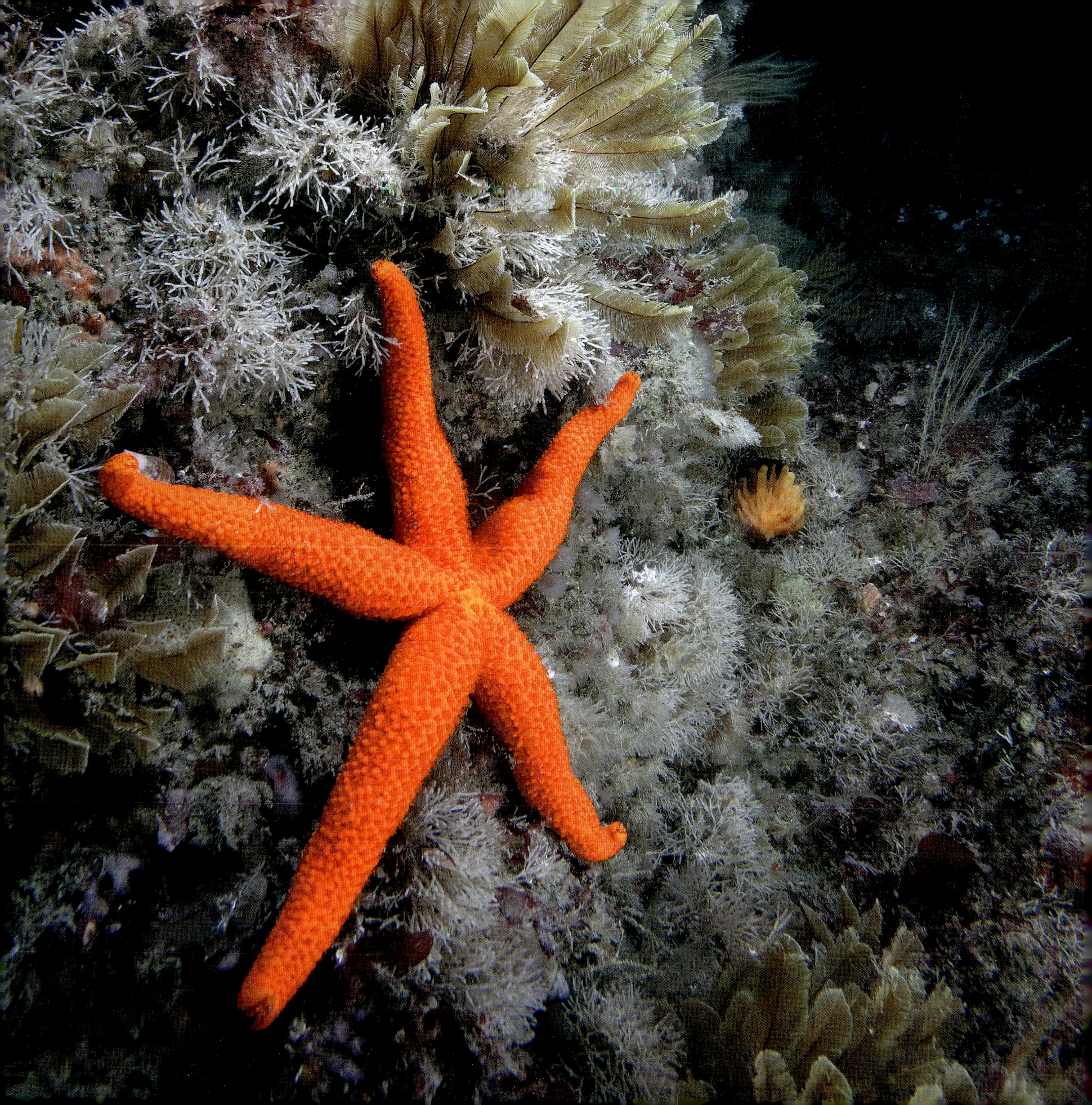

CONTENTS

LES GIBBON > HIGHLY COMMENDED

Deer Fly Reflection
(Deer Fly, *Chrysops relictus*)
Arglam Farm, East Yorkshire, England
See page 90

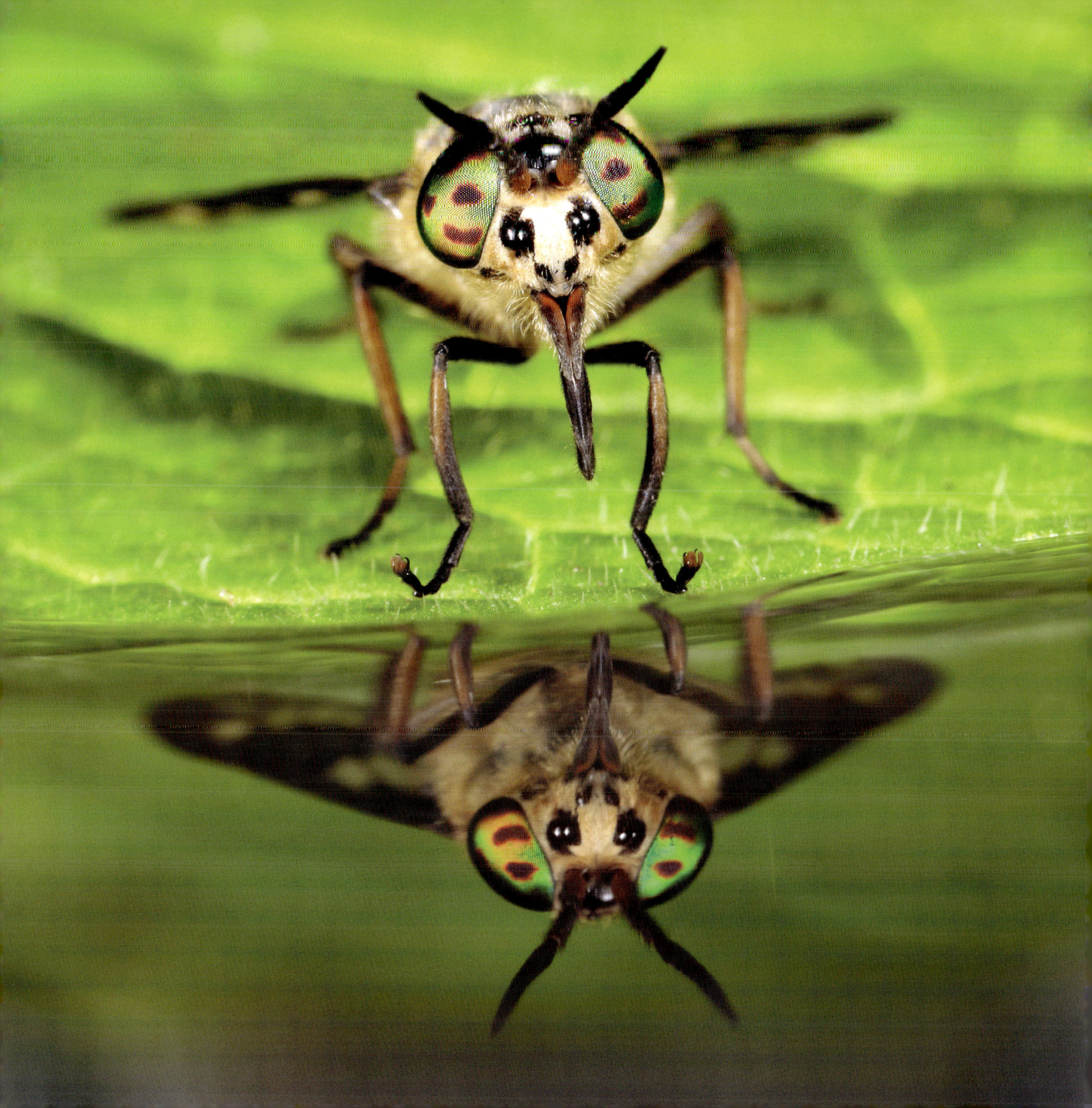

THE BRITISH WILDLIFE PHOTOGRAPHY AWARDS

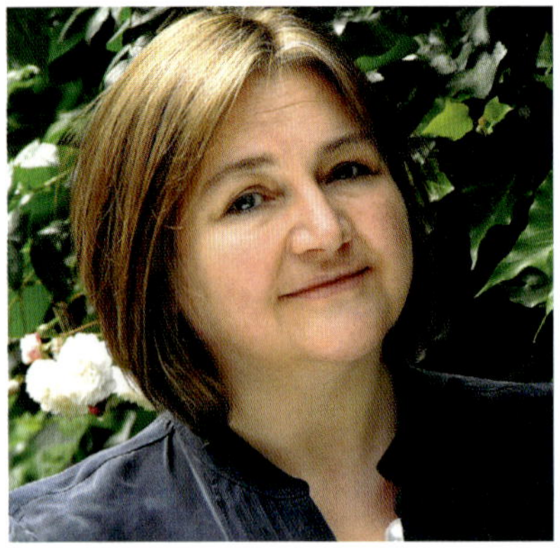

MAGGIE GOWAN
BWPA DIRECTOR

The British Wildlife Photography Awards were established to recognise the talents of photographers practicing in Britain, while at the same time highlighting the great wealth and diversity of Britain's natural history. The driving motivation to set up the Awards evolved through the nation's growing awareness of the local environment and the need for its protection. With a nationwide touring exhibition of the best entries, the Awards aim to:

- celebrate British wildlife, in all its beauty and diversity, through a collection of inspirational photographs
- recognise the talents of photographers (of all nationalities) practising in Britain
- showcase the very best of British nature photography to a wide audience
- engage all ages with evocative and powerful imagery
- raise awareness about British biodiversity, species and habitats

THE CATEGORIES

1. Animal Portraits

In this category, the judges were looking for images that capture the character or spirit of the subject in an imaginative way, giving a sense of the animal's 'personality'.

2. Animal Behaviour

Entries showing unusual or remarkable behaviour in an original and compelling composition were sought in this category, especially those showing the judges something they don't know or haven't seen before.

3. Urban Wildlife

Increasingly, wildlife can be found in our towns and cities. In this category, the judges were looking for an original image that shows animal adaptation to urban environments.

4. Hidden Britain

Images in this category reveal the secret universe that is life in the undergrowth – a life that is all around us but rarely seen.

5. Coast and Marine

Imaginative photographs that reveal the behaviour of marine animals or create a sense of place or occasion. This includes marine animals near the sea, underwater, at the sea shore and within the coastal zone.

6. Wildlife in My Backyard

Gardens represent a vital habitat for some of our most threatened species. In this category we were looking for stunning photos of animals in the garden, backyard, balcony, or any space where people do something to encourage wildlife.

7. Wild Woods

The judges were looking for a winning image that celebrates the beauty and importance or sheer splendour of our British woods. Entries could be portraits of woodland wildlife, breathtaking wooded landscapes, seasonal scenes, intricate details of plants, or show the relationships between species and habitats that occur within our woods and forests. There are plenty of fantastic photographs of woodland wildlife and seasonal landscapes full of mood and atmosphere waiting to be taken.

8. Habitat

This category celebrates animals in their environment and plants and their partners. The judges sought original images portraying the relationship between wildlife and its environment.

9. British Seasons

A portfolio of four images portraying British wildlife at its best in each of the four seasons; or a portfolio of four scenes from just one season. Each image should capture the essence of the season along with the wildlife subject.

10. Living Landscape: Connectivity

Being connected to the natural world is good for our health, our happiness and for the future. It is The Wildlife Trusts' vision to create a resilient environment, rich in wildlife and great for people, called a 'Living Landscape'. This category was purposely left open to interpretation to encourage fresh and imaginative responses. Connectivity might be physical, sensory, seasonal or otherwise. Positive images illustrating how habitats, species, communities and individuals interconnect within rural and urban settings, were encouraged.

11. Documentary Series

In this category, we aimed to showcase a sequence of up to six images on any British wildlife, conservation or environmental issue, showing innovation in storytelling.

12. Wildlife in HD Video

This category invited an inspirational and dynamic sequence (up to 90 seconds long) illustrating the unique power of HD video as a medium for capturing British wildlife. Some stills from the film are featured in this book. The whole film and commended entries can be viewed at www.bwpawards.co.uk

13. *Outdoor Photography* Editor's Choice

Each month the competition was running, the editor of *Outdoor Photography* magazine chose his favourite image for publication in future editions of the magazine.

14. Young People's Awards WildPix/BWPA

In the Young People's Awards there were two age categories. There were no specific subjects; we were looking for striking and memorable images of any British wildlife species taken by under 12s and 12–18 year olds.

For further information about the annual competition and touring exhibition, please visit **www.bwpawards.co.uk**

BWPA

BRITISH WILDLIFE PHOTOGRAPHY AWARDS

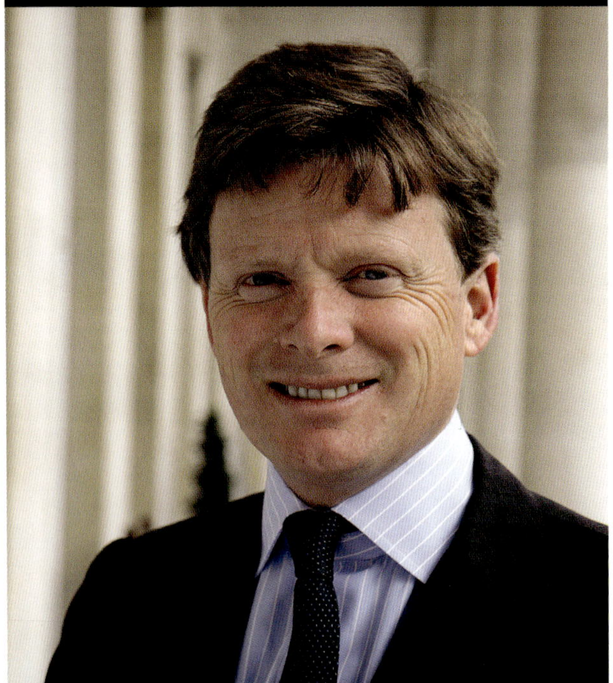

FOREWORD

RICHARD BENYON MP

ENVIRONMENT MINISTER

It's a real pleasure to introduce you to this delightful book. These photographs, as well as being aesthetically beautiful in their own right, are also a vital reminder of this nation's greatest wealth.

As a child I was captivated by the natural world. I spent hours looking under stones for fascinating creepy-crawlies, or gazing up at the signs of changing seasons in the trees. I continue to feel that rapture as an adult, and I spend as much time as I can getting close to nature. But the truth is that many of our ecosystems and much of our wildlife is in danger. We are still losing species at an alarming rate.

In Government we are working hard to halt and reverse this trend, both internationally and at home. The actions needed to achieve this target range from habitat conservation to species recovery, from reducing wildlife crime to battling invasive species.

'Engaging people' is one of the four fundamental themes running through our national biodiversity strategy. We know from our research that the more we can connect people with nature, the more people will want to protect it. There is already a lot of great collaborative work going on across the country, with the support of many committed organisations, individuals and charities, including those who have sponsored this competition (see pages 10–13). For example, a project to reintroduce the short-haired bumble bee, funded by Natural England and backed by the RSPB, Hymettus and the Bumblebee Conservation Trust with support from local farmers, is a real cause for celebration.

This is a great example of the type of dedicated partnership between farmers, scientists and conservation organisations that can make a real difference for wildlife in this country and is absolutely vital for the ongoing pollination of crops.

These partnerships are also helping a range of previously endangered farmland birds and butterflies. Agreements between farmers and the RSPB in the South West are boosting numbers of cirl buntings to ensure that these distinctive birds have the appropriate uncropped arable field margins, hedges and orchards which are a rich source of insects. And in the Cotswolds, local nature lovers are working with Butterfly Conservation to produce flower-strewn hillsides that will bring back from extinction the Large Blue butterfly in an area made famous by Laurie Lee's famous novel Cider with Rosie*.*

By working together, we can protect our wildlife and stop the degradation of habitats, leaving our natural environment in a better state for our children and future generations.

Richard Benyon MP
Environment Minister

ALEX SABERI > HIGHLY COMMENDED

Fairytale 1
(Red Deer, *Cervus elaphus)*
Richmond Park National Nature Reserve, Surrey, England
See page 139

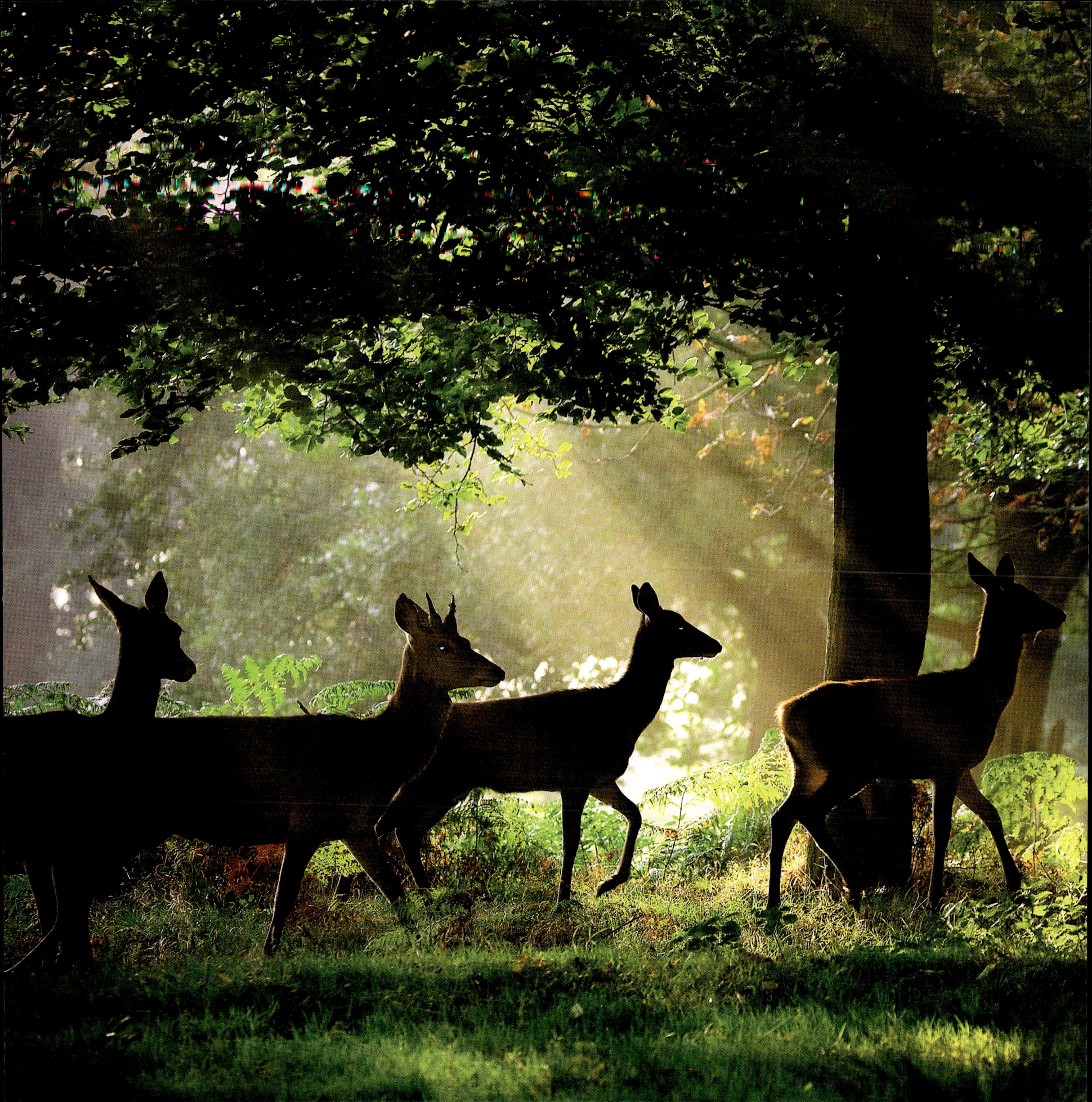

THE SPONSORS

Canon
The British Wildlife Photography Awards 2012 are held in association with Canon. Canon is a world leader in imaging products and solutions for the digital home and office.

Canon is passionate about providing consumers with imaging solutions that empower creativity and innovation, ranging from digital compact and SLR cameras, through broadcast lenses and portable X-ray machines to multi-function and production printers, all supported by a range of value-added services.

To learn more about Canon products, please visit Canon's website:
www.canon.co.uk

Sky+HD
Sky+HD is proud to sponsor the Wildlife in HD Video category of the British Wildlife Photography Awards 2012.

Sky+HD is the UK's leading high definition TV service, with over four million customers now enjoying their favourite TV with up to five times more picture detail. With up to 60 HD channels to choose from, we offer a stunning range of entertainment across all genres, enabling our customers to get immersed in the action, emotion and detail of their favourite programmes whatever their passion.

It is our privilege to be sponsoring the Wildlife in HD Video category of the British Wildlife Photography Awards. This category puts into clear focus the beauty and richness of nature which is so superbly brought to life in HD.

To find out more about Sky's products and services, please visit
www.sky.com/products

Countryside Council for Wales

The Countryside Council for Wales is the government's statutory advisor on sustaining natural beauty, wildlife and the opportunity for outdoor enjoyment in Wales and its inshore waters. CCW champions the environment and landscapes of Wales and its coastal waters as sources of natural and cultural riches, as a foundation for economic and social activity, and as a place for leisure and learning opportunities. CCW aims to make the environment a valued part of everyone's life in Wales.

www.ccw.gov.uk

RSPB

The RSPB speaks out for birds and wildlife, tackling the problems that threaten our environment. With over a million members, it is a powerful voice for nature.

www.rspb.org.uk

WWF

WWF is building a future where people and nature thrive together. That's why WWF is as passionate about tackling global climate change, and finding ways to share the planet's resources more sustainably, as it is about protecting endangered wildlife.

www.wwf.org.uk

The Wildlife Trusts

In May 1912, Charles Rothschild formed the Society for the Promotion of Nature Reserves (SPNR), which had one main objective – to save Britain's finest wildlife sites – a revolutionary concept at the time. In the 1950s, the SPNR was adopted by emerging Wildlife Trusts as their national association and would play a key role in their development. A hundred years on, The Wildlife Trusts are supported by 800,000 members and manage 2,300 nature reserves. All work with landowners, local communities and partners to aid nature's recovery on a wider scale and create A Living Landscape. They also work to protect the UK's marine environment and secure Living Seas, surveying and collecting vital data and celebrating marine wildlife. Find out how to join your local Wildlife Trust at

www.wildlifetrusts.org/joinus

THE SPONSORS

Buglife

Buglife is a registered charity and the only organisation in Europe devoted to the conservation of all invertebrates – everything from bees to beetles, woodlice to worms and jumping spiders to jellyfish! Invertebrates are vital for the health of the planet. They pollinate our crops and wild flowers, recycle nutrients back into the ground, and are a vital source of food for other animals such as birds and mammals. The food we eat, the fish we catch, the birds we see, the flowers we smell and the hum of life we hear simply would not exist without wonderful, amazing, fascinating bugs.

www.buglife.org.uk

Countryside Jobs Service

Countryside Jobs Service is a small ethically focused company providing information on countryside careers including jobs, volunteers, professional training and more. Established in July 1994, CJS is run by a small, dedicated team of ex-rangers, smallholders and ecologists working on a co-operative basis to ensure they publish the widest range of relevant information for readers.

www.countryside-jobs.com

***Outdoor Photography* magazine**

Outdoor Photography's intriguing balance of features, tests, techniques, travel and conservation stories make it the most relevant magazine in the dynamic world of nature photography. Each issue features an inspiring array of landscape, wildlife and travel features, with regular, intelligent contributions by leading nature photographers from around the world.

www.thegmcgroup.com

VisitWoods

VisitWoods aims to inspire everyone to enjoy woods and the benefits they can offer. Our website hosts an online database of the UK's woods and forests which are open to the public. VisitWoods celebrates all that is special about UK woods and lets you search from more than 10,000 woods to visit using a postcode or town name. We celebrate the beauty, importance, fun and sheer splendour of woods with new ideas and activities. VisitWoods is a partnership project led by the Woodland Trust to promote woods to all.

www.visitwoods.org.uk

Wildlife Extra – the web's wildlife magazine

Wildlife Extra is for people who like watching, conserving and photographing wildlife and hearing the latest wildlife news.

Wildlife Extra brings you a guide to UK nature reserves and National Parks, a selection of wildlife and conservation jobs, lets you know the best places to watch whales or go on a safari holiday and provides wildlife photography tips, competitions and articles. Free subscription.
www.wildlifeextra.com

Sparsholt College Hampshire

With a heritage of training for land-based and rural industries spanning one hundred years, Sparsholt College Hampshire offers specialist courses in conservation and wildlife, animal management, woodland management and arboriculture, fishery management, equine studies, horticulture and agriculture, among others.

Set on a 200-hectare campus in the heart of rural Hampshire, students from the age of 16 through to degree level benefit from outstanding practical facilities and expert tuition, shaping the future of rural industries.
www.sparsholt.ac.uk

RSPB Wildlife Explorers

RSPB Wildlife Explorers is the junior membership of the RSPB. With over 200,000 members, it is the world's biggest club for young people who love wildlife and want to make a difference.

The RSPB speaks out for birds and wildlife, tackling the problems that threaten our environment. More than a million members provide a powerful voice for nature. Nature is amazing. Help us to keep it that way.
www.rspb.org.uk/youth

THE JUDGES

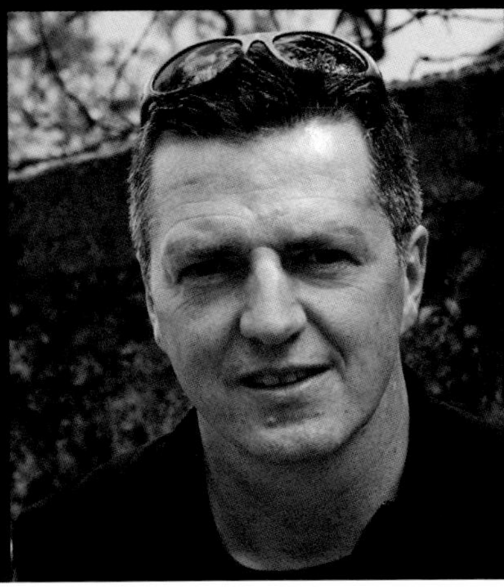

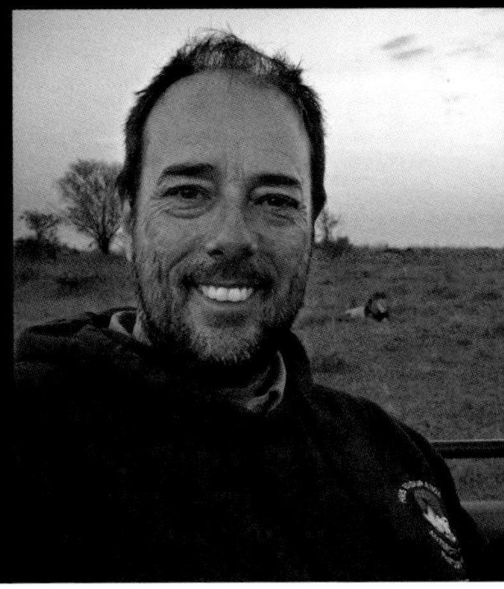

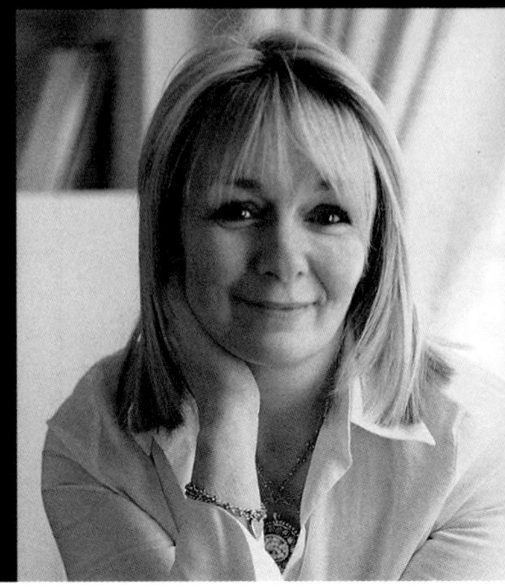

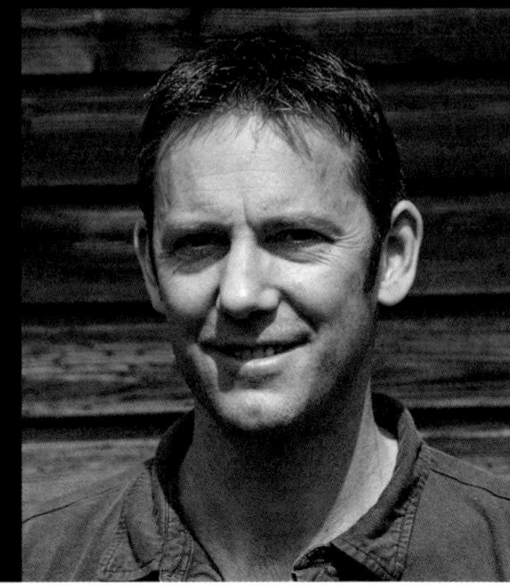

Steve Watkins

Editor
Outdoor Photography magazine

Steve Watkins is an award-winning travel photographer, and writer and editor of *Outdoor Photography* magazine. He has worked closely with the BBC in creating travel programmes and has written BBC books including *Unforgettable Journeys to Take Before you Die*. Steve has walked on the wild side for 20 years as an adventurer and travel photographer. In the past eight years he has shot 120,000 images in 65 countries.

Andy Rouse

Photographer

Andy Rouse is one of the best-known British wildlife photographers, with 15 books to his name and regular contributions to newspapers and magazines. For seven years in a row he has been commended and won nine prizes in the BBC Wildlife Photographer of the Year competition, a feat not achieved by any other British wildlife photographer and few worldwide. He is considered one of the top wildlife photographers and has a huge following in almost every country of the world. Andy presented a 13-part series on wildlife photography for Channel 5/Animal Planet. He regularly gives theatre shows, workshops and takes clients on expeditions to all corners of the globe. His last two books, on tigers and mountain gorillas, raised considerable sums of money for conservation and he is currently involved in the 2020Vision project to showcase the UK's fragile ecosystems.

Sue Herdman

Editor
The National Trust magazine

Sue Herdman is a journalist who started her career as a feature writer working for broadsheets including *The Daily Telegraph*, *The Sunday Telegraph*, *The Times* and the *Guardian*. In 1998 she moved to magazine production, working as a Features and Commissioning Editor for BBC Worldwide magazines. She is currently the editor of *National Trust Magazine*, which this year marks 78 years in production and holds the sixth-highest magazine circulation figure in Britain, with 3.6 million readers.

James Tims

Picture Library Manager
AA Publishing

James Tims has been a travel photographer for 18 years. Published in books, magazines and websites across five continents, he has an unfaltering passion for inspirational and emotive photography. His relationship with AA Publishing spans almost his entire career, and at the beginning of 2011 he took over the role to manage the picture library and reprographics department. Committed to delivering the very best imagery at the highest possible standard, upholding the AA's ethos of publishing market-leading leisure titles is a core value. Embracing new technologies in digital image delivery and publishing, James is excited about the new challenges and demands that face the industry, and is proud to be part of the BWPA competition.

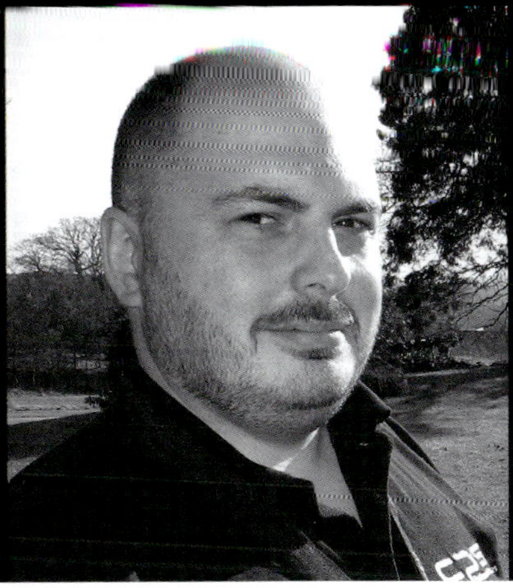

David Fidler

Pro Imaging Account Manager
Canon UK

David Fidler has worked in both the commercial and retail sections of the photographic industry for over 13 years. For the past eight years he has worked for Canon UK Ltd as their Professional Imaging Account Manager. Within this role, he is responsible for driving business in the nature and wildlife market and working with some of the UK's top photographers.

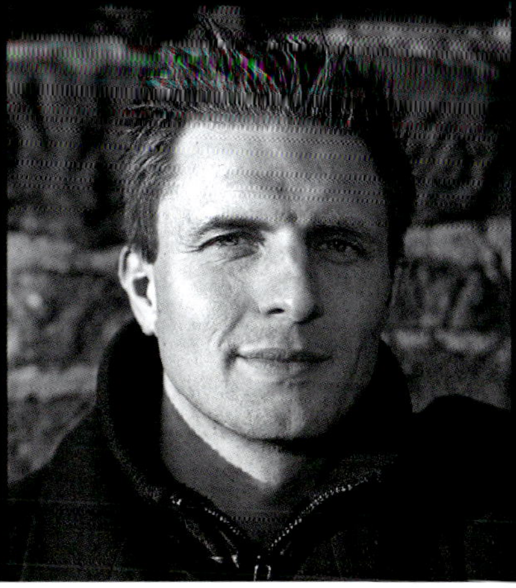

Mark Ward

Editor
RSPB *Birds* magazine

Mark Ward is the editor of the RSPB's *Birds* magazine. He has been writing about birds and nature from a young age – his first feature appeared in *Birdwatch* magazine when he was 15. Mark has enthused many people about wildlife through his writing in titles including *Bird Watching, Natural World* and *Country Walking*; his newspaper columns, and books including *Bird Identification and Fieldcraft: A Birdwatcher's Guide*. Passionate about nature, his ambition is to see as many species of wildlife in the UK as possible in his lifetime.

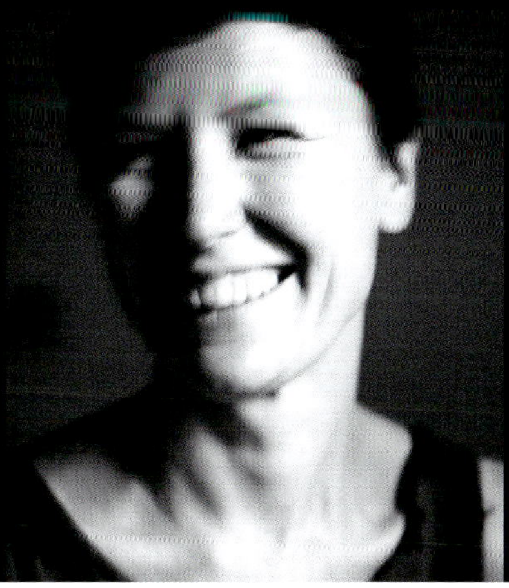

Diane Smyth

Deputy Editor
British Journal of Photography

Diane Smyth is the deputy editor of the *British Journal of Photography*, one of the oldest photography magazines in the world. She has also written on photography for Aperture, Photomonitor, Guardian.co.uk, Thetimes.co.uk, The Telegraph's Telephoto site, *Creative Review* and *Philosophy of Photography*. Diane has reviewed portfolios at the Rhubarb-Rhubarb International Photographic Review, The Whitechapel Gallery, the National Media Museum and University College Falmouth, and helped judge the National Portrait Gallery's Taylor Wessing Photographic Portrait Prize in 2009.

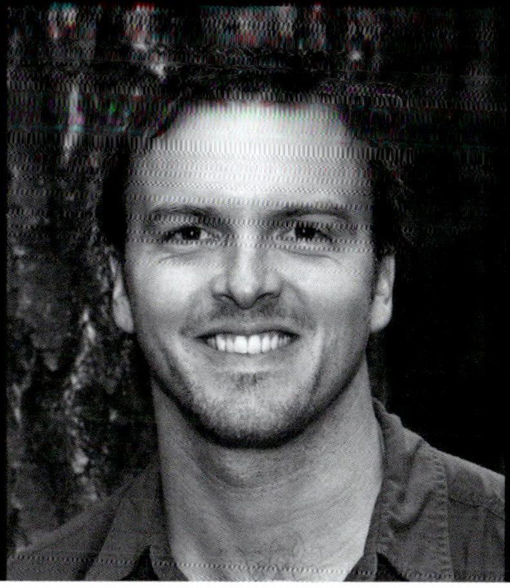

Neil Aldridge

Conservation Photojournalist

Neil Aldridge is the winner of the 2011 British Wildlife Photography Awards 'Documentary Series' category. As a photojournalist, Neil works on conservation stories that showcase controversial and endangered wildlife and debate humankind's relationship with these creatures and their environments. His work is influenced by his years growing up in South Africa and by his time working in the UK conservation sector. He is also a professional wildlife guide and runs photo safaris to Africa and Europe. The field skills used in guiding play a central role in the success and enjoyment of his photography. In 2011, after three years of writing and photographing, he published his first book: *Underdogs: the Fight to Save South Africa's Wild Dogs*.

THE JUDGES

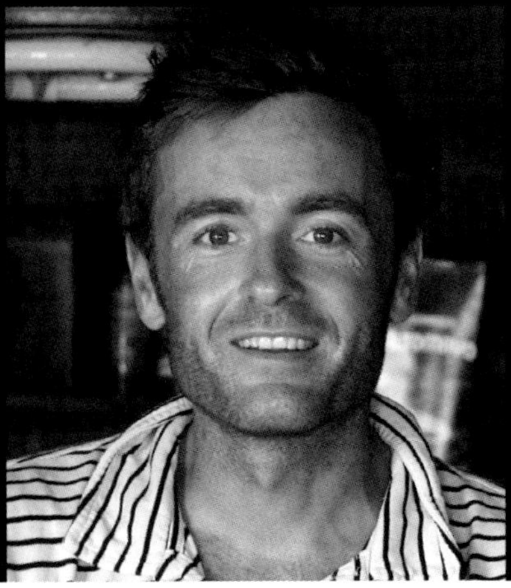

Greg Armfield

Photographic and Film Manager
WWF-UK

Greg is photographic and film manager for WWF-UK. He has worked in the photographic industry for over 12 years, combining both the NGO sector (National Trust and WWF) and private sector (Emap and Construction Photography). Greg has worked in all facets of the industry, from artistic director at Construction Photography image library to photographer, editor and commissioner. He has worked for WWF for the last six years, originally heading up the photographic department but now also leading on film, with responsibility for all photo commissions.

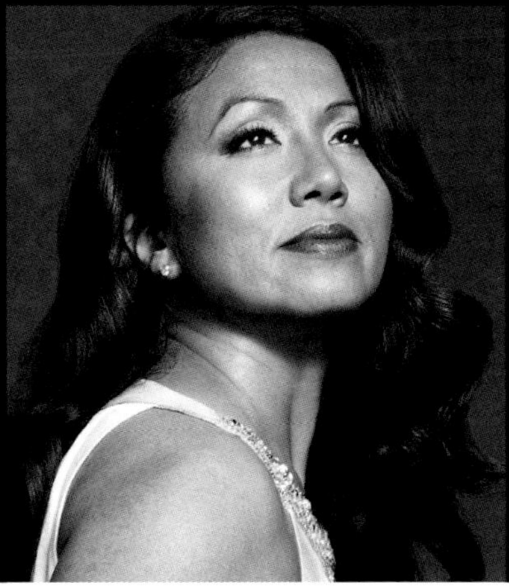

Christina Vaughan

Founder and Chief Executive Officer
Image Source

Christina Vaughan brings more than 22 years of experience in the creation, distribution and marketing of commercial imagery to Image Source, an online fully digitised company that trades directly through the internet as well as through an extensive network of more than 200 photo agencies in over 80 countries worldwide. Christina sets the vision for Image Source as well as managing investor relationships with the institutional shareholders in the business. She has extensive experience as a panellist and industry commentator and has played an active role in leadership and direction on several industry trade bodies including PACA, ASPP and BAPLA. Christina is the current president of the European Trade Body CEPIC, which represents over 1,000 picture sources in Europe.

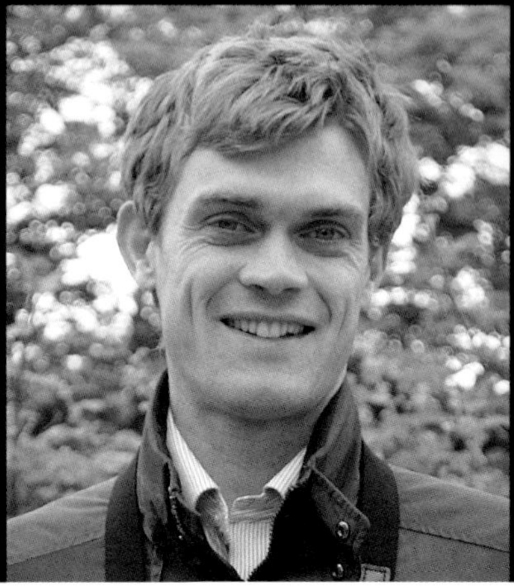

Adam Cormack

Communications Manager
The Wildlife Trusts

Adam is communications manager for The Wildlife Trusts, where he has worked for seven years. He has edited, written for and picture researched a range of the Trusts' publications including *Natural World*, the junior membership magazine *Wildlife Watch* and a number of books including wildlife guides and a forthcoming history of The Wildlife Trusts. His work brings him into contact with some of the UK's leading wildlife photographers. A self-professed amateur naturalist, he can often be found rambling in his local woods and meadows in search of butterflies, birds and wildflowers, and occasionally trying to photograph them.

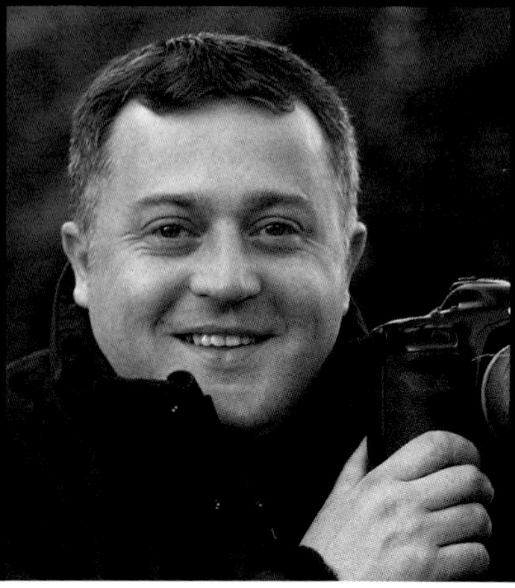

Danny Green

Wildlife Photographer

Danny is a multi-award-winning photographer based in the UK. Although Danny travels widely to many locations worldwide in search of images, he has a deep passion for the natural history of the British Isles. Danny is represented by leading agencies including Nature Picture Library, RSPB Images and NHPA. Over the past seven years he has been working on a book about the fauna and flora of northern Europe. Danny also leads workshops and trips for Nature's Images, one of the country's leading wildlife photography tour operators.

Stephen Moss

Chairman
BWPA Judging Panel

A lifelong naturalist, Stephen is one of Britain's leading nature writers, broadcasters and natural history television producers. During a BBC career spanning 30 years, based mainly at the Natural History Unit in Bristol, his series included the BAFTA award-winning *Springwatch, Big Cat Diary*, *Birding with Bill Oddie*, *The Nature of Britain* and *Birds Britannia*. He has written more than 20 books, including *A Bird in the Bush* (a social history of birdwatching), *This Birding Life*, *A Sky Full of Starlings* and *The Bumper Book of Nature*. His latest book is *Wild Hares and Hummingbirds: the Natural History of an English Village*, and he also writes a regular monthly column on birdwatching for the *Guardian*. Stephen appears frequently on BBC Radio Four, including the Today programme, and has presented several radio series on British birds and their songs.

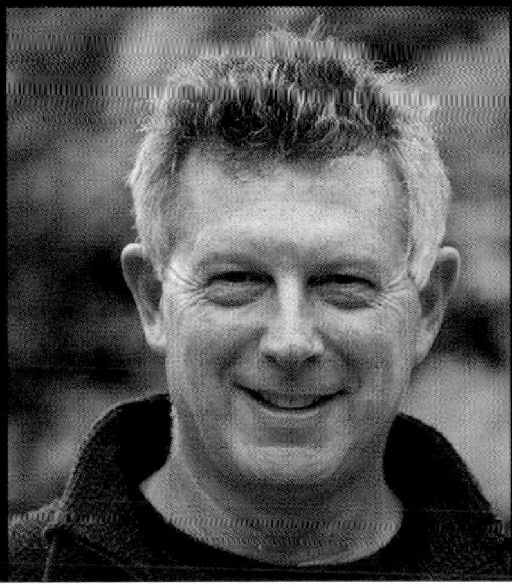

WITH THANKS TO

All the photographers who have participated in the awards

The sponsors, supporters and judges

AA Publishing

Natural England

The National Trust and these properties:

- Calke Abbey, Derbyshire
- Nunnington Hall, Yorkshire
- Penrhyn Castle, North Wales

Horniman Museum

Nature in Art

Moors Valley Country Park

Luton Museums

The National Trust for Scotland

The RSPB's Wildlife Explorers team

British Society of Underwater Photographers

Made By Hippo Ltd

Picture This

Pinkeye Graphics

Exhibitions on Tour

Matthew Chatfield

Carl Crawley

Ralph Tribe

Clare Webb

Victoria Skeet

Emma Louise Clarke

Chris Hart

Kate Foreshew

Jennie Hart

Charlotte Salaman

Babs Gowan

Rebecca Blanchard

Karen Price

Dympna Marsh

Kat Dixon

Venetia Khan

Phil Gowan

BRITISH WILDLIFE
PHOTOGRAPHER OF THE YEAR

BRITISH WILDLIFE PHOTOGRAPHER OF THE YEAR 2012

OVERALL WINNER

DR MATT DOGGETT

Gannet Jacuzzi
(Northern Gannet, *Morus bassanus)*
Scotland

I wanted to capture gannets beneath the waves. Using dead mackerel we attracted a number of gannets as they searched for the fish amongst a maelstrom of bubbles. I like this image as almost every stage of the action is captured – gannets entering the water, gannets eyeing up the fish, gannets taking the fish, gannets eating the fish and then finally leaving. It looks like chaos but the gannets know exactly where the other birds are.

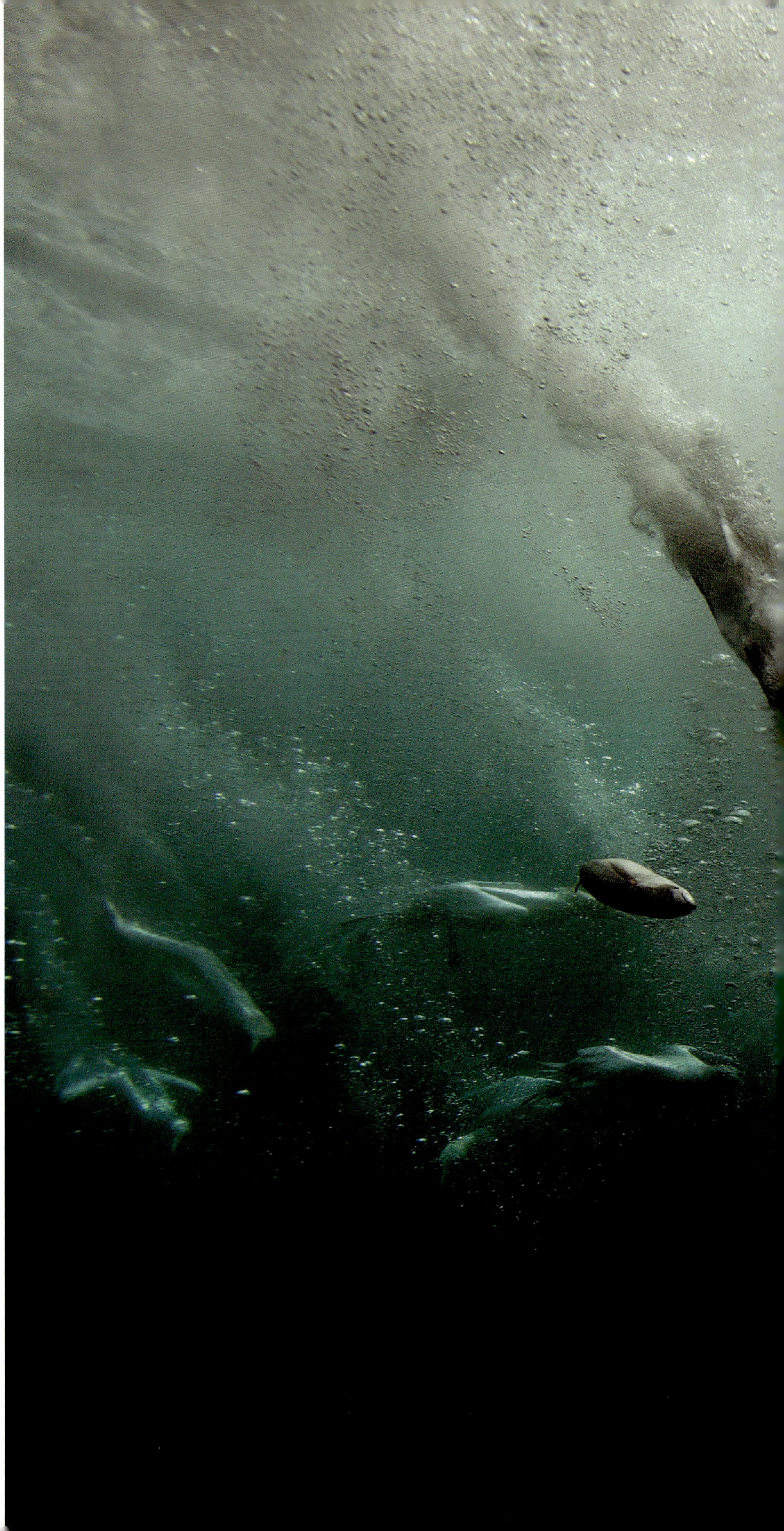

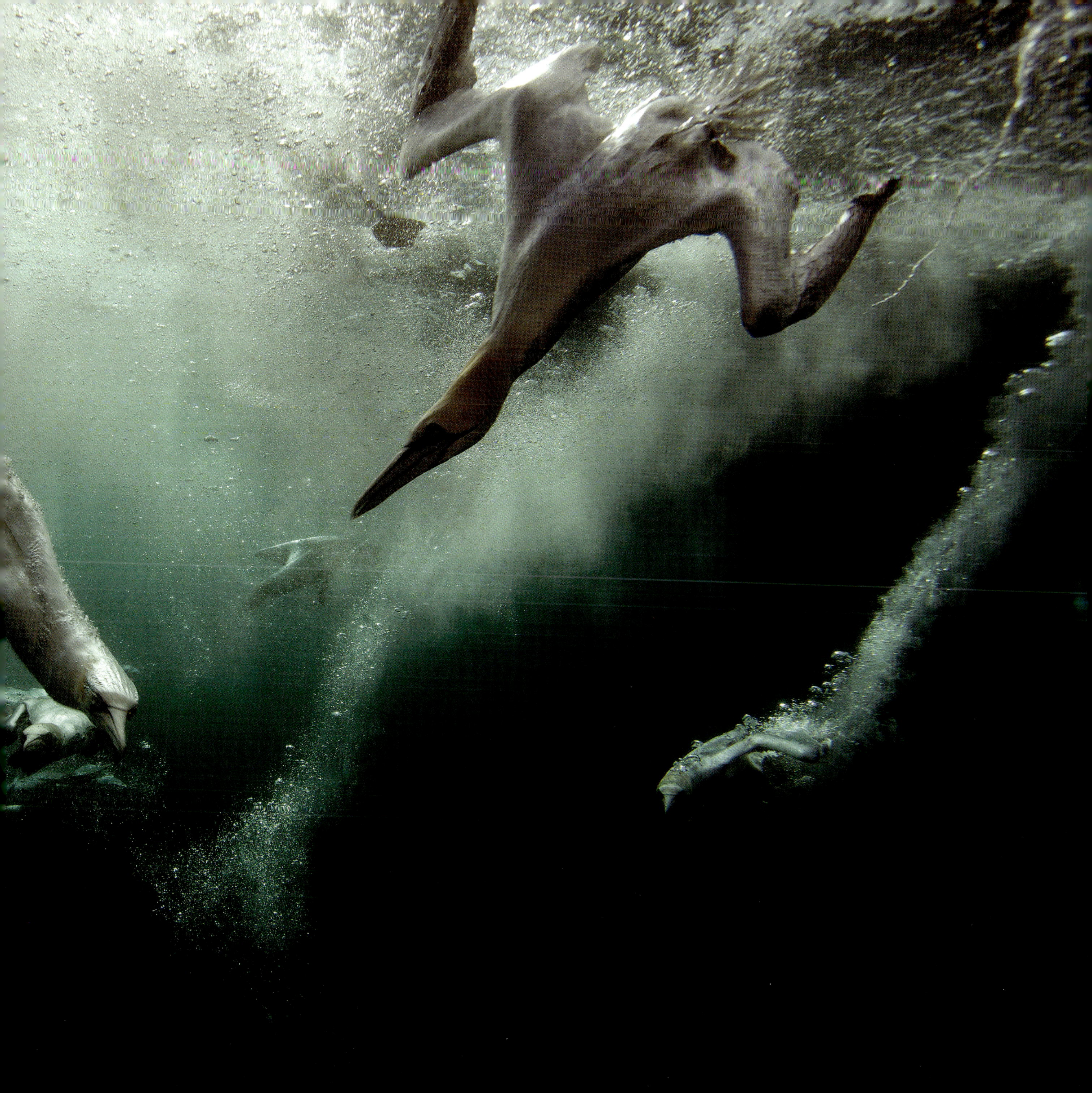

ANIMAL PORTRAITS

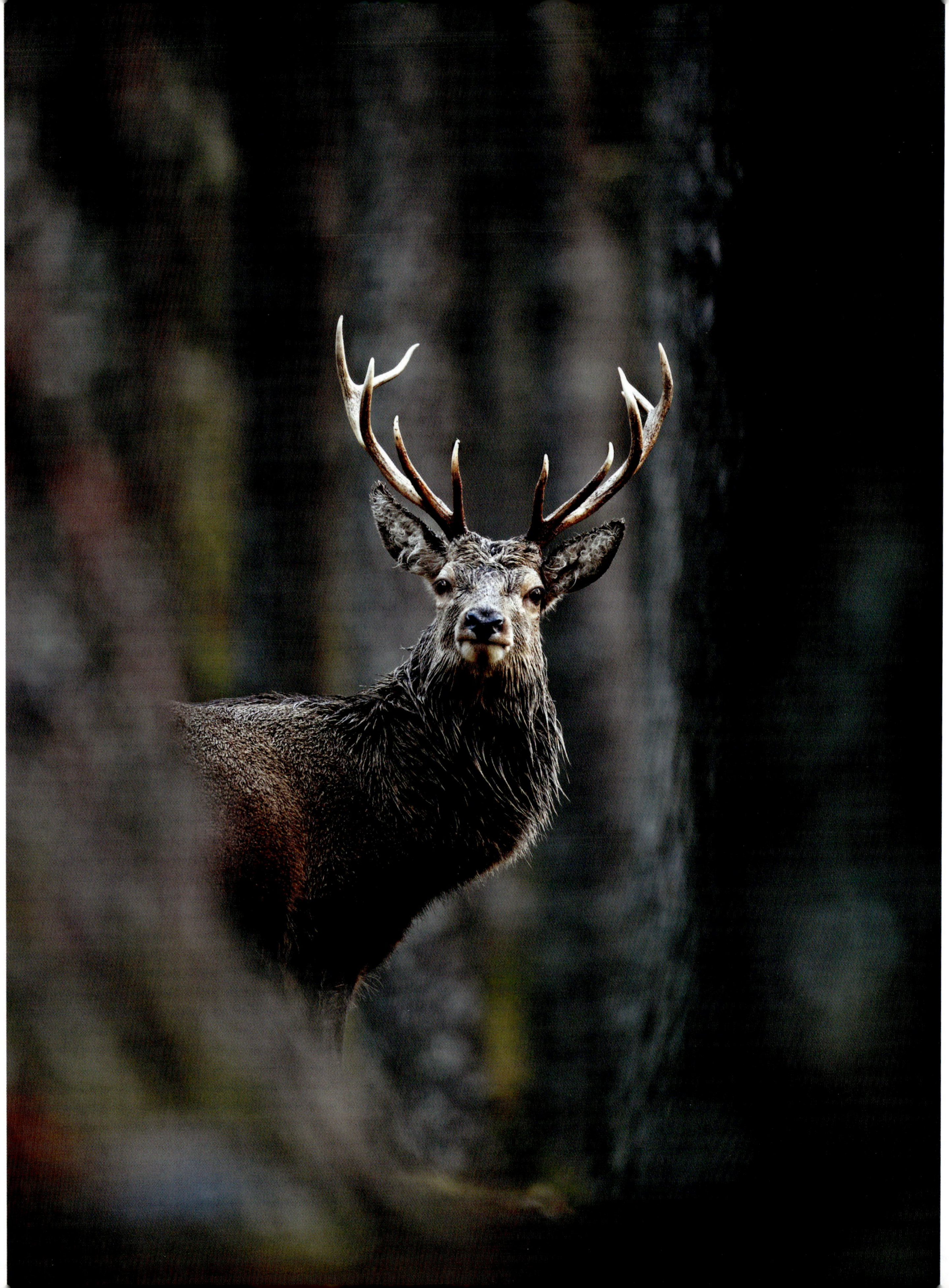

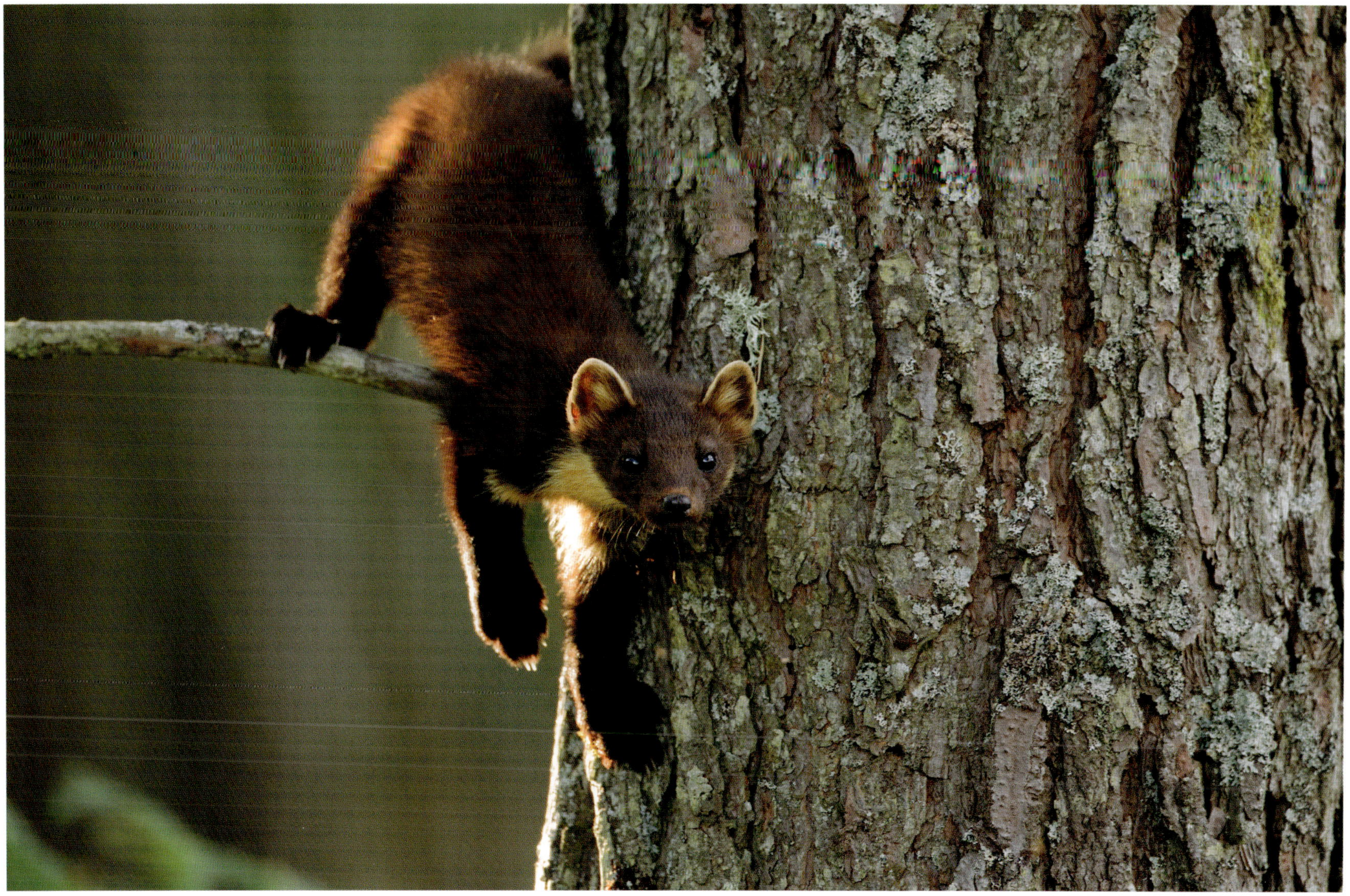

ANIMAL PORTRAITS WINNER

NEIL McINTYRE <

Red Deer Stag in Pine Forest
(Red Deer, *Cervus elaphus)*
Cairngorms National Park, Highland, Scotland

Over the winter months some estates subsidise the feeding of the red deer to help them over the hard times. The estate where this shot was taken feed them in a pine wood, perfect for the shot I was after. After a few near misses, eventually one stag stood in the perfect position and I got the shot I wanted, with the out-of-focus tree trunks framing the stag.

MARK HAMBLIN/2020VISION Λ HIGHLY COMMENDED

Pine Marten in Scots Pine
(Pine Marten, *Martes martes)*
Kinlochewe, Wester Ross, Scotland

I'd been given access to a site where a family of martens was visiting a feeding station during the day. A hide was set up, so all I had to do was wait. Twelve hours later at 8pm I was still waiting when suddenly I heard strange playful noises just out of sight. After several minutes two young martens burst into the clearing quickly followed by two more, and I was richly rewarded with a fantastic 15-minute encounter.

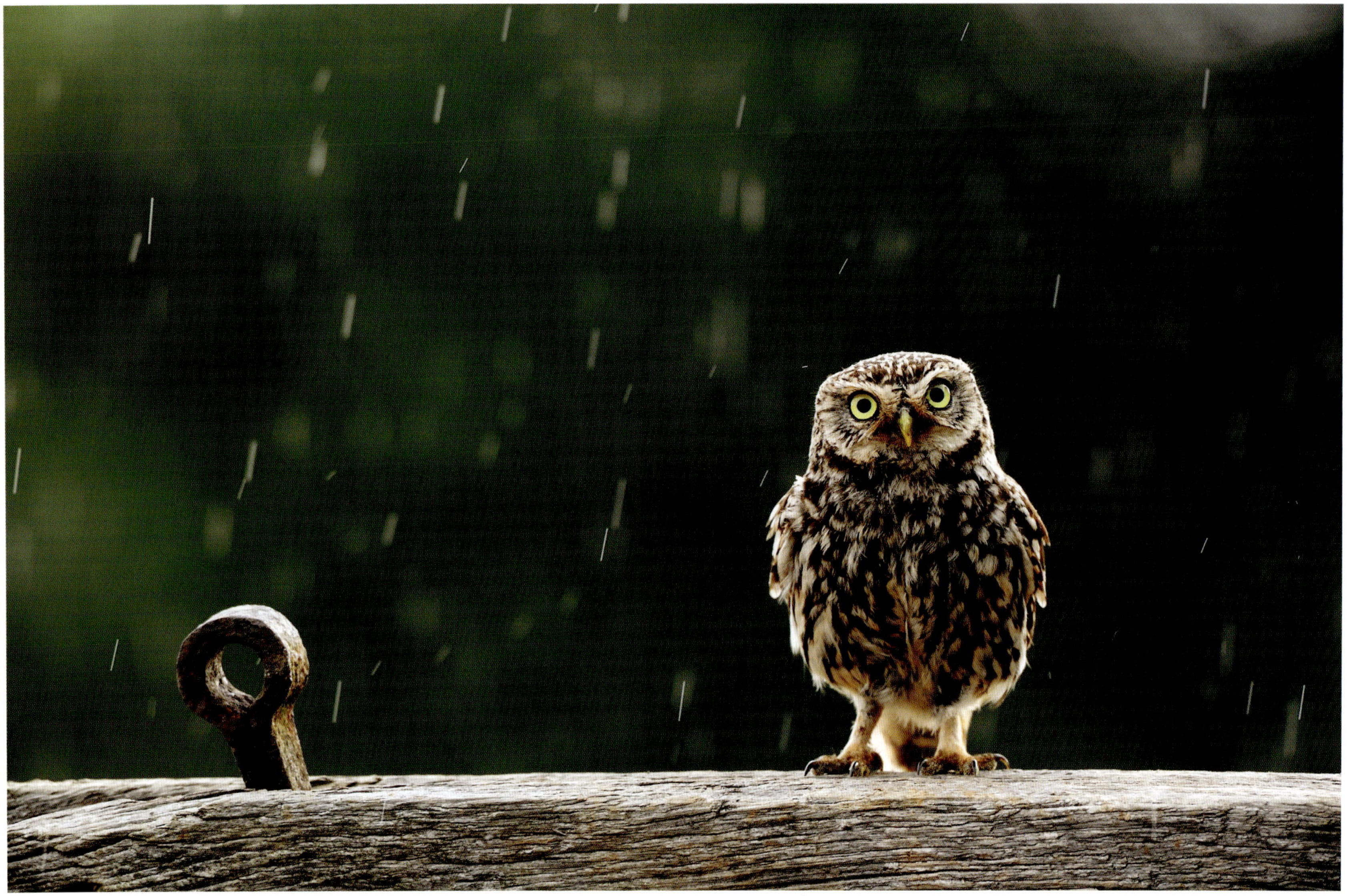

CRAIG CHURCHILL ⋀ HIGHLY COMMENDED

Adult Little Owl in the Falling Rain

(Little Owl, *Athene noctua)*
Adderbury, Oxfordshire, England

I had watched a pair of little owls feeding from the top of a small conifer. They would drop down onto a post then to the surface of a ploughed field to feed. I placed a post on its side and erected a hide next to it. The owls visited on several occasions but I had to wait for a week for the rain, which lasted just five minutes, so was I lucky to have it visit during that period.

GERALD ROBINSON > HIGHLY COMMENDED

A Swan Like You Have Never Seen One Photographed Before

(Mute Swan, *Cygnus olor)*
Kingsmill Reservoir, Mansfield, Nottingham, England

I wanted to capture a different view of a swan other than the usual swan photos. I had the idea of the type of photo I wanted. I sat on the ground with about 15 swans to let them get used to me. I was surprised how at ease they were with me after a while, though some were very inquisitive. Some seemed like they almost wanted to pose for me.

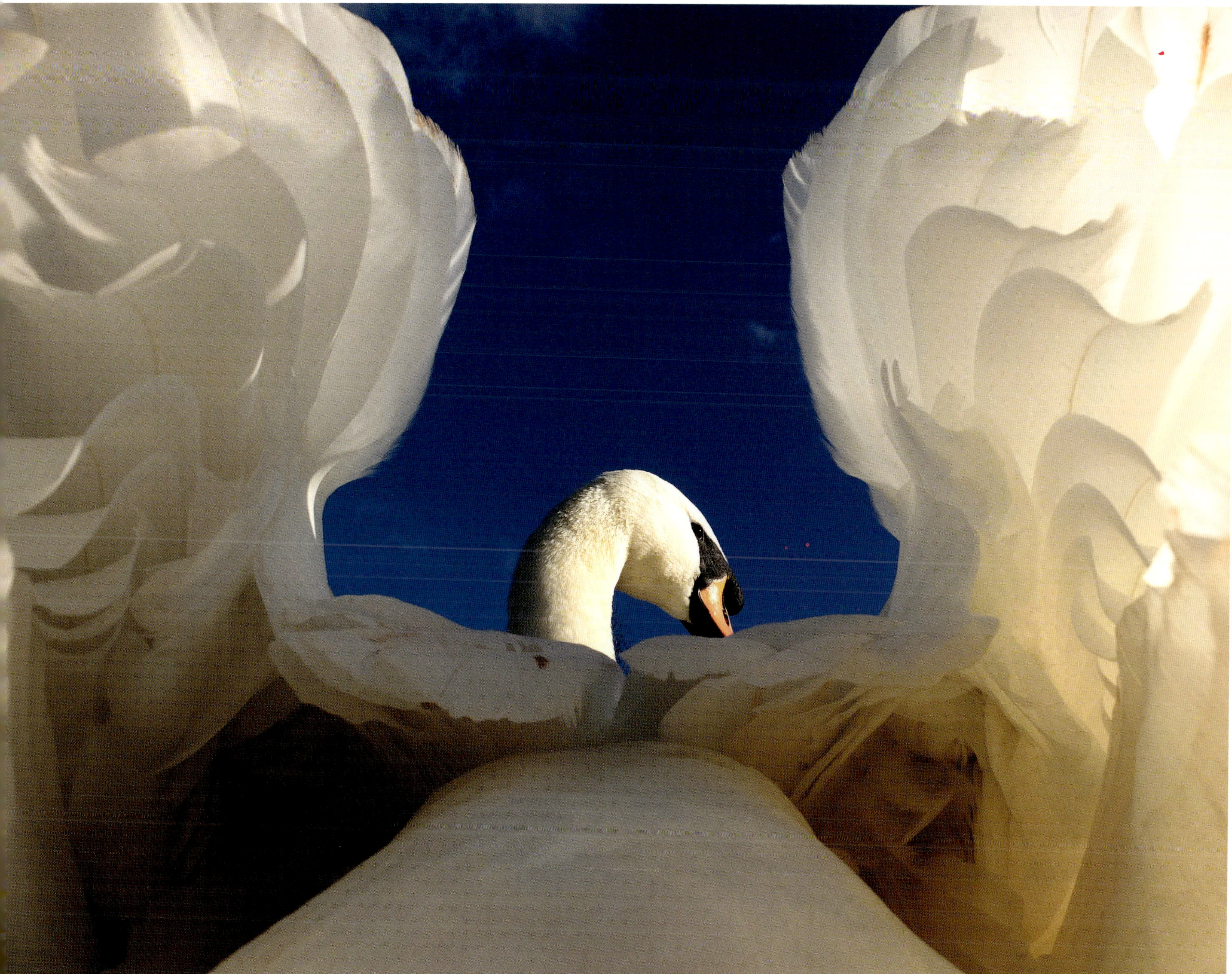

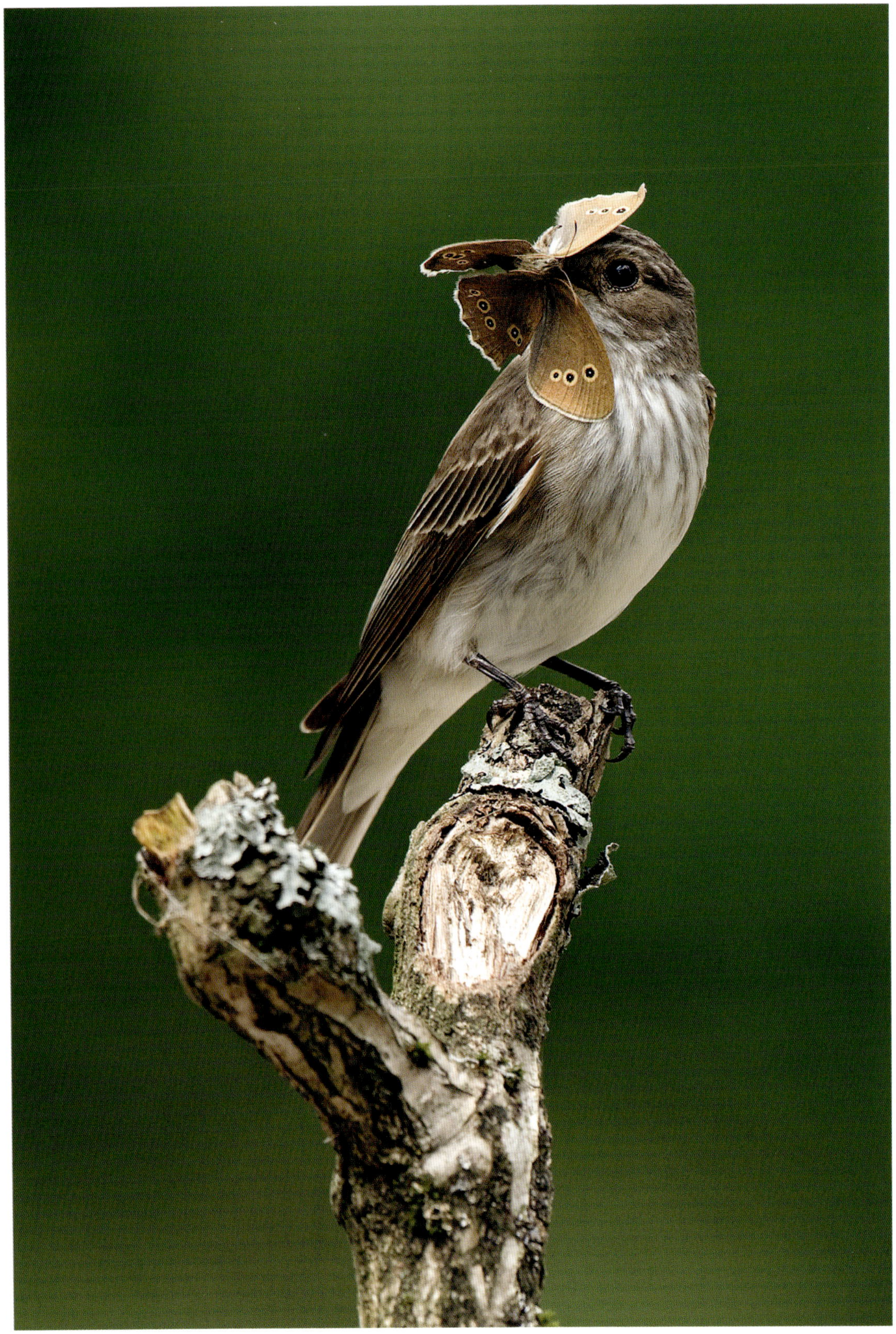

ANDY SANDS HIGHLY COMMENDED

Spotted Flycatcher
(Spotted Flycatcher, *Muscicapa striata*;
Ringlet Butterfly, *Aphantopus hyperantus)*
West Sussex, England

After several sessions photographing a pair of spotted flycatchers as they returned to feed their young in a friend's back garden, I embarked on my final attempt. Two hours later this bird turned up with a ringlet butterfly. It perched for a while and encouraged the young to fledge. This was the last image I took of these birds and the only one where it turned its head sufficiently to see the eye!

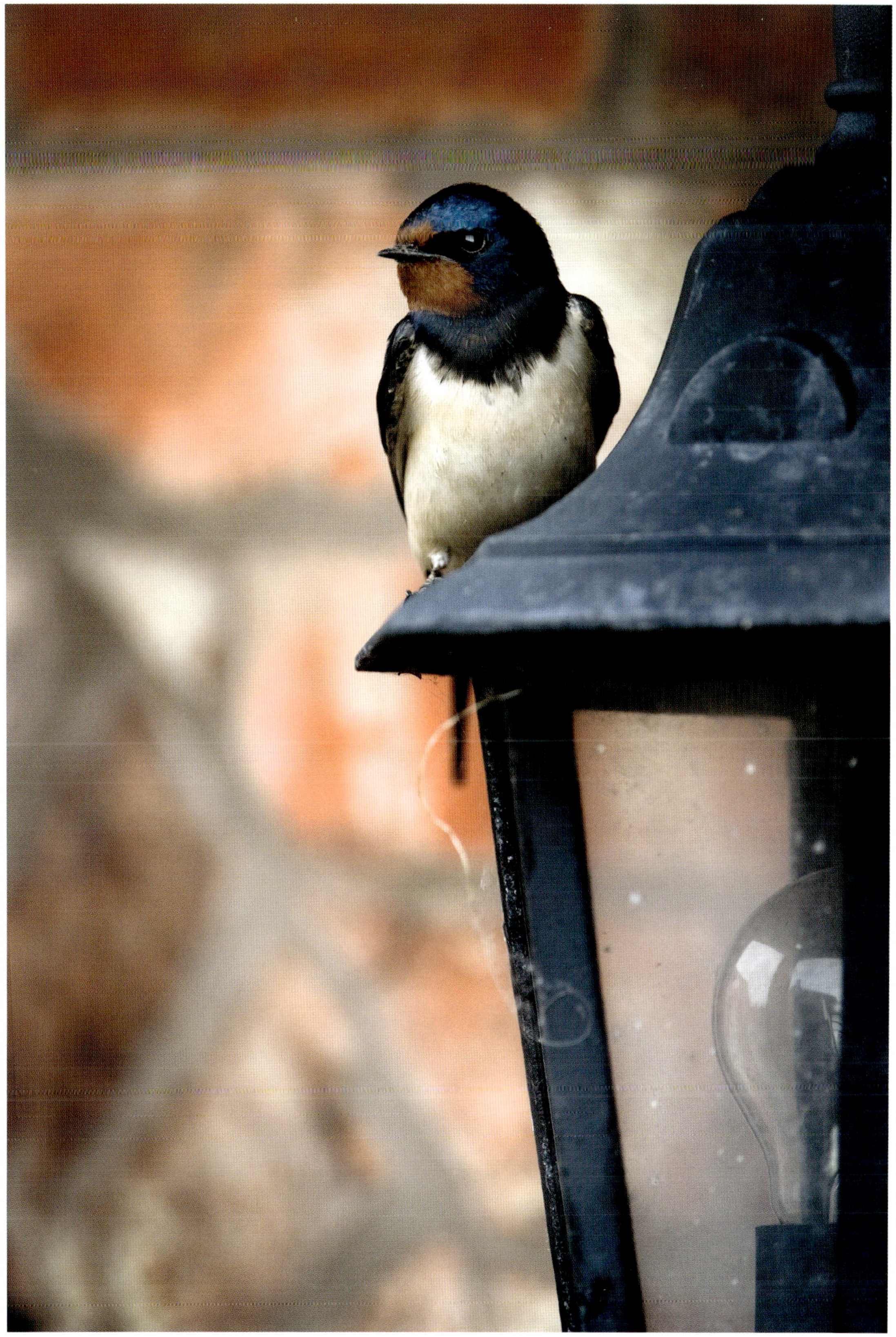

STEVE MAGENNIS HIGHLY COMMENDED

Sitting Pretty
(Swallow, Hirundo rustica)
Northamptonshire, England

A friend's country garden is full of wildlife. As I don't get a wide variety in my garden I spent time photographing a number of species including this swallow, which nested above the back door. Due to it being used to people I was able to photograph it without a hide. The pair loved to sit on this lamp, adding to the atmosphere of the image.

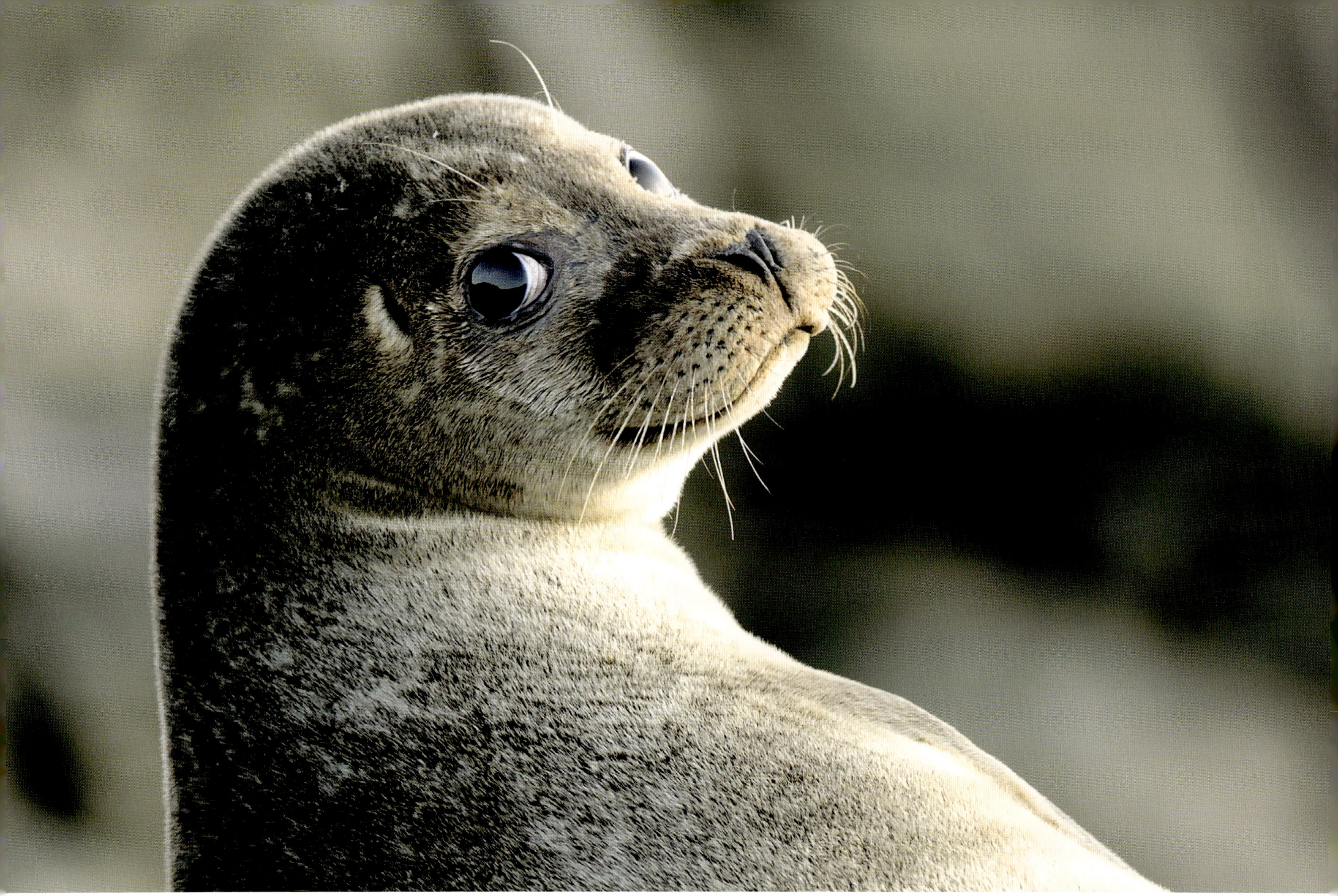

JOHN MONCRIEFF HIGHLY COMMENDED

Over the Shoulder Seal
(Common Seal, *Phoca vitulina)*
Troswick, Shetland Isles, Scotland

I initially went out early in the morning looking for otters. I had no luck with the otters, but came across a small group of common seals hauled up on rocks as the sun began to rise. They remained at ease as I crept closer for several photographs, and this particular one rewarded me with an 'over the shoulder' pose.

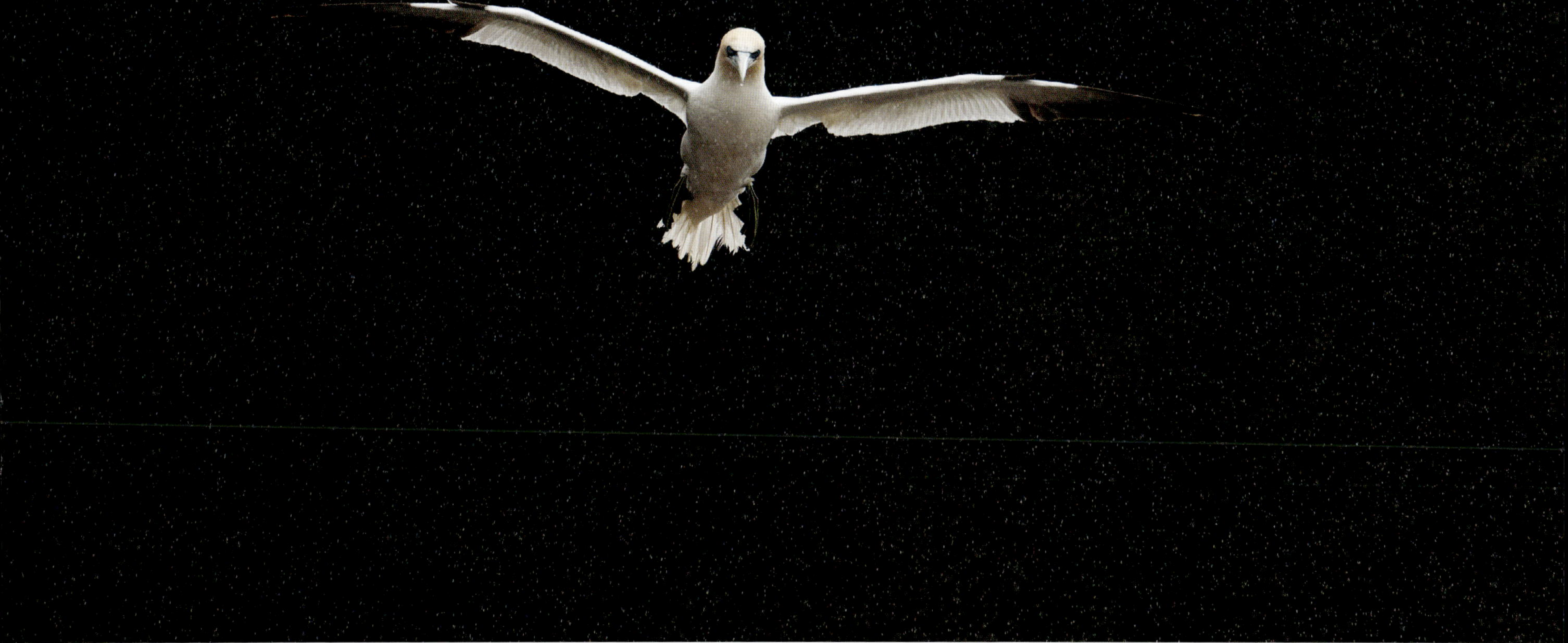

ANDREW PARKINSON HIGHLY COMMENDED

Gannet in the Pouring Rain
(Northern Gannet, *Morus bassanus*)
Hermaness National Nature Reserve, Shetland Isles, Scotland

After a 90m abseil and a treacherous scramble across a tidal boulder field I was able to work in a gannet colony that no other photographer had ever worked in. As yet another rainstorm blew in from the North Atlantic, I was able to isolate this lone gannet against the dark cliffs and show the extent of the rain. Hours later I was cut off by the tide and had to negotiate the ascent in near darkness.

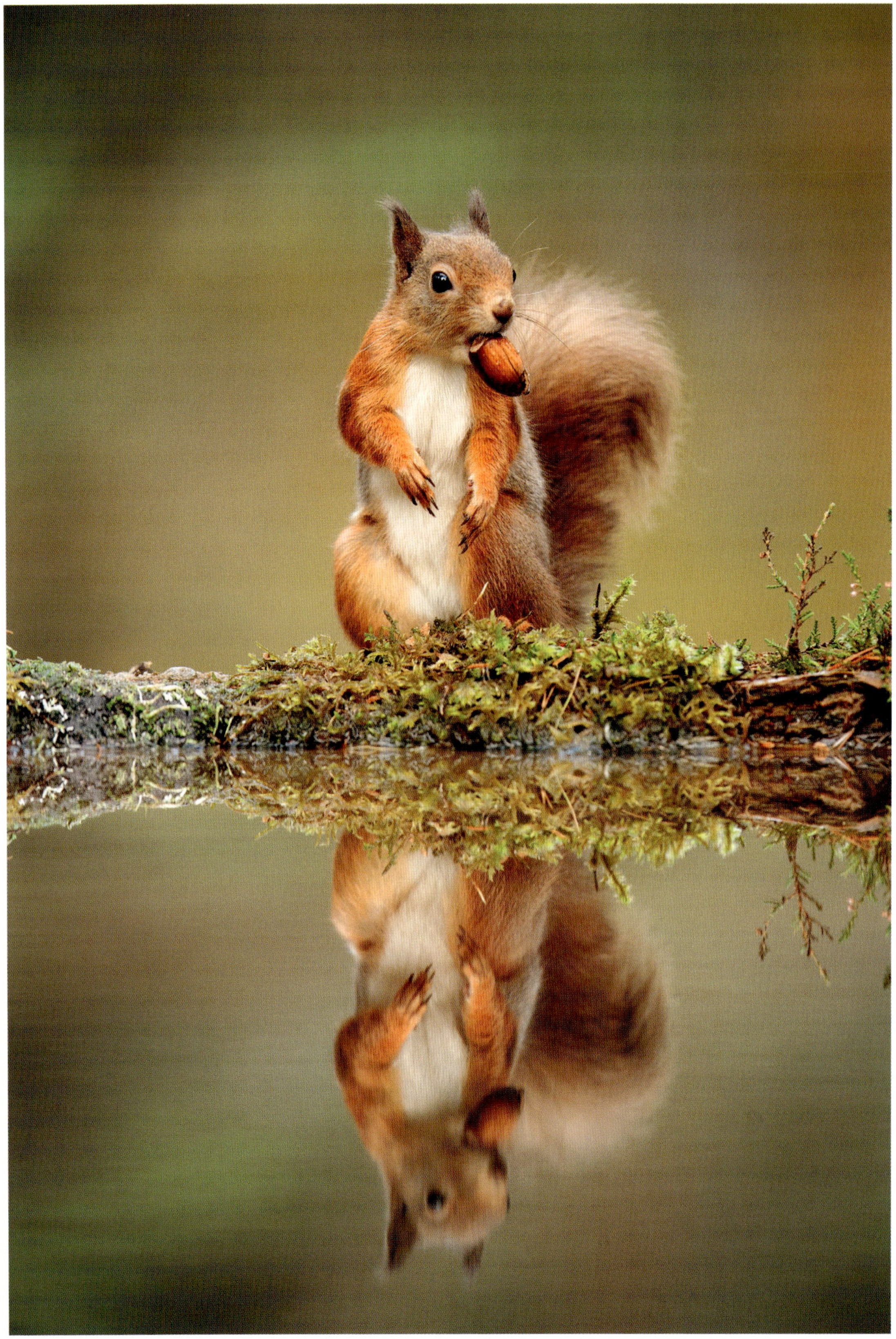

MARK HAMBLIN < HIGHLY COMMENDED

Red Squirrel at Woodland Pool
(Red Squirrel, *Sciurus vulgaris)*
Cairngorms National Park, Highland, Scotland

Red squirrels frequent the pine woodland near my home where I have been feeding them at a regular spot for the past five years. In an attempt to get something different I began to hide nuts along the edge of my woodland pool that was specifically constructed for low-level shooting from a hide that is sunk into the ground. This shooting angle allows for a perfect reflection when conditions are still.

TOM LANGLANDS > HIGHLY COMMENDED

Sparrowhawk Preening
(Sparrowhawk, *Accipiter nisus)*
Near Kirkcudbright, Dumfries and Galloway, Scotland

This was taken from a hide in a wood near Kirkcudbright, Dumfries and Galloway. It had been reported that a sparrowhawk was appearing regularly. I was photographing small birds when it flew in, landing on a dead branch close to the hide. There it preened itself for several minutes before flying off. The light had been poor but as the sparrowhawk arrived it improved enough to allow me to grab these shots.

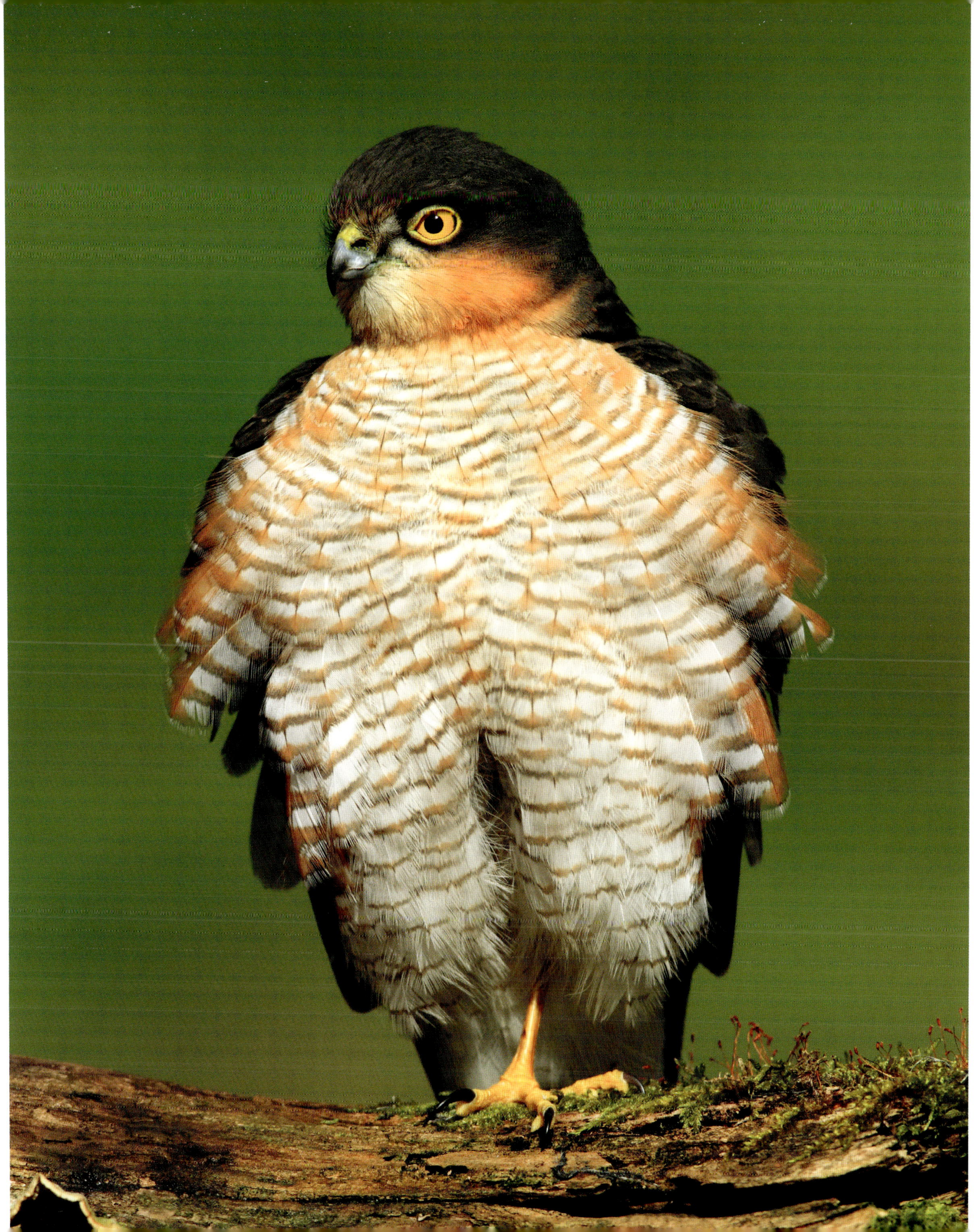

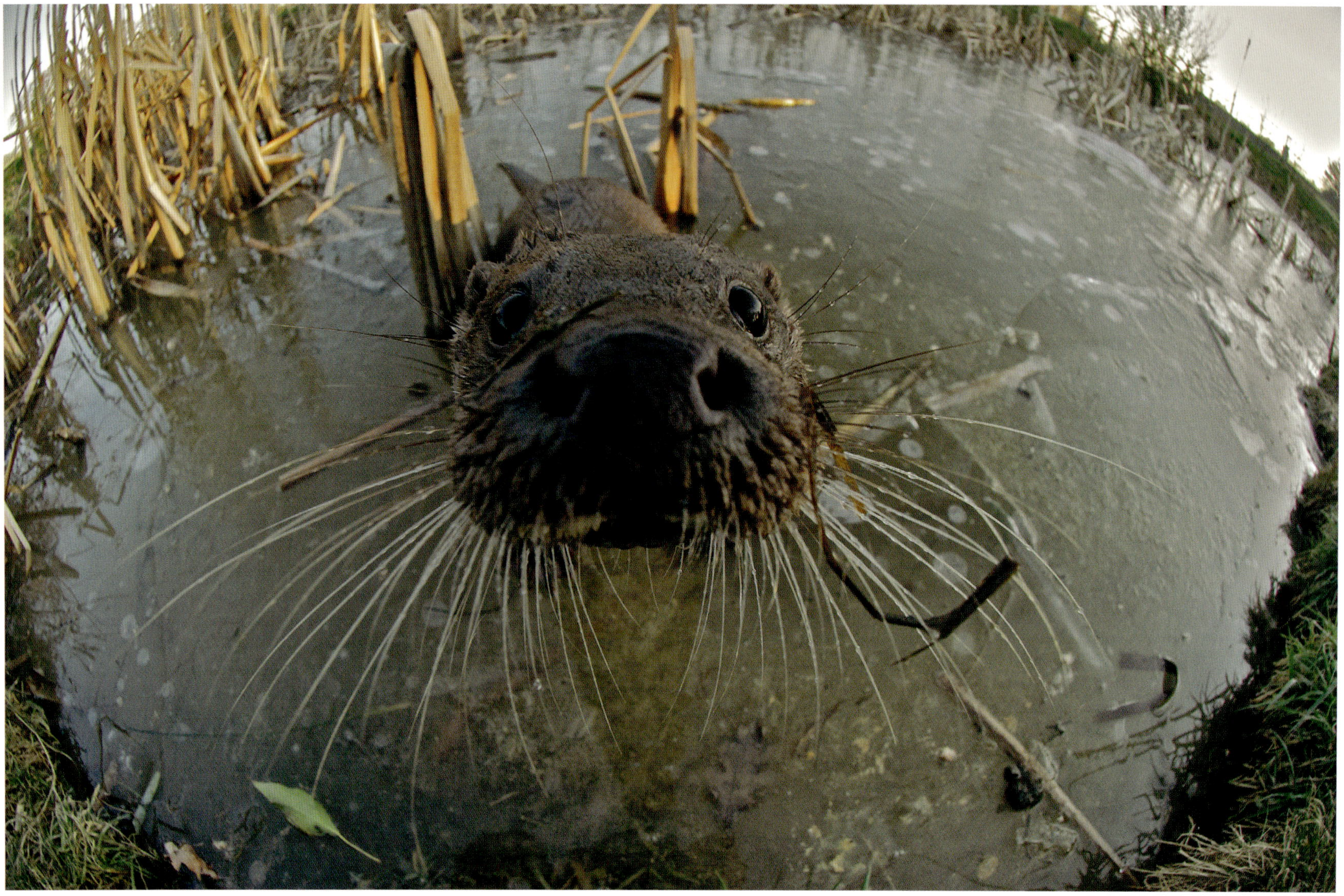

MATT BINSTEAD HIGHLY COMMENDED

Otter on Ice
(European Otter, *Lutra lutra*)
British Wildlife Centre, Surrey, England

Working with captive animals allows me to get closer than I ever could in the wild. This opens up new possibilities such as this close-up. I think the fisheye lens can be used to great effect with wildlife, not just for quirky portraits, but also for telling a story. This otter looks like she is on top of her own little world with the curve the lens has created.

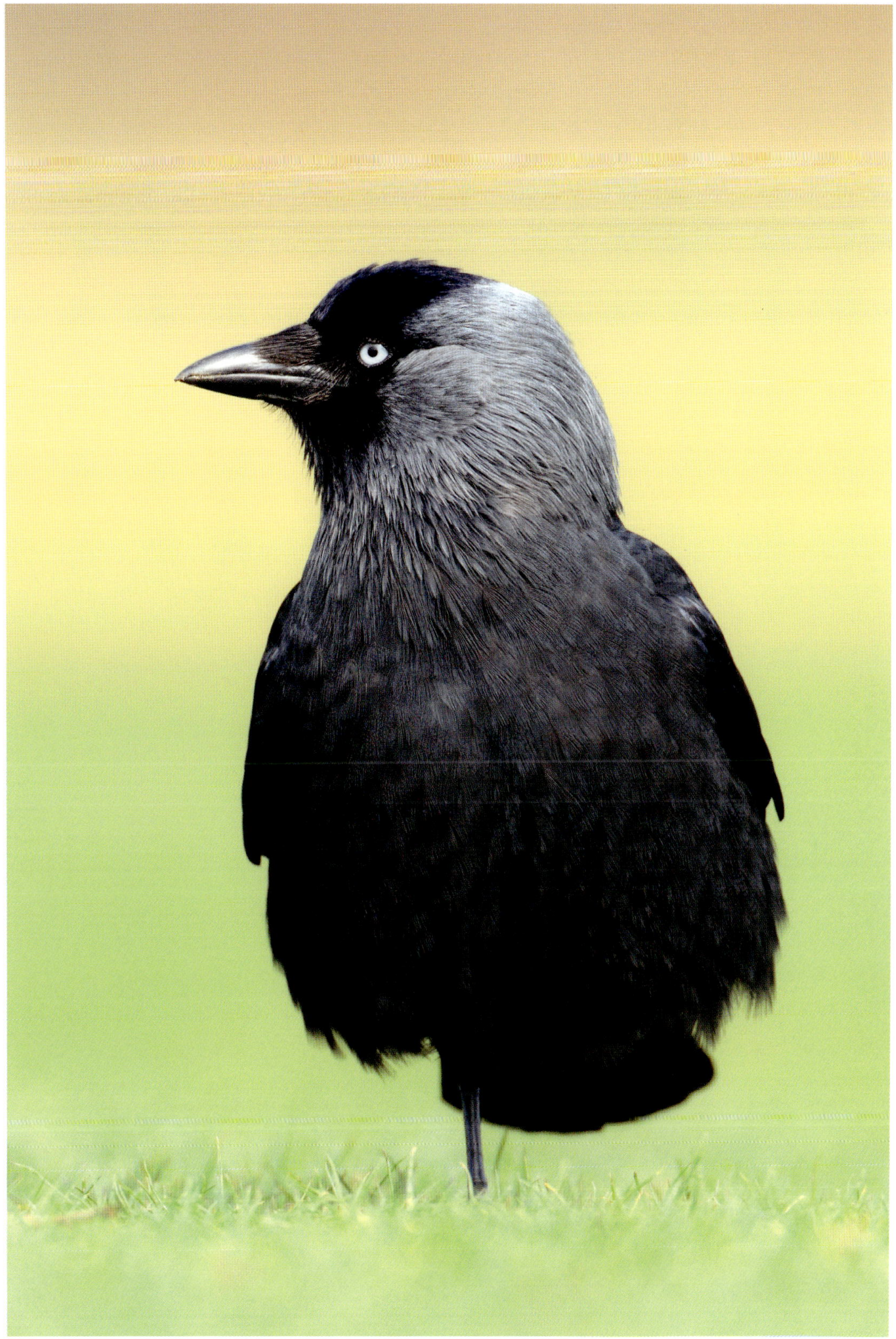

RICHARD PETERS HIGHLY COMMENDED

Balancing Act
(Western Jackdaw, *Corvus monedula*)
Bushy Park, Surrey, England

I was lying low on the ground trying to photograph some grebes when I spotted this jackdaw getting ever closer. I slowly twisted myself around so as not to startle the bird, and as I got him in the frame he stopped moving, turned to show me his profile and balanced on one leg, almost as if he knew I wanted to take his picture.

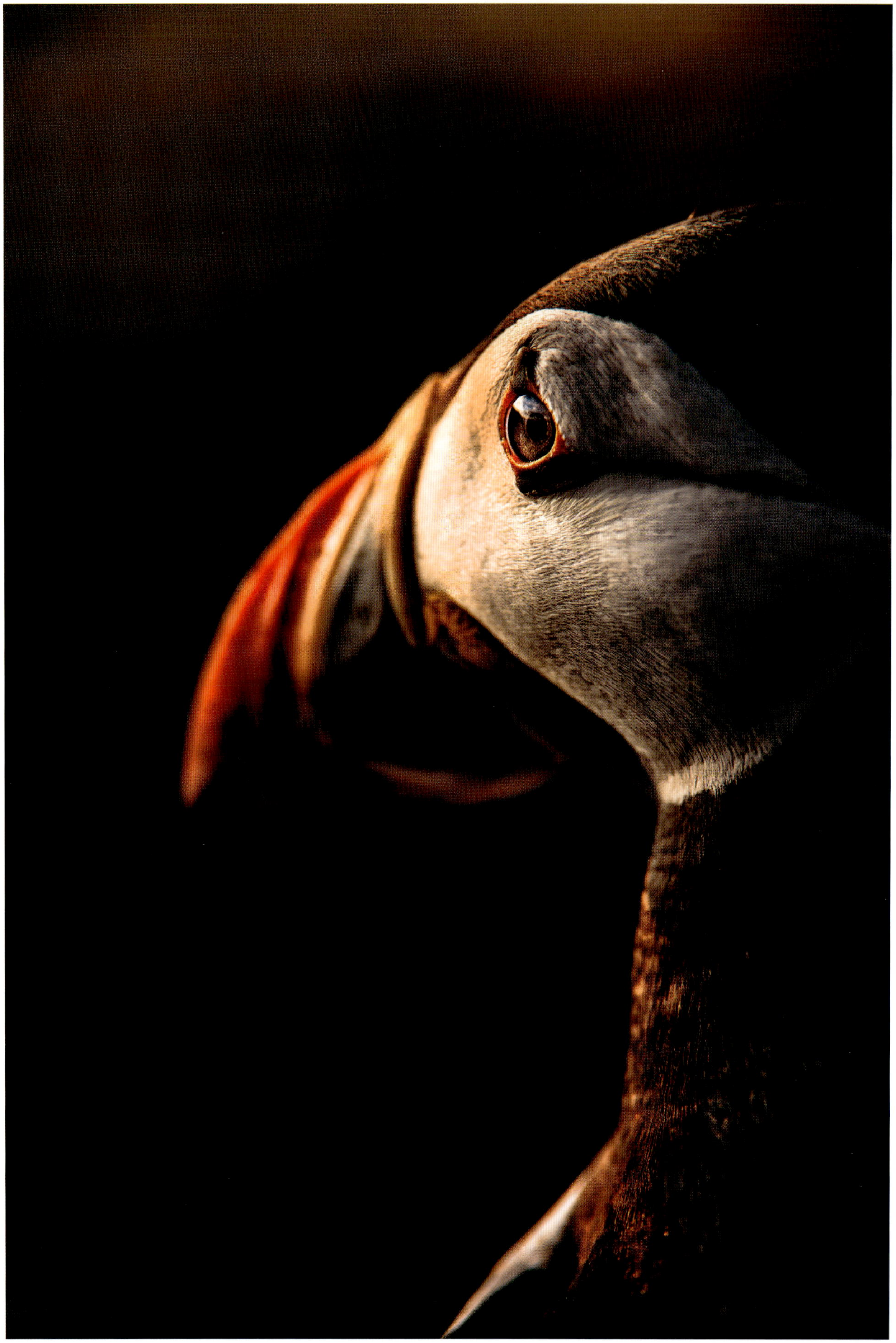

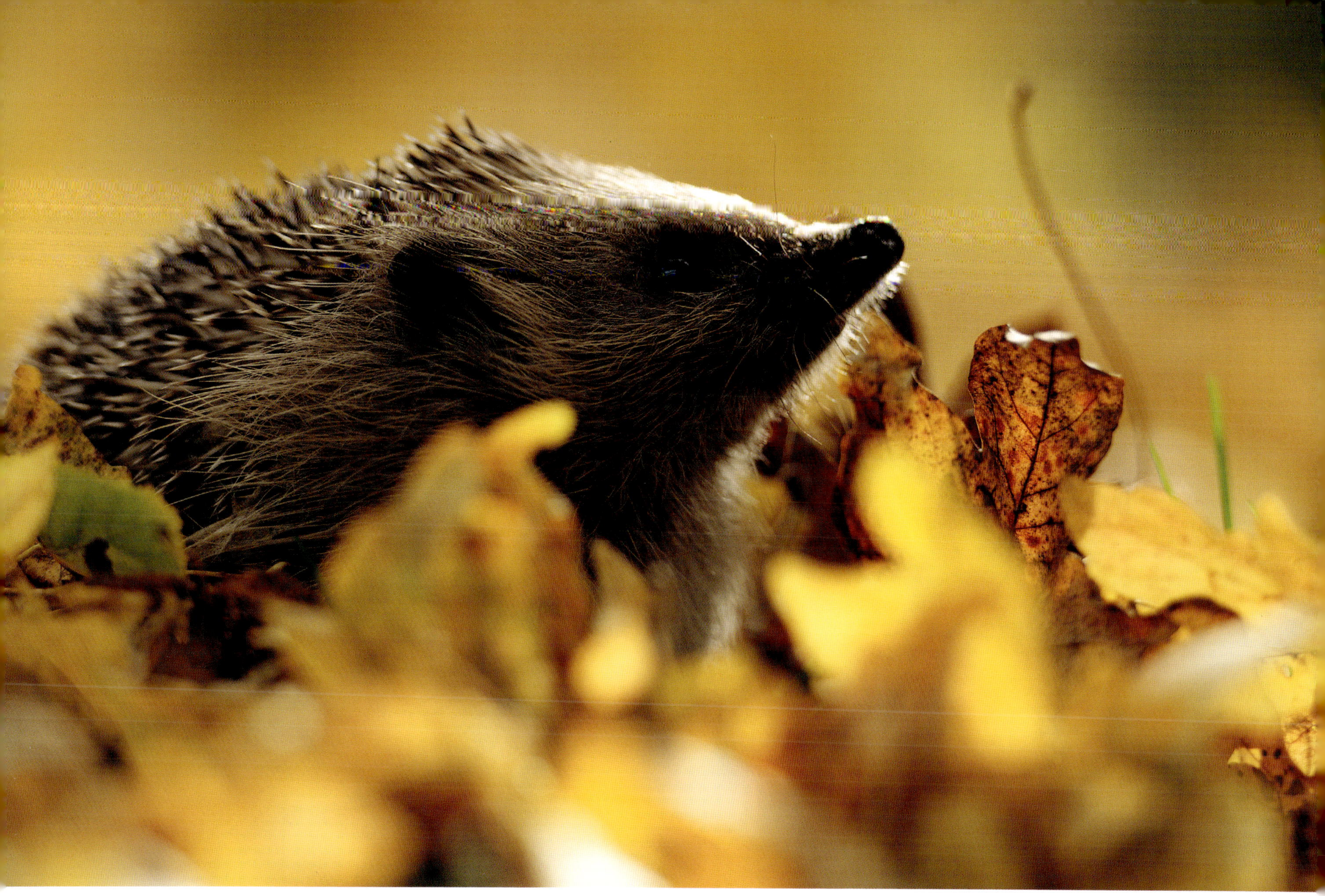

RICHARD COSTIN < HIGHLY COMMENDED

Unsure Puffin
(Atlantic Puffin, *Fratercula arctica)*
Skomer Island, Wales

I have worked a lot with puffins and on an amazing evening when the light is a gift, you need to work swiftly. I spotted this chap perfectly placed in the setting sun. I quickly yet cautiously made my way close enough for some portrait shots. Puffins are a photographer's dream for portraits as they have faces of such character.

MATT BINSTEAD Λ

Autumn Hedgehog
(Hedgehog, *Erinaceus europaus)*
British Wildlife Centre, Surrey, England

I caught the lovely golden colours of autumn lighting a patch of leaves one day and wanted to capture it in an unusual way. Using a tame hedgehog, I set him in the leaves and positioned myself so he was nicely backlit with the leaves glowing and almost transparent. I waited until he had finished rooting around, then as soon as he lifted his nose I took this picture.

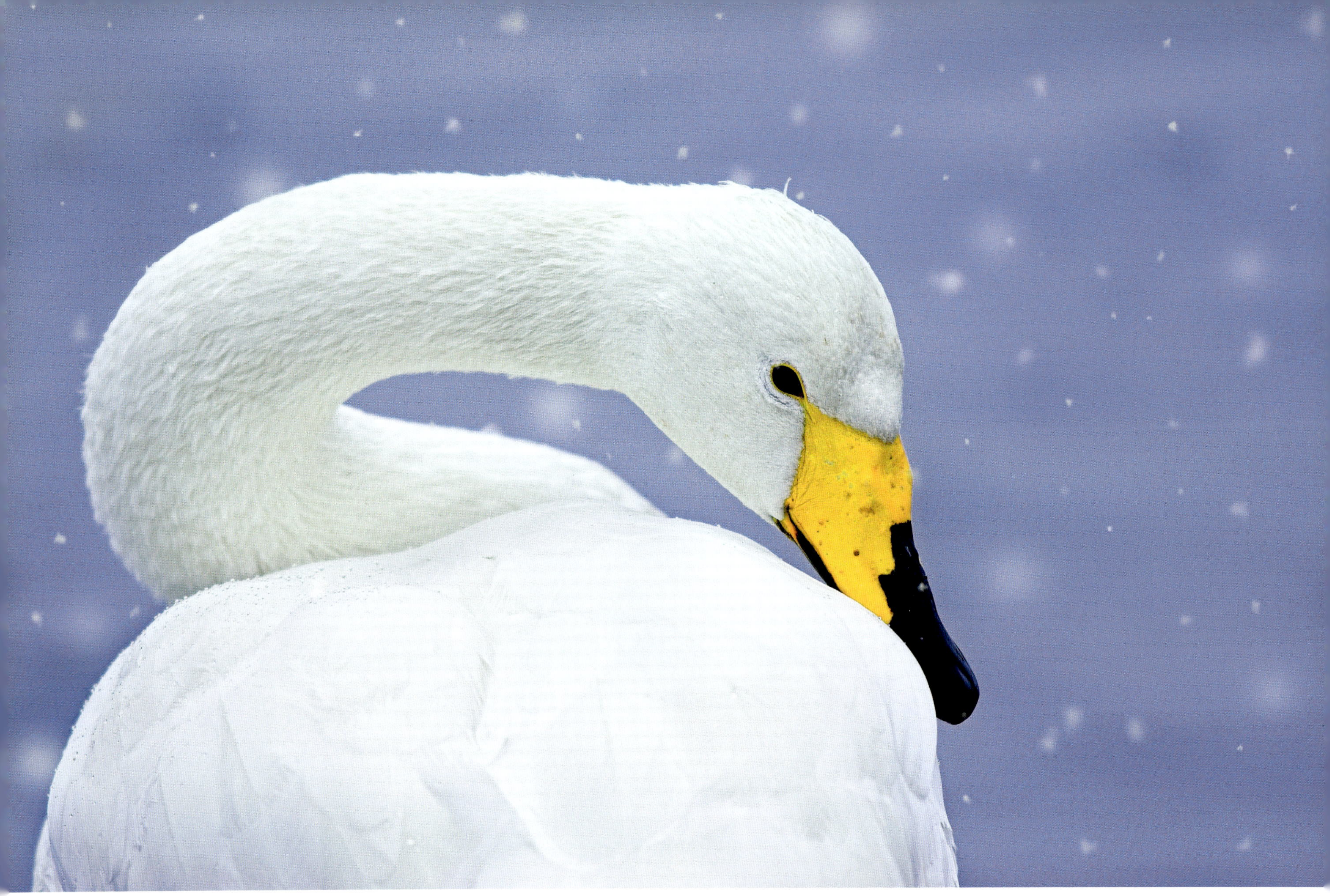

BEN HALL

Whooper Swan in Snowfall
(Whooper Swan, *Cygnus cygnus)*
Martin Mere Wildfowl & Wetlands Trust Reserve, Lancashire, England

I love photographing wildlife in adverse weather conditions. Falling snow in particular can conjure a special kind of atmosphere. I composed the image relatively tightly in order to highlight the elegant shape of the swan's neck.

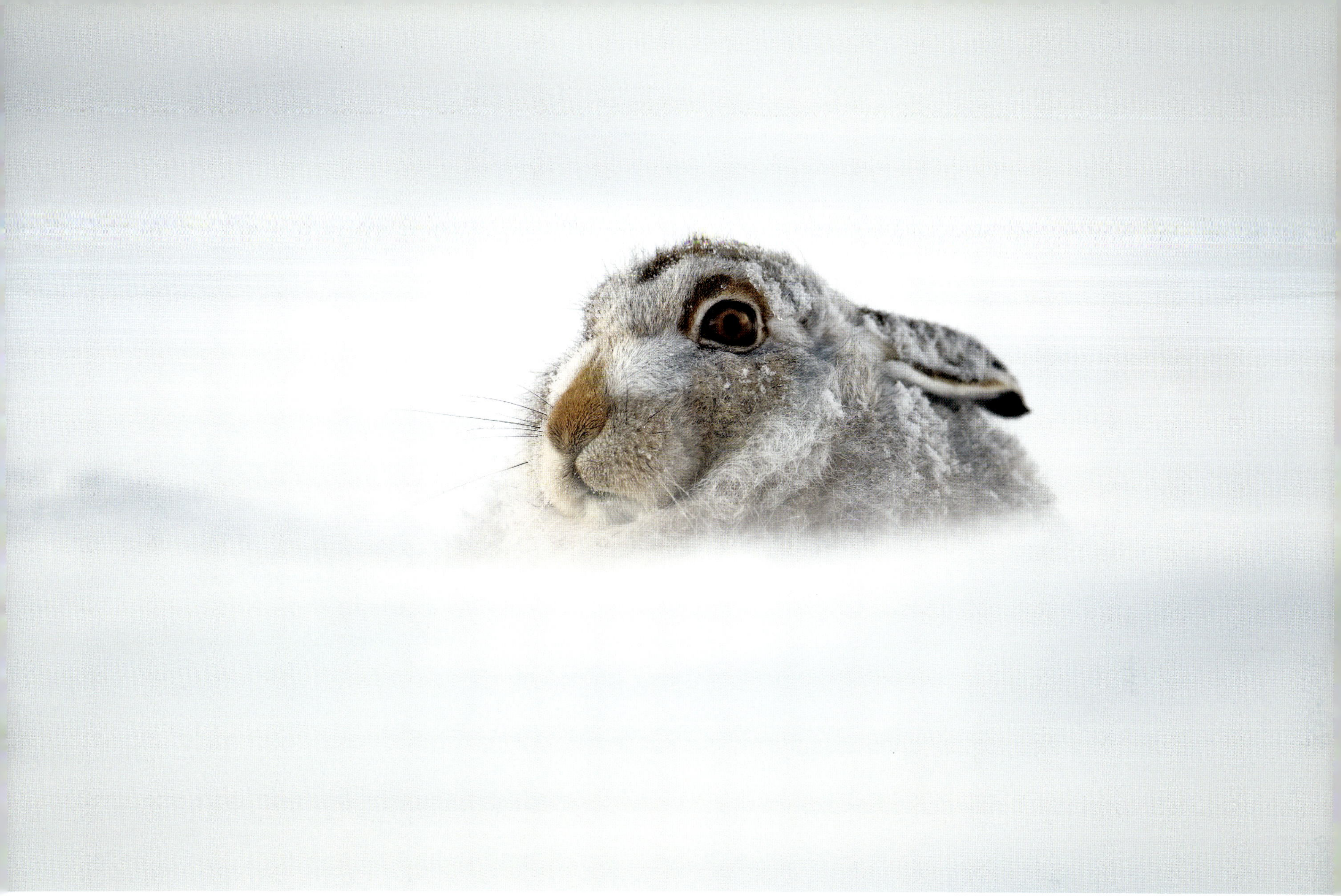

JULES COX

Snowed In
(Mountain Hare, *Lepus timidus)*
Cairngorms National Park, Highland, Scotland

A mountain hare shelters in a snow blizzard, perfectly camouflaged in her pure white winter pelage. I found this old female hare high on a Scottish hillside not far from the path, sitting out the adverse weather conditions. She was perfectly relaxed, allowing me to photograph her even as the snow threatened to bury her completely. As I made my way back down the hill out of the blizzard I looked back and could just about see her still sitting there through the snowfall. The encounter left me marvelling at what hardy mammals mountain hares are.

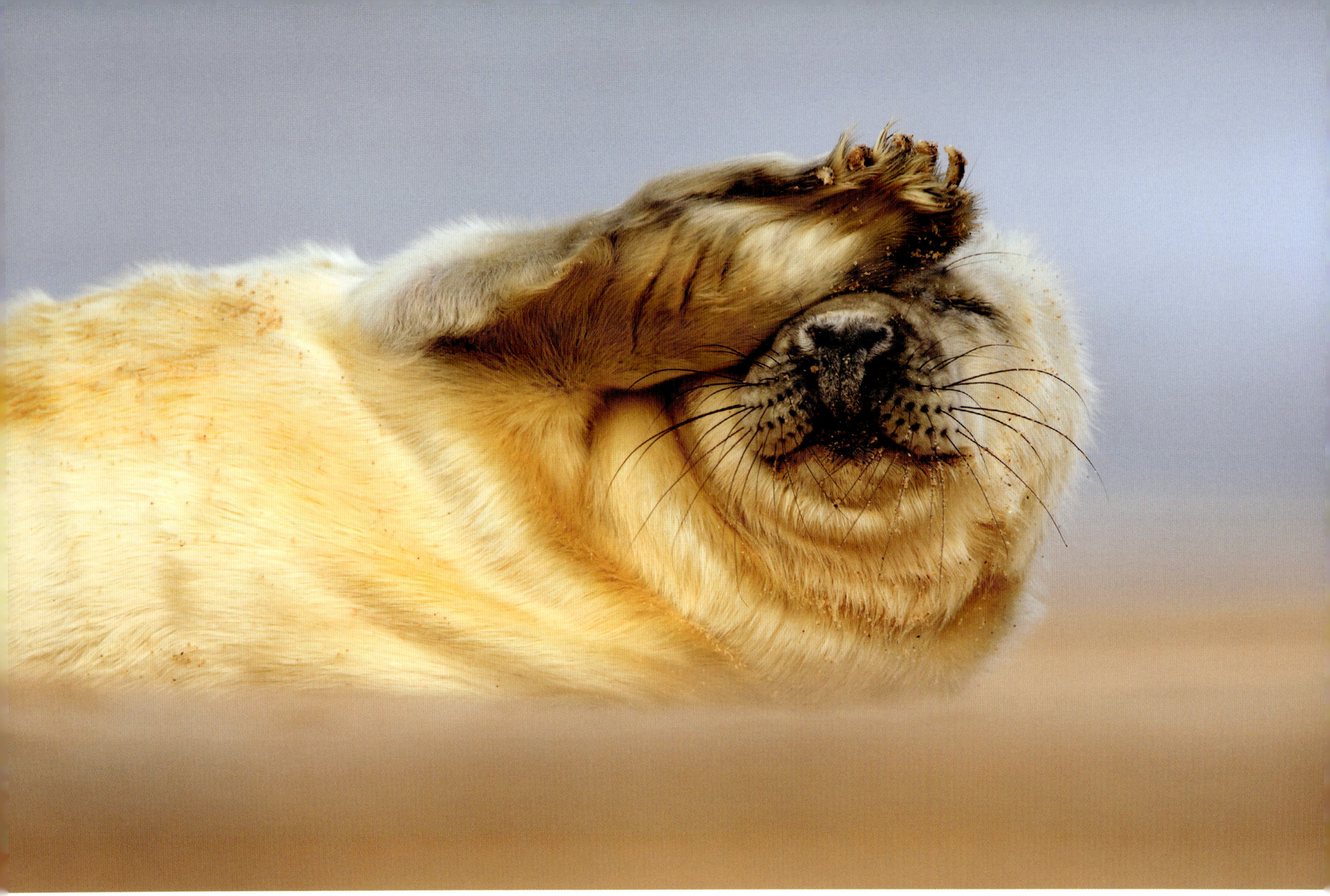

SIMON LITTEN

The Morning After
(Grey Seal, *Halichoerus grypus*)
Horsey, Norfolk, England

Grey seals give birth during the winter months so I positioned myself at the base of the Horsey dunes on a cold November morning. Using my longest lens to avoid any disturbance I lay on the sand for several hours and watched the seal pup. Mostly it slept but occasionally it would wake up and give me a brief chance to exercise my frozen shutter finger!

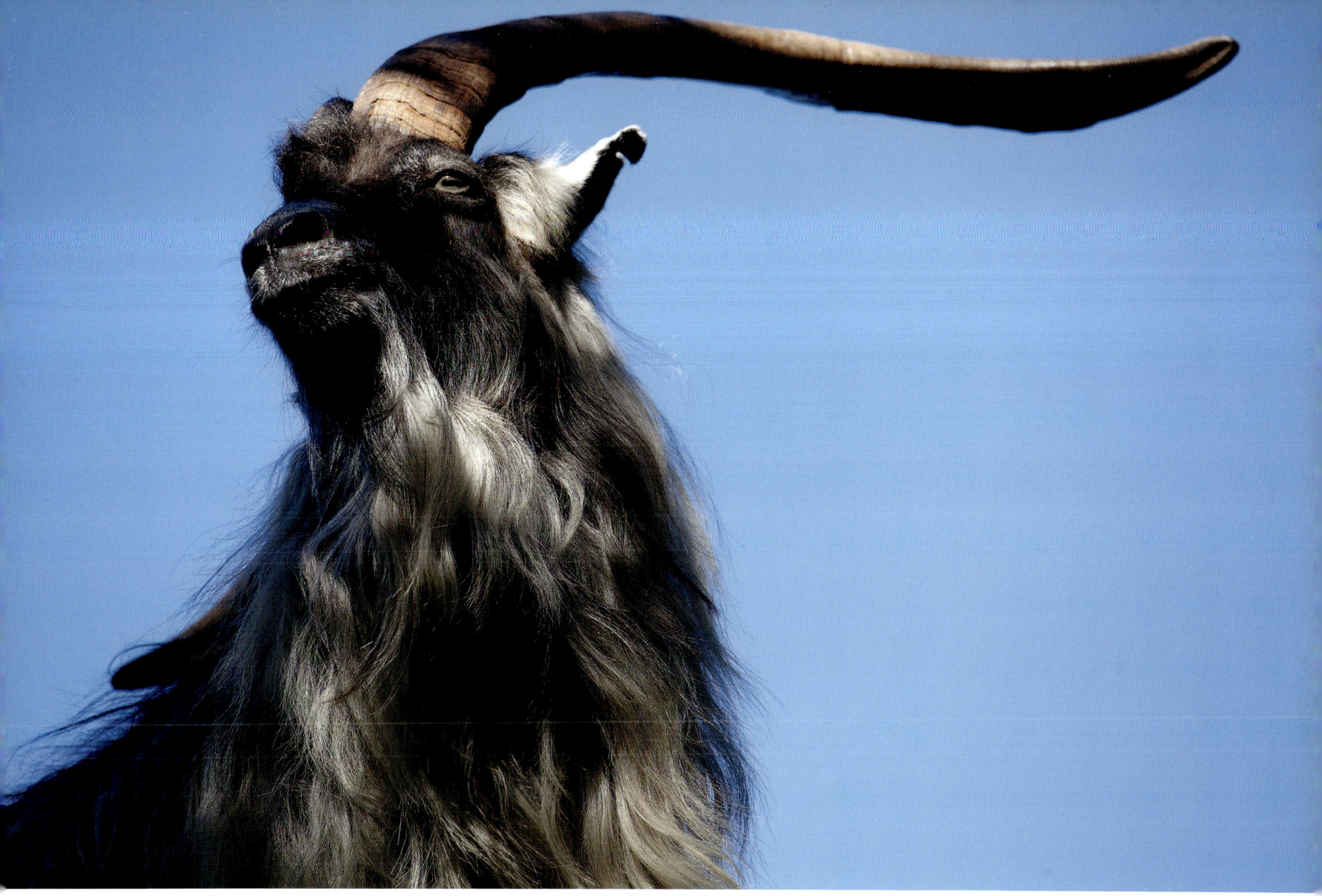

JAN DAVIES

Geifr Gwyllt
(Wild Goat, *Capra aegagrus*)
Snowdonia National Park, Gwynedd, Wales

Many years ago I lived in an old farmhouse that was only accessed by a rickety footbridge. As we crossed I often recalled the tale of the Three Billy Goats Gruff. One unusually balmy day in late autumn, I decided to find the goat that was captured by my imagination. I came across a small herd of *geifr gwyllt* – wild goats seeking shade amongst the oak trees. In a languid manner, this magnificent creature tilted his head as if to salute me.

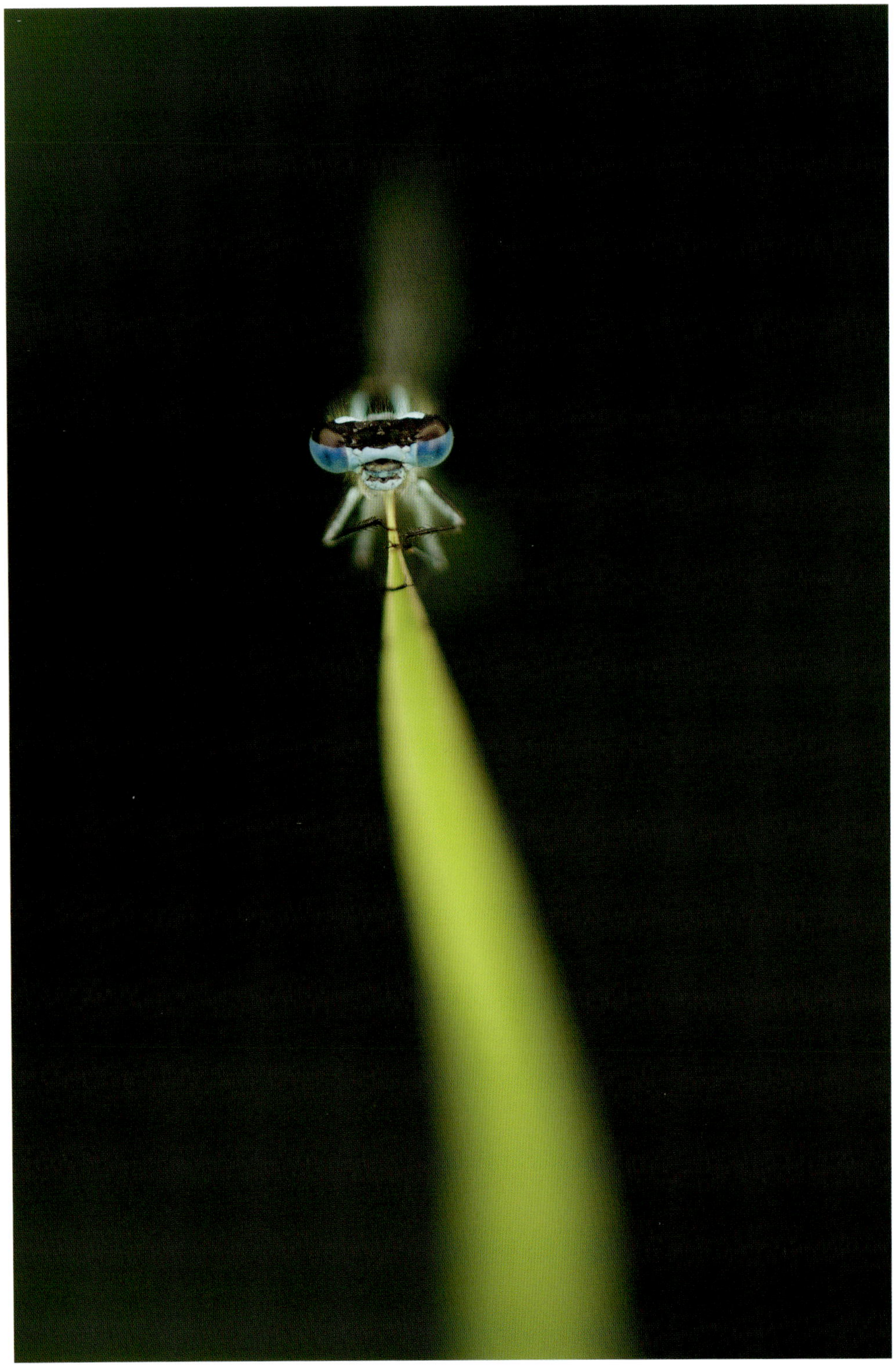

PAUL HOBSON <

Azure Damselfly Roosting on Grass Stalk

(Azure Damselfly, *Coenagrion puella)*
Pleasley Pit Country Park, Derbyshire, England

On damp, cool days damselflies roost in the grasses around their breeding ponds. I carefully searched for one and photographed it from above, hand holding the camera because a tripod would have crushed too many grasses. I wanted a very shallow depth of field so chose F3.5.

JOAN O'HAGAN >

Monster or Cutie?

(Puss Moth, *Cerura vinula)*
Worcestershire, England

Monster or cutie or just an ordinary animal going about its business? You decide! My partner found an abandoned willow branch by the roadside, and brought it home. It grew! And it became covered with hungry puss moth caterpillars. Windy conditions meant that to photograph this one I had to tear off a twig and put it on the ground.

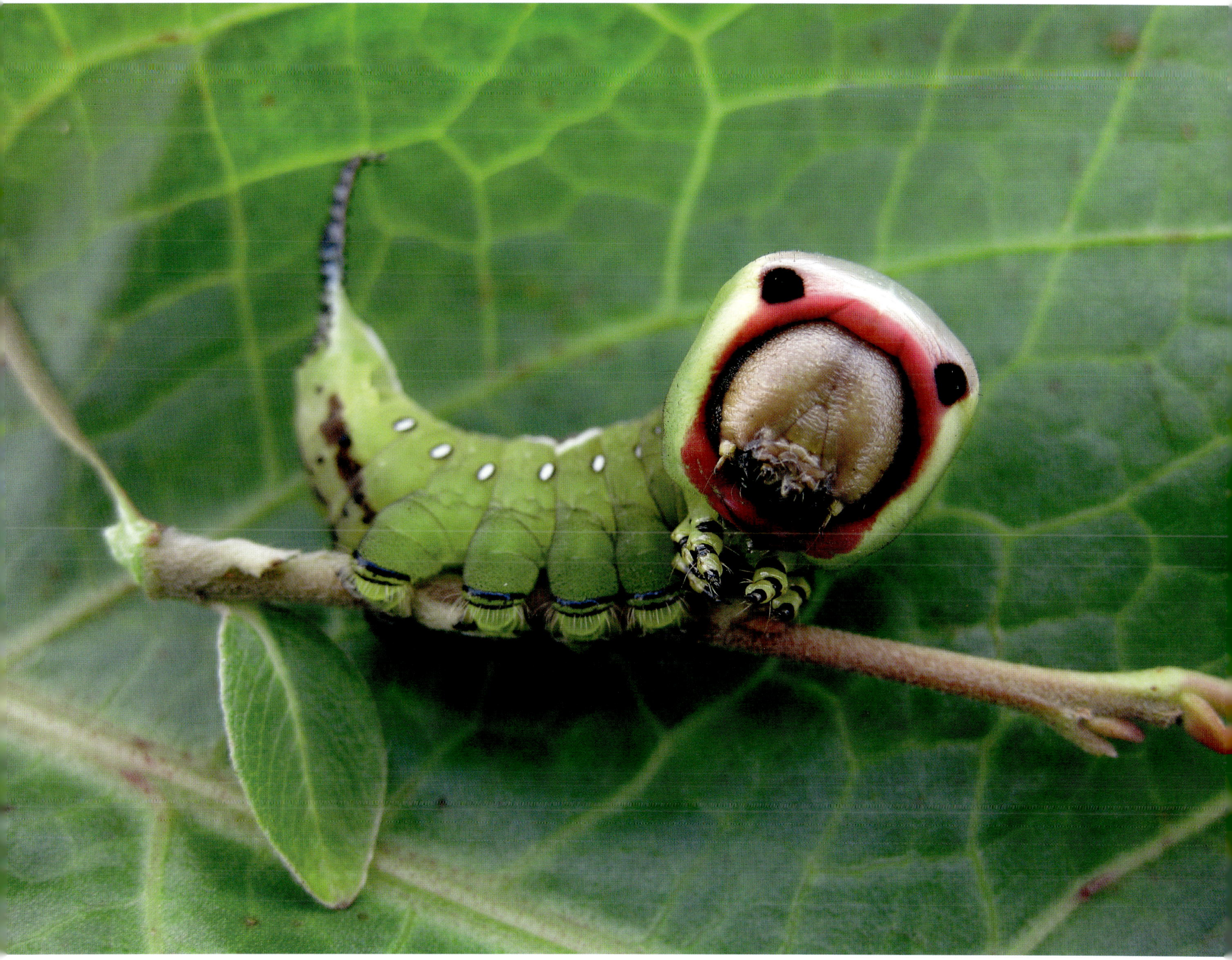

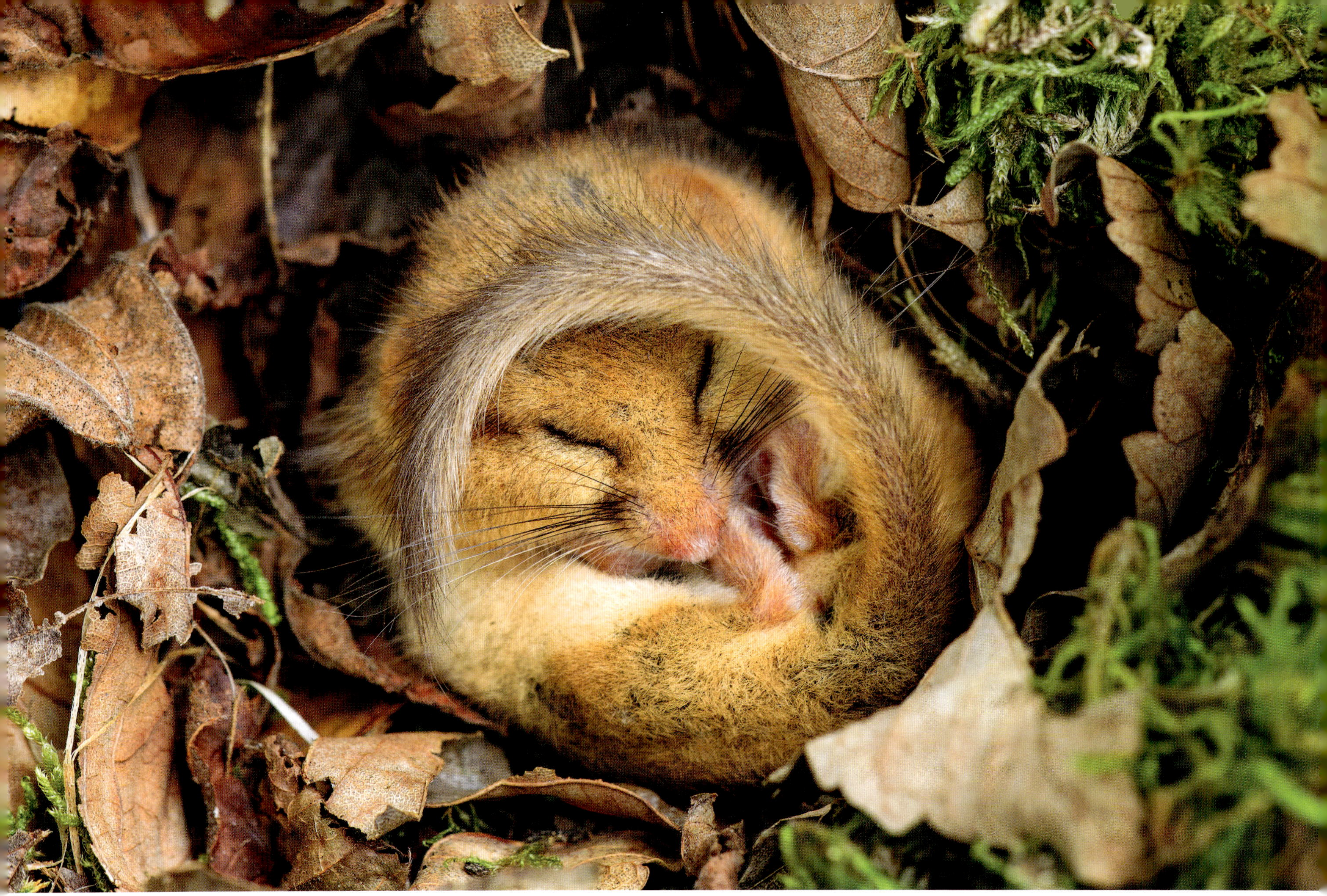

SIMON PHILLPOTTS

Snug as a Bug
(Common Dormouse, *Muscardinus avellanarius)*
Yorkshire Dales National Park, North Yorkshire, England

Dormice have been successfully reintroduced into the Yorkshire Dales National Park after an absence of 100 years. Populating a coppiced woodland, their nesting boxes are regularly checked by licensed National Park staff to record numbers, weight and health of the population. By attending several official counts I was able to obtain a view into a rarely seen world and capture this image of a torpid (hibernating) dormouse.

LUKE MASSEY

Eyelashes
(Wild Boar, *Sus scrofa*)
Forest of Dean, Gloucestershire, England

After following the sound of boar for much of the day I was eventually accepted by the adults and the piglets soon came to investigate me. Wild boar have poor eyesight but a great sense of smell. Being downwind of them meant they weren't able to detect me until they were very close. The piglets actually sat on my legs at one point!

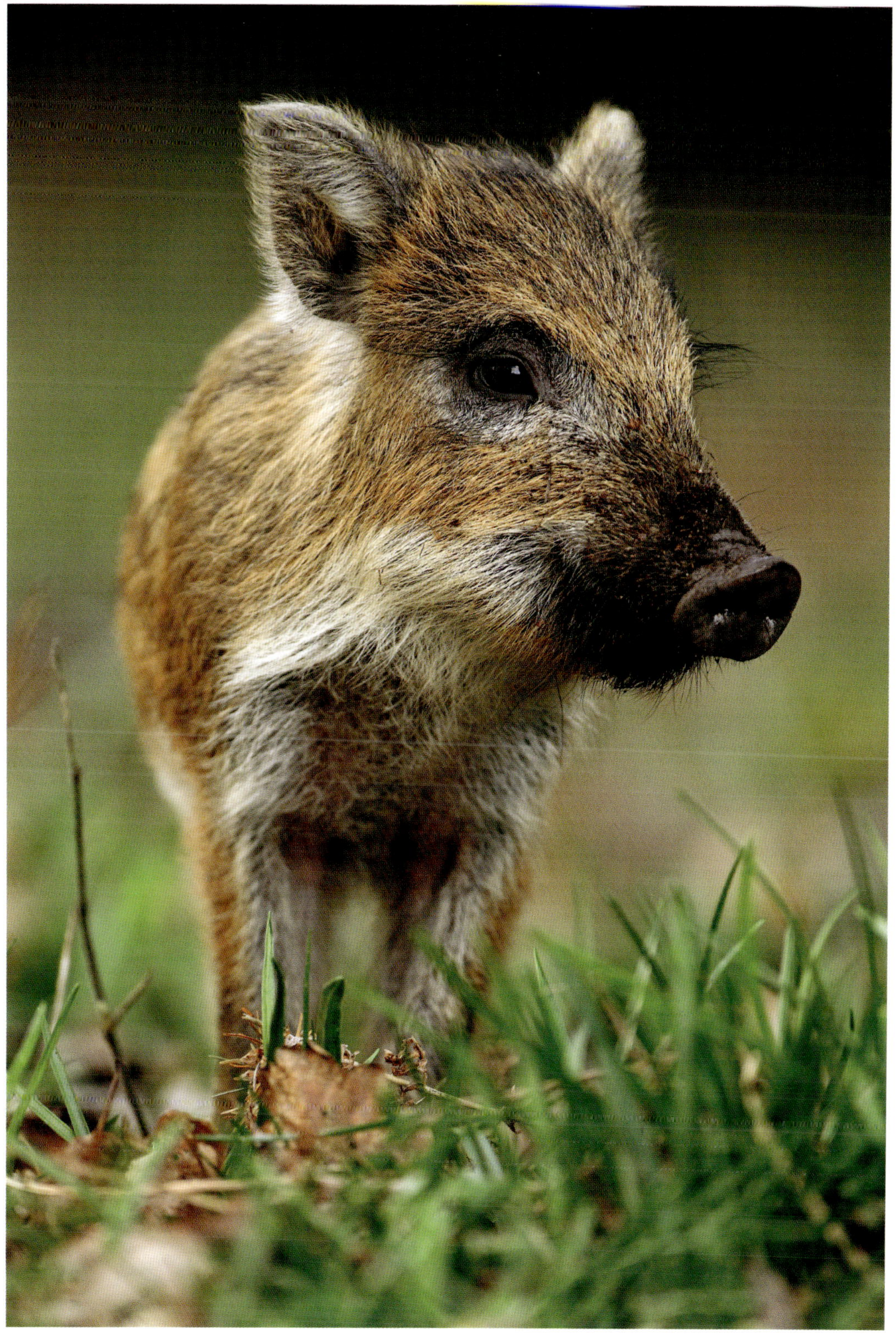

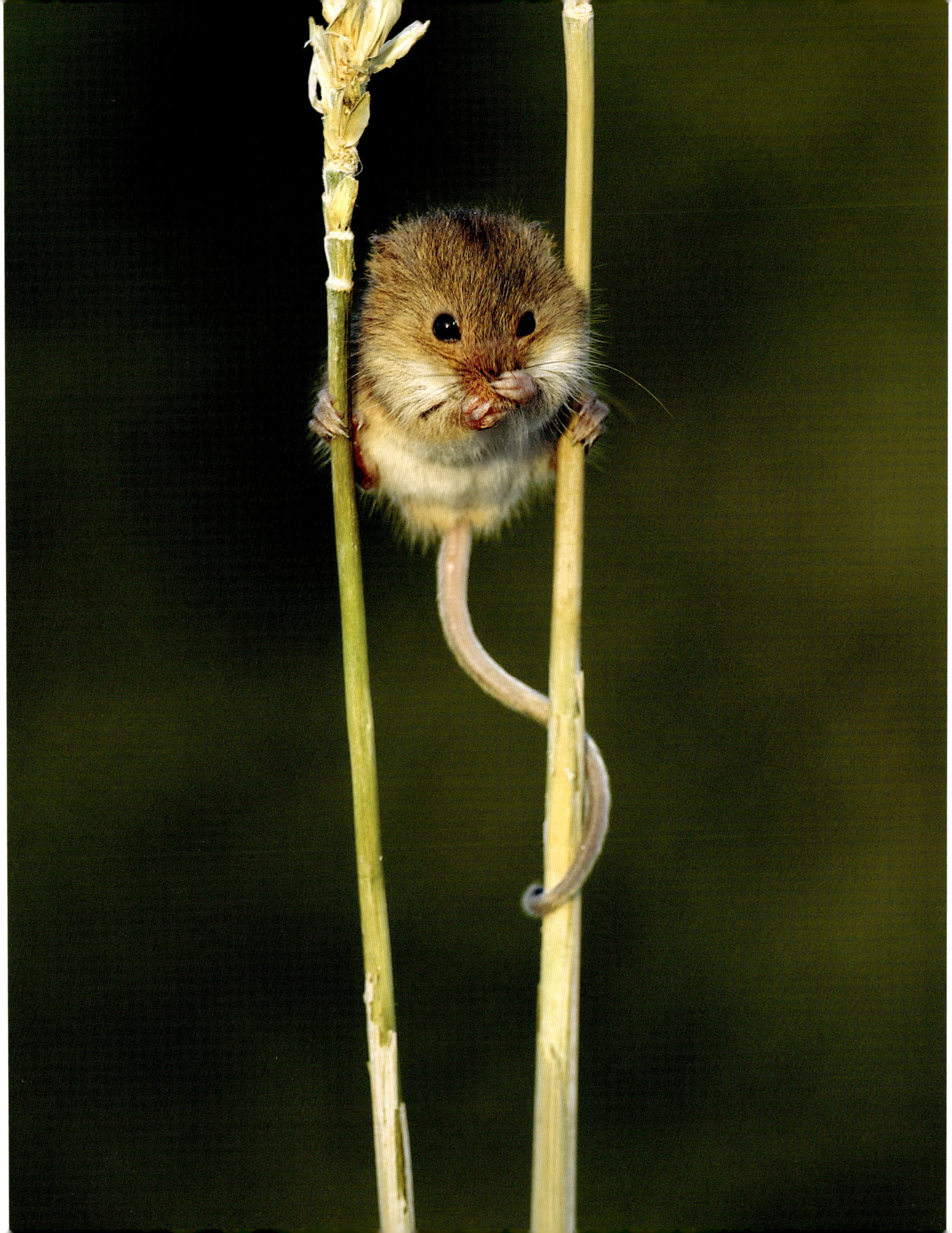

MATT BINSTEAD <

Shh... Keep Quiet
(Harvest Mouse, *Micromys minutus*)
British Wildlife Centre, Surrey, England

It's moments like this that can make a good image into something special. I was photographing this mouse carefully balancing between two corn stalks when he started to have a wash. I fired off a few shots and was lucky enough to capture this image. I like to think he has a big secret to keep and is desperately holding his mouth shut so it doesn't escape.

PENNY DIXIE >

Crested Tit in the Evening Sun
(Crested Tit, *Lophophanes cristatus*)
Caledonian pine forest, Strathspey, Scotland

I had the chance to photograph near a feeder that had regular visits from crested tits. I wanted an image that really showed off the bird's iconic crest. Facing towards the low sun and making sure I had a dark background, I waited until the tiny bird stuck its crest up so that the sun shone through its feathers.

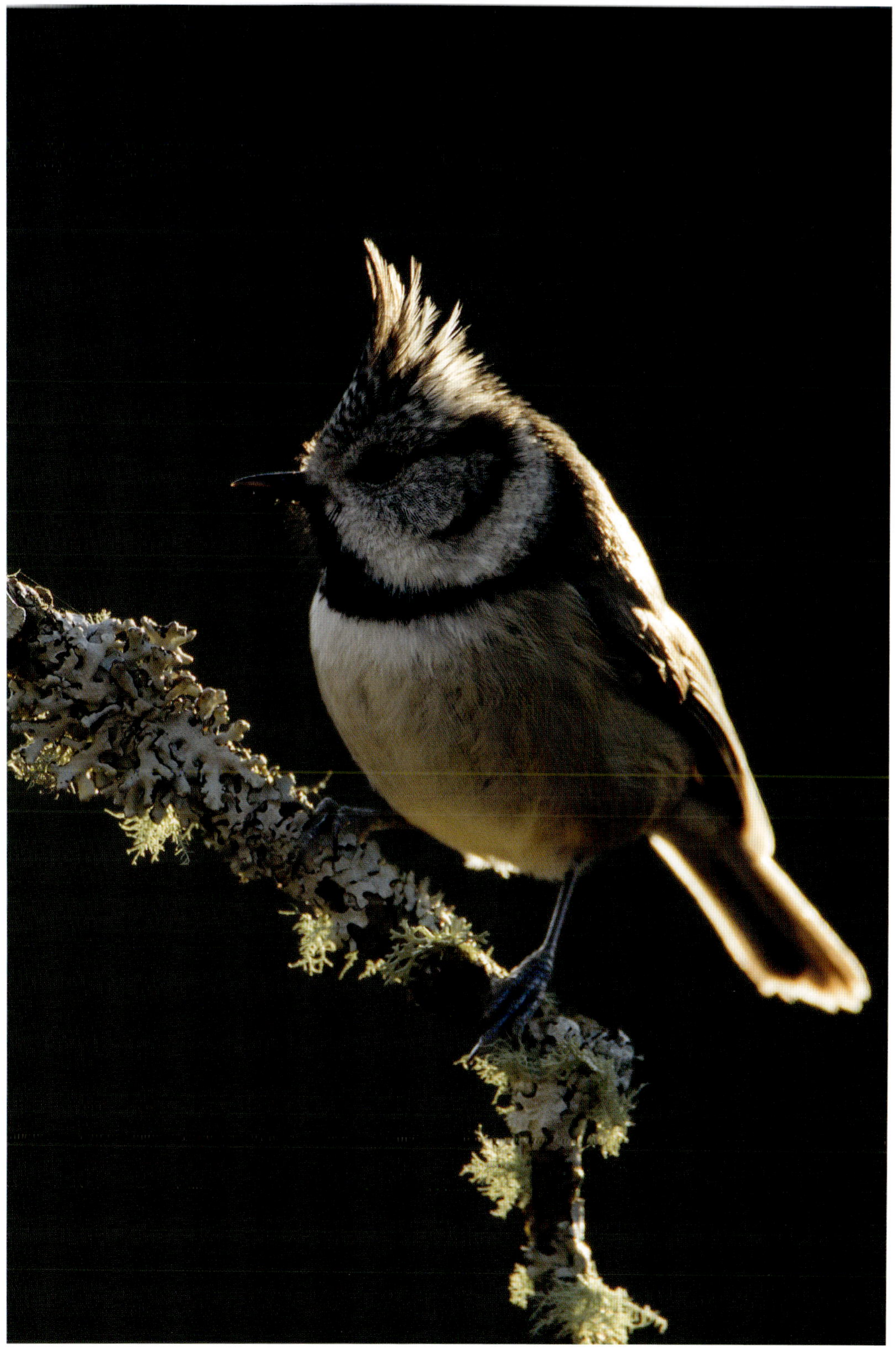

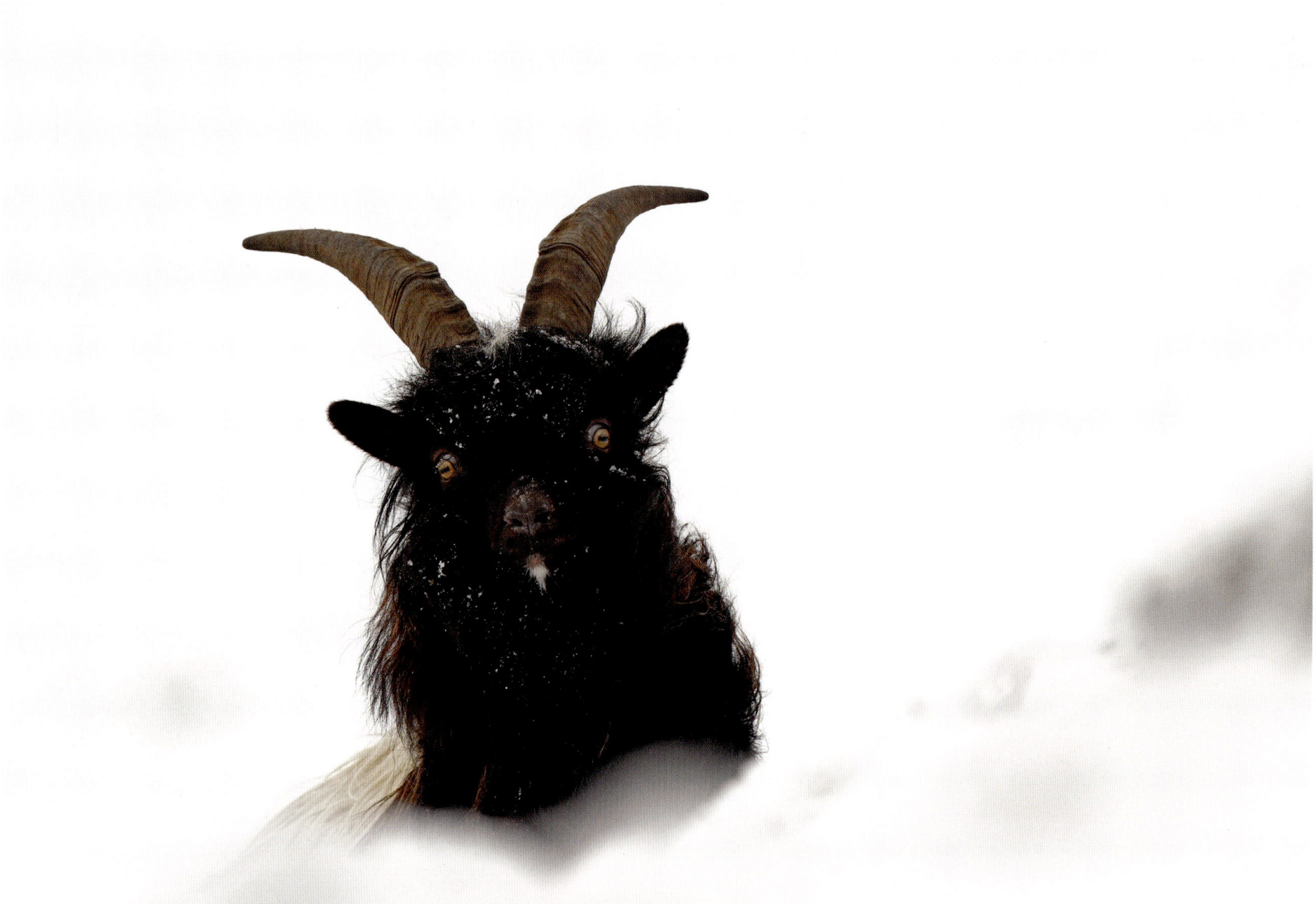

NEIL McINTYRE

Wild Goat in Snow
(Feral Goat, *Capra hircus)*
Highland, Scotland

With the first fall of snow of the winter I headed to the hills in search of mountain hares. I was making my way up the hill towards some hares I could see in the distance when, out of the blue, this wild feral goat appeared before me. We both looked at each other at the same time. The look on its face suggests it was just as surprised as I was.

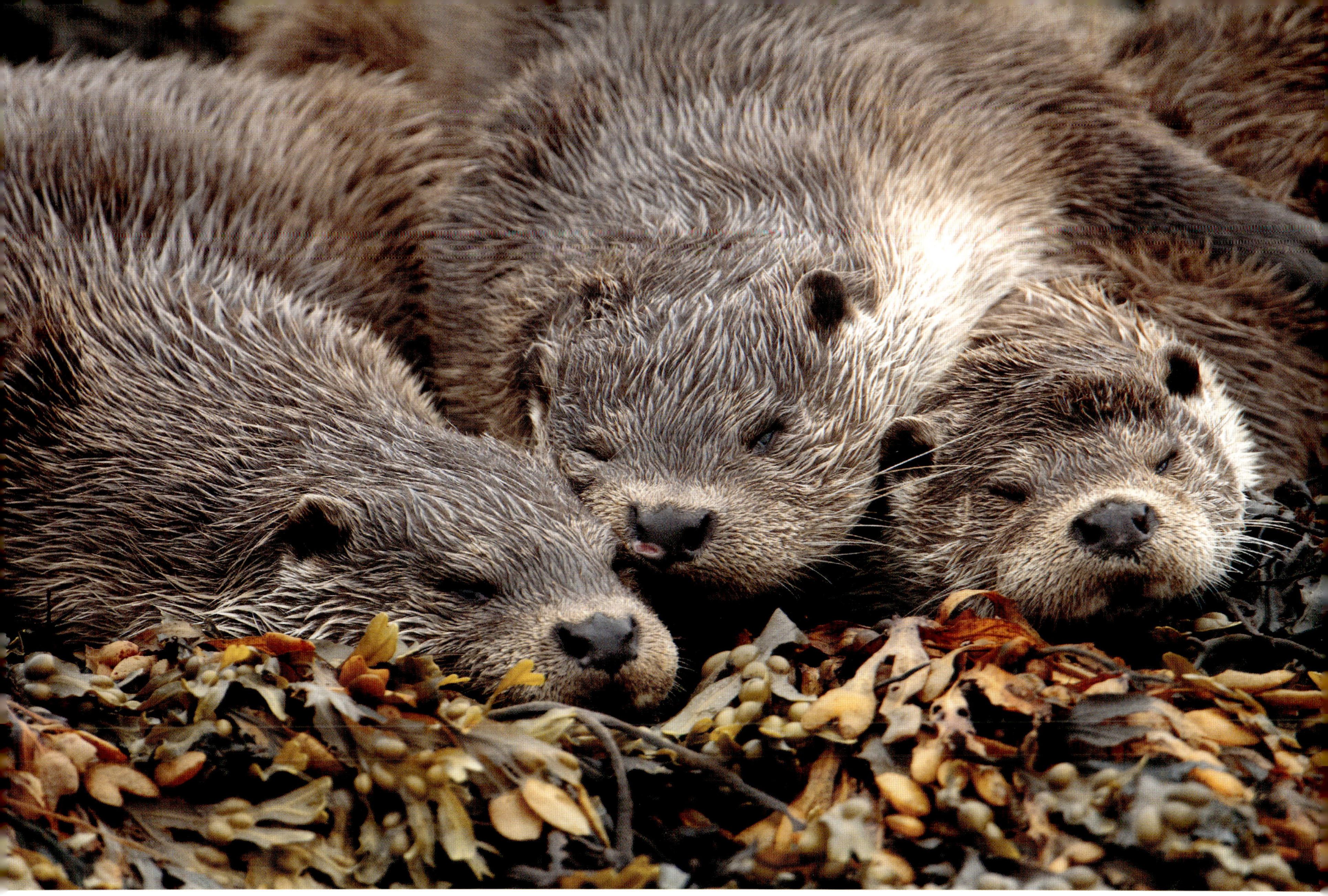

RICHARD SHUCKSMITH

Oh So Sleepy
(European Otter, *Lutra lutra*)
Shetland Isles, Scotland

I had been following this family of otters: a mum and two cubs. They had been fishing all morning and then fell into a heap to get some much-needed sleep. Otter families always want to stay as close as possible to each other and seem to crave contact with one another.

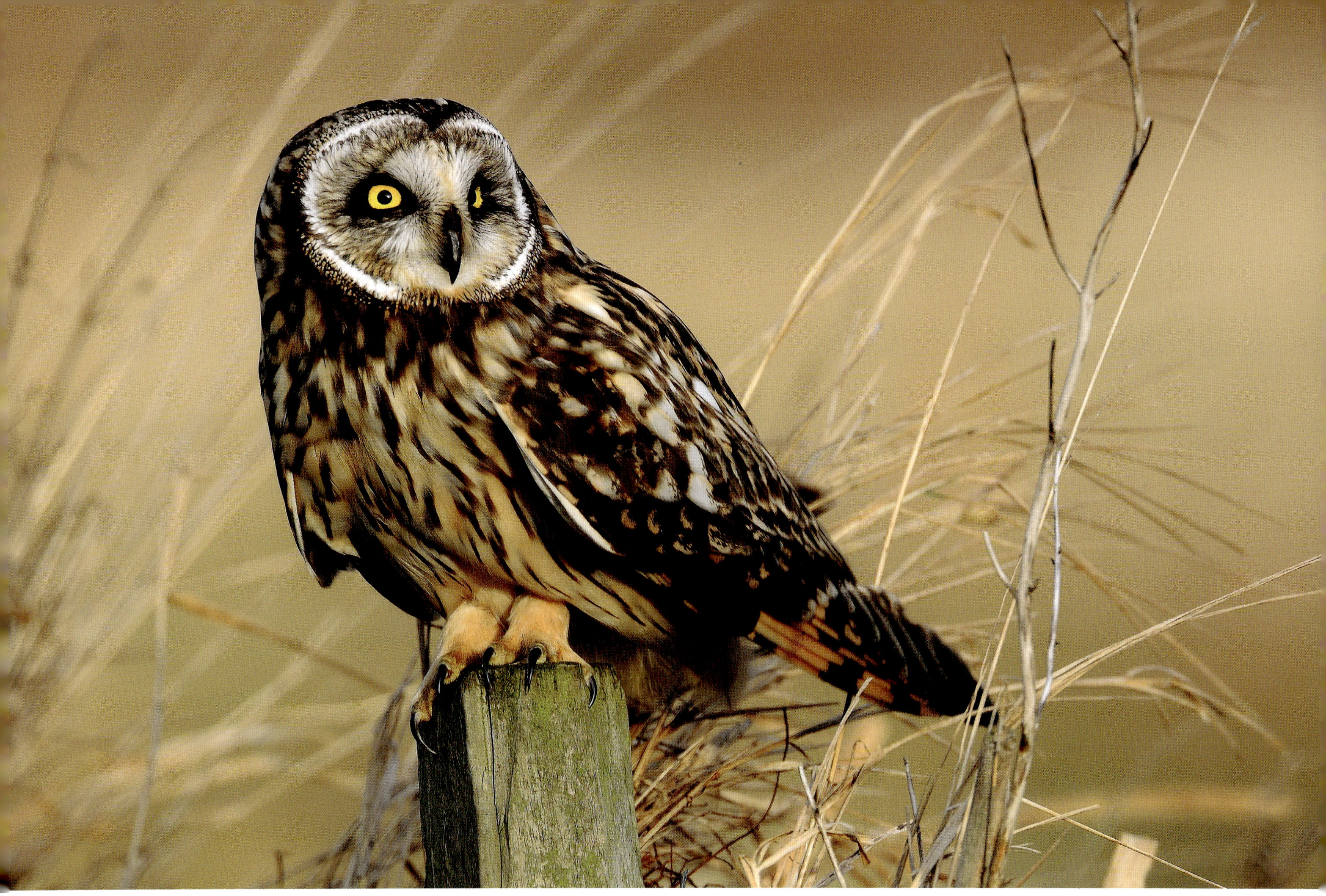

JOHN TAYLOR

Short-eared Owl
(Short-eared Owl, *Asio flammeus)*
Cresswell, Northumberland, England

On a photography trip I encountered this bird hunting on the roadside, flying from post to post.

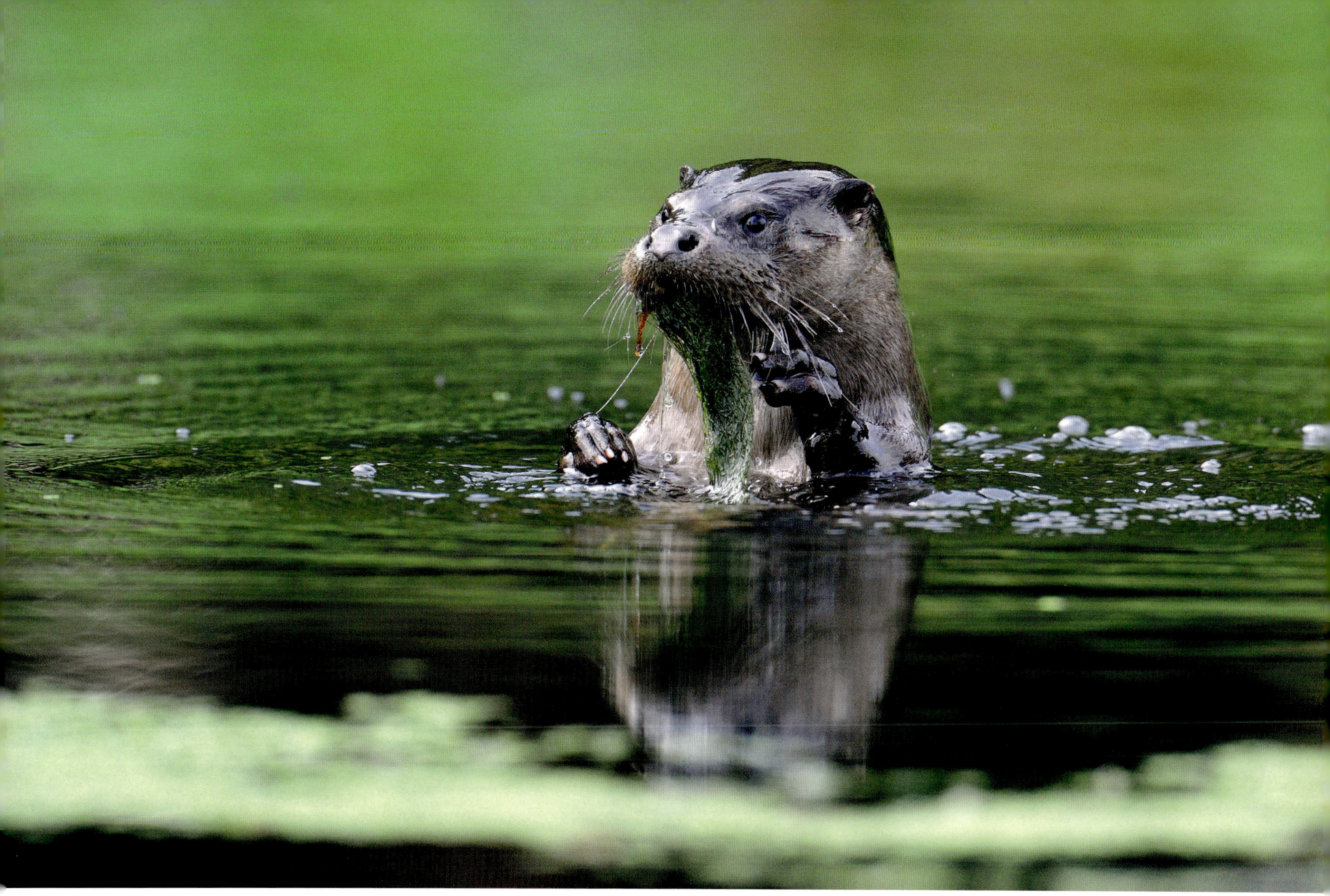

RICHARD PETERS

Otter Portrait
(European Otter, *Lutra lutra*)
Dorset, England

I spent the morning lying flat on the riverbank watching this otter swim about at some distance. Eventually my patience paid off when I finally had a close encounter. The otter dived, then popped up in front of me, and contentedly floated in the water, tucking in to some tasty morsels from the riverbed.

ANIMAL BEHAVIOUR

SPONSORED BY
SPARSHOLT COLLEGE
HAMPSHIRE

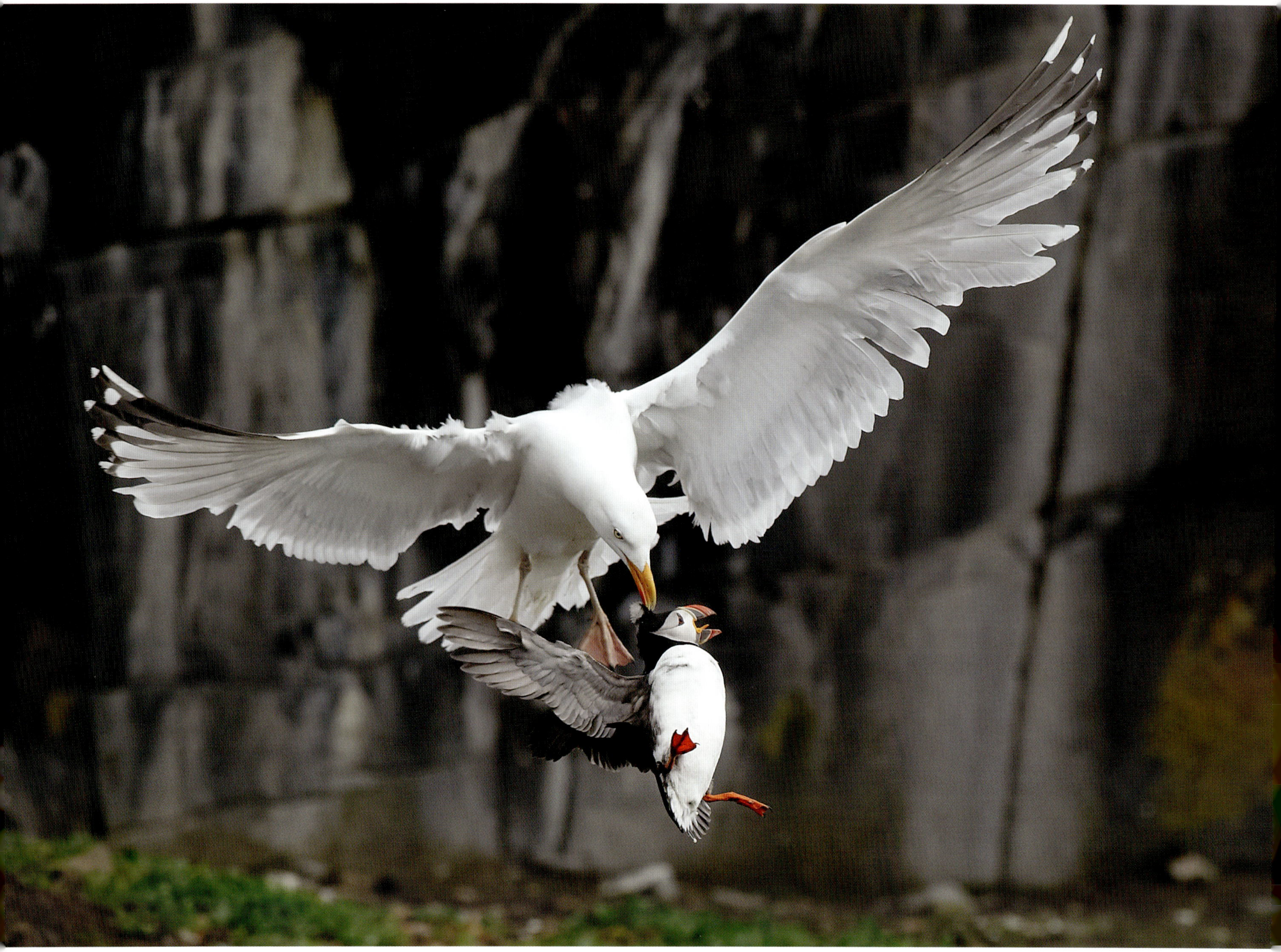

ANIMAL BEHAVIOUR WINNER

AMANDA HAYES <

Herring Gull Plucking Puffin from the Ground
(Herring Gull, *Larus argentatus*;
Atlantic Puffin, *Fratercula arctica)*
Inner Farne, Northumberland, England

Over a week I'd attempted to photograph gulls stealing puffins' sand-eels as they landed. This proved difficult as the puffins often outwitted the gulls. One day a puffin landed and a herring gull grabbed the back of the puffin's neck, lifting it high off the ground. Although sudden, I managed to obtain some shots. Fortunately for the puffin, it wriggled free and escaped. I'm not sure whose heart was beating faster – the puffin's or mine!

ROBERT LIND > HIGHLY COMMENDED

Rudd Catches Red-eyed Damselfly Midair
(Common Rudd, *Scardinius erythropthalmus*;
Red-eyed Damselfly, *Erythromma najas)*
River Stour estuary near Brantham, Suffolk, England

The image is of the common rudd leaping out of the water to catch a red-eyed damselfly that was perched on a *Phragmites* reed stem. The fish did this repeatedly and learned to check a damselfly was there and in the correct position so that it would be able to catch its abdomen. Not every attempt was successful and the event happened every 10–15 minutes so I had to be ready to capture the moment!

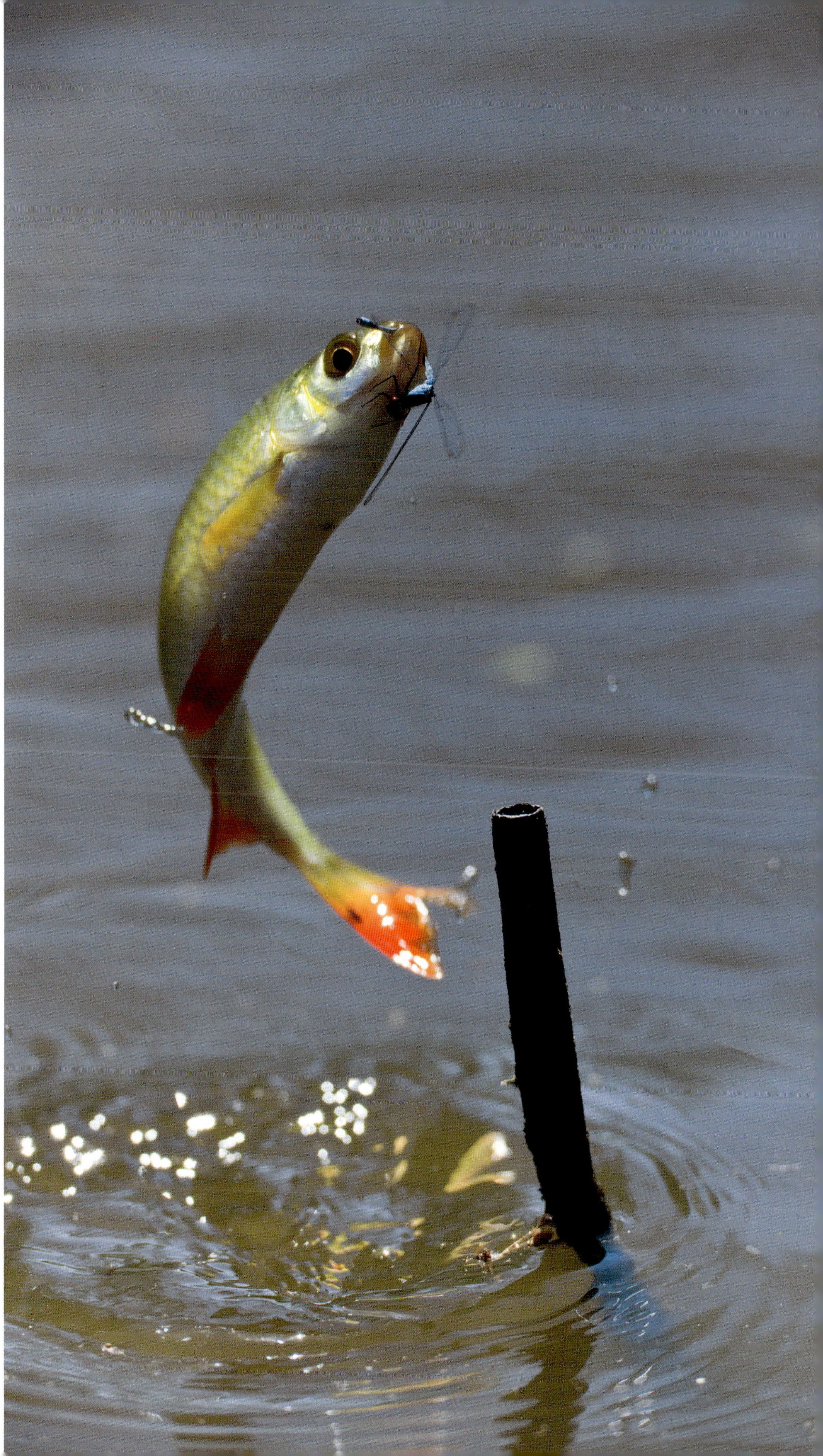

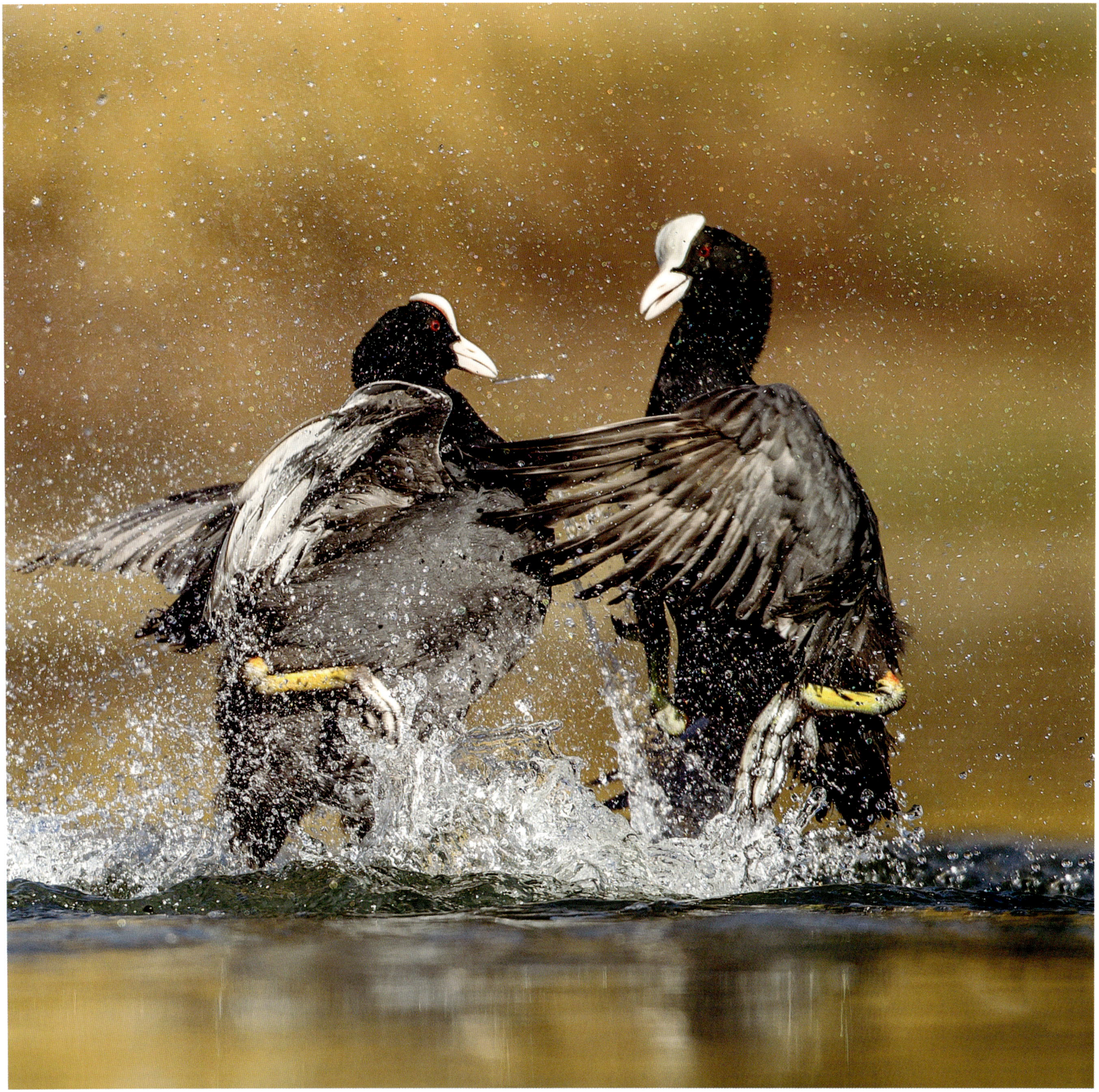

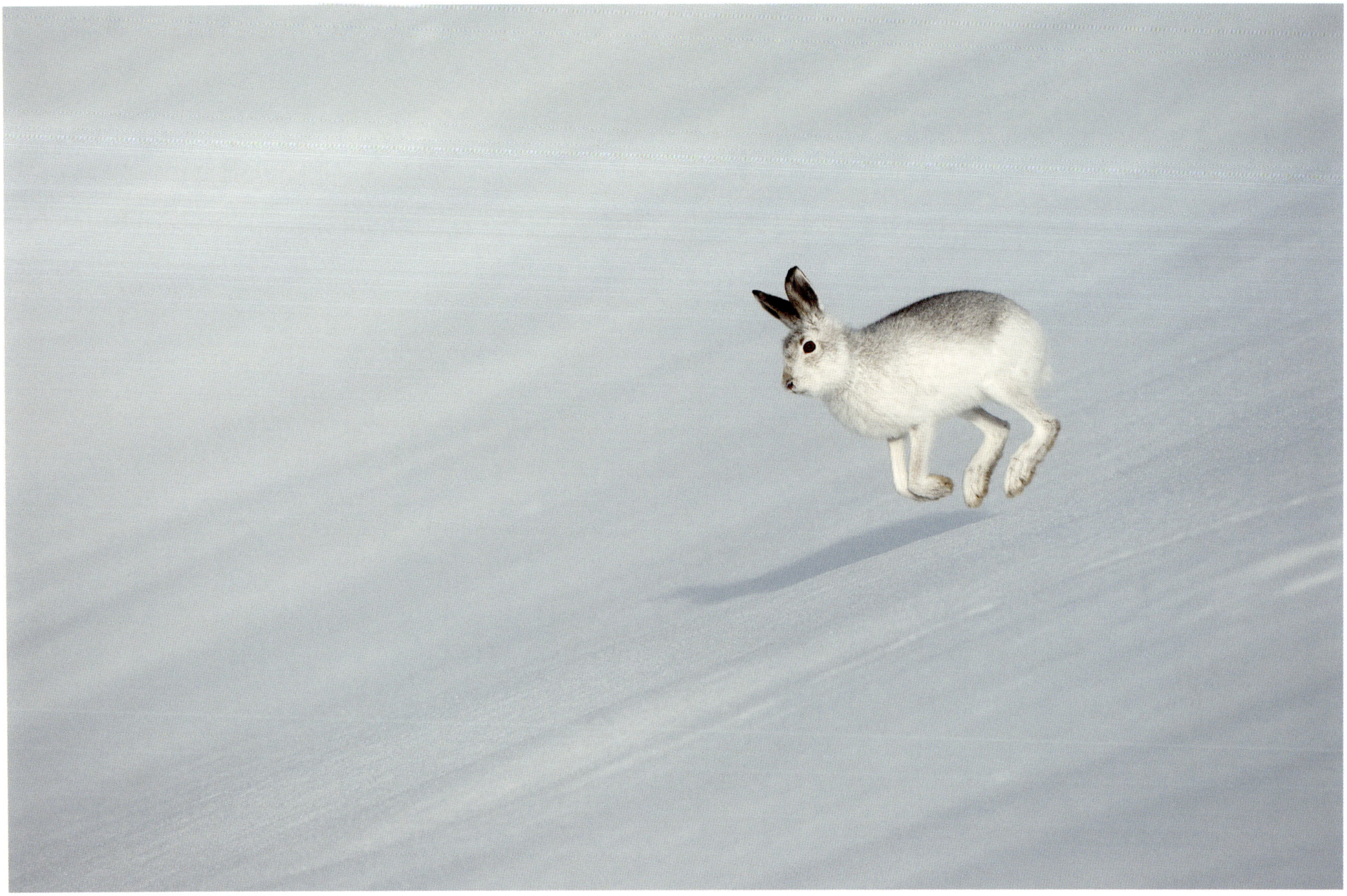

MARK SMITH < HIGHLY COMMENDED

Fighting Coots

(Eurasian Coot, *Fulica atra*)
Local pond, London, England

For the coot, its patch or territory is everything and dictates every part of its behaviour. Outbursts of fury and anger erupt if an intruder dares to stray over poorly defined borders. To me, this is the most defining feature of a coot's behaviour and every fight excites me, whether it's a two-second squabble or a minute-long wrestling match.

MARK HAMBLIN Λ HIGHLY COMMENDED

Mountain Hare Running Across Snow-covered Mountain

(Mountain Hare, *Lepus timidus*)
Monadhliath Mountains, Highland, Scotland

Mountain hare photography is a roller-coaster of emotions. This encounter was no different. Frustration at first, having fought my way for an hour through thigh-high snow only to see several hares disappear from view, but then exhilaration as this individual appeared from over the ridge in front of me and proceeded to run in my direction across virgin snow. And for once I got it sharply in focus!

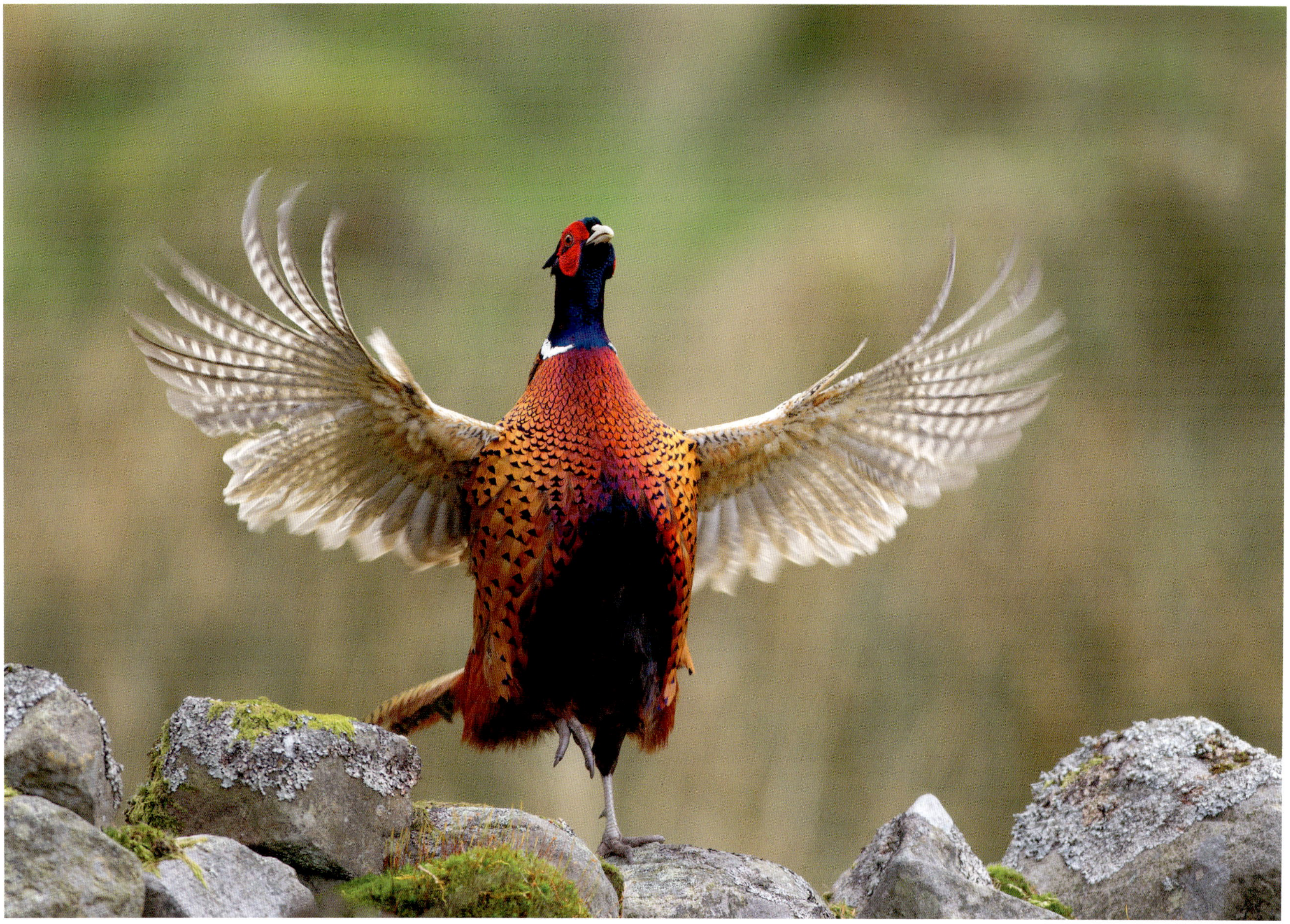

SIMON PHILLPOTTS HIGHLY COMMENDED

The Guardian

(Pheasant, *Phasianus colchicus)*

Yorkshire Dales National Park, North Yorkshire, England

I spent much of the spring in the company of this handsome and dominant pheasant. Over time he became accustomed to my presence and would let me work close to him. I also learned to anticipate his display calls. The dry-stone wall overlooked his territory and allowed him a good vantage point to spot both hens and any rival males (who would generally beat a hasty retreat for fear of a pasting!)

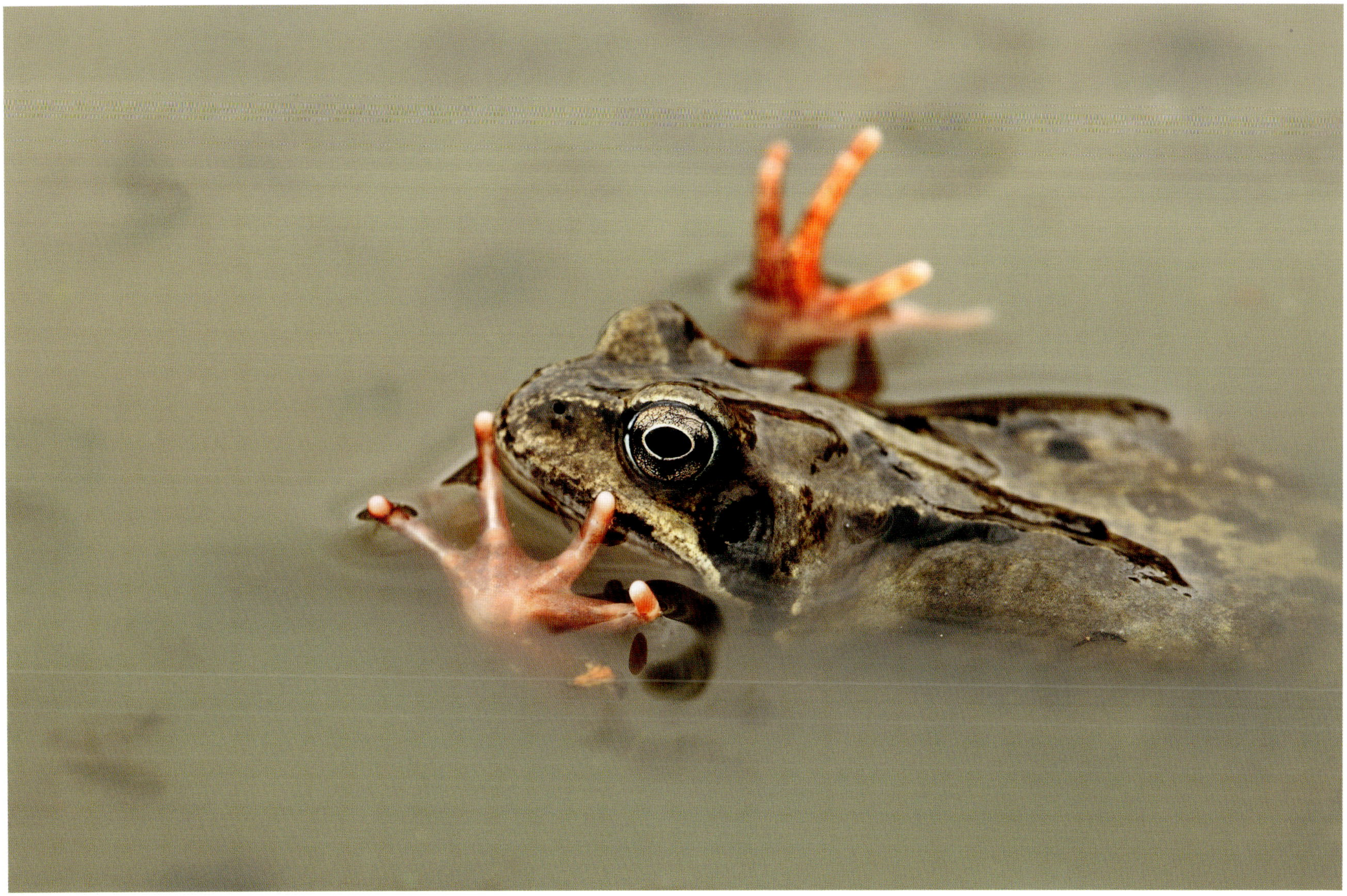

JAMES YAXLEY HIGHLY COMMENDED

Hands Up
(Common Frog, *Rana temporaria*)
Vinegar Pond, Mousehold Heath, Norwich, Norfolk, England

In this image of frogs mating, the male is the main subject. All that can be seen of the female are her front feet and toes, which look like four-fingered hands held up in an act of surrender.

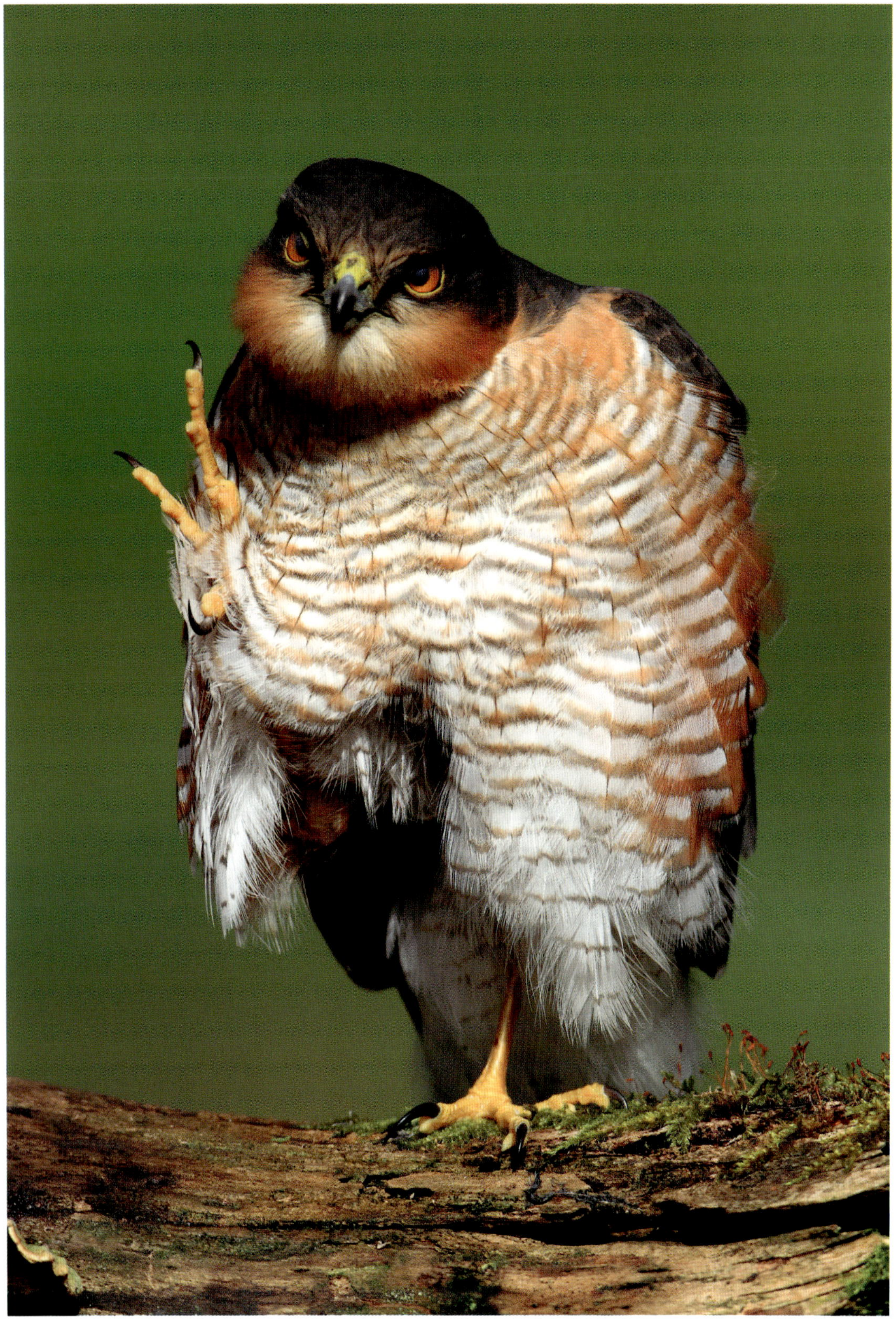

TOM LANGLANDS < HIGHLY COMMENDED

Sparrowhawk Preening

(Sparrowhawk, *Accipiter nisus)*
Near Kirkcudbright, Dumfries and Galloway, Scotland

This photograph was taken from a hide in a wood near Kirkcudbright. It had been reported that a sparrowhawk was appearing regularly. I was photographing small birds when it flew in, landing on a dead branch close to the hide. There it preened itself for several minutes before flying off. The light had been poor, but as the sparrowhawk arrived it improved enough to allow me to grab these shots.

ROBERT HOWARD > HIGHLY COMMENDED

Corn Bunting Singing at Dawn

(Corn Bunting, *Miliaria calandra)*
Great Waldingfield, Suffolk, England

Corn buntings are in serious decline across the UK, but I am fortunate enough to have a good population close to my home. I set out that morning after some dawn chorus shots. I had just got into a position where I know the buntings perch. Just as the sun rose, one came to the sapling and began to sing, the cold air showing off his breath and beautifully backlit.

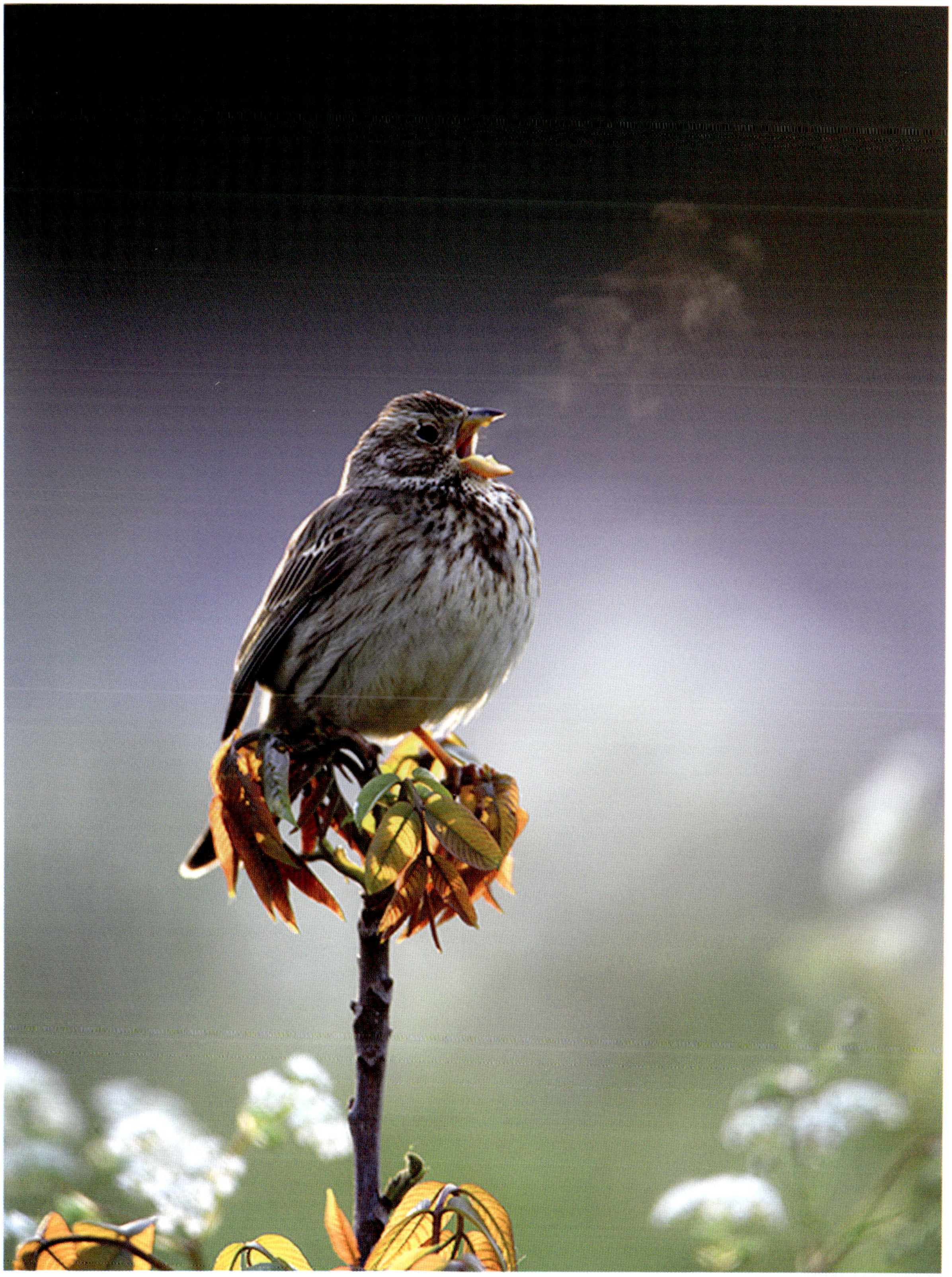

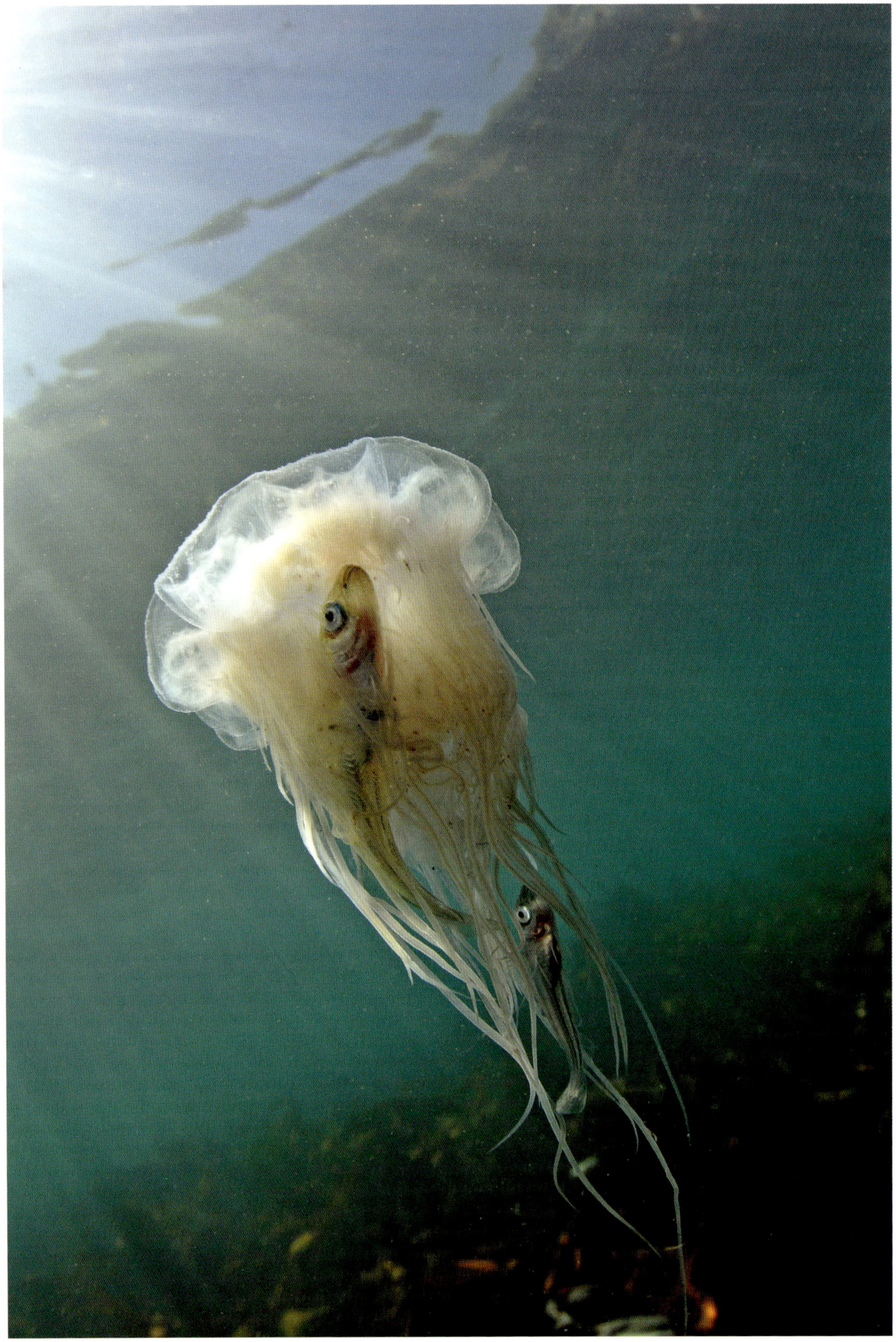

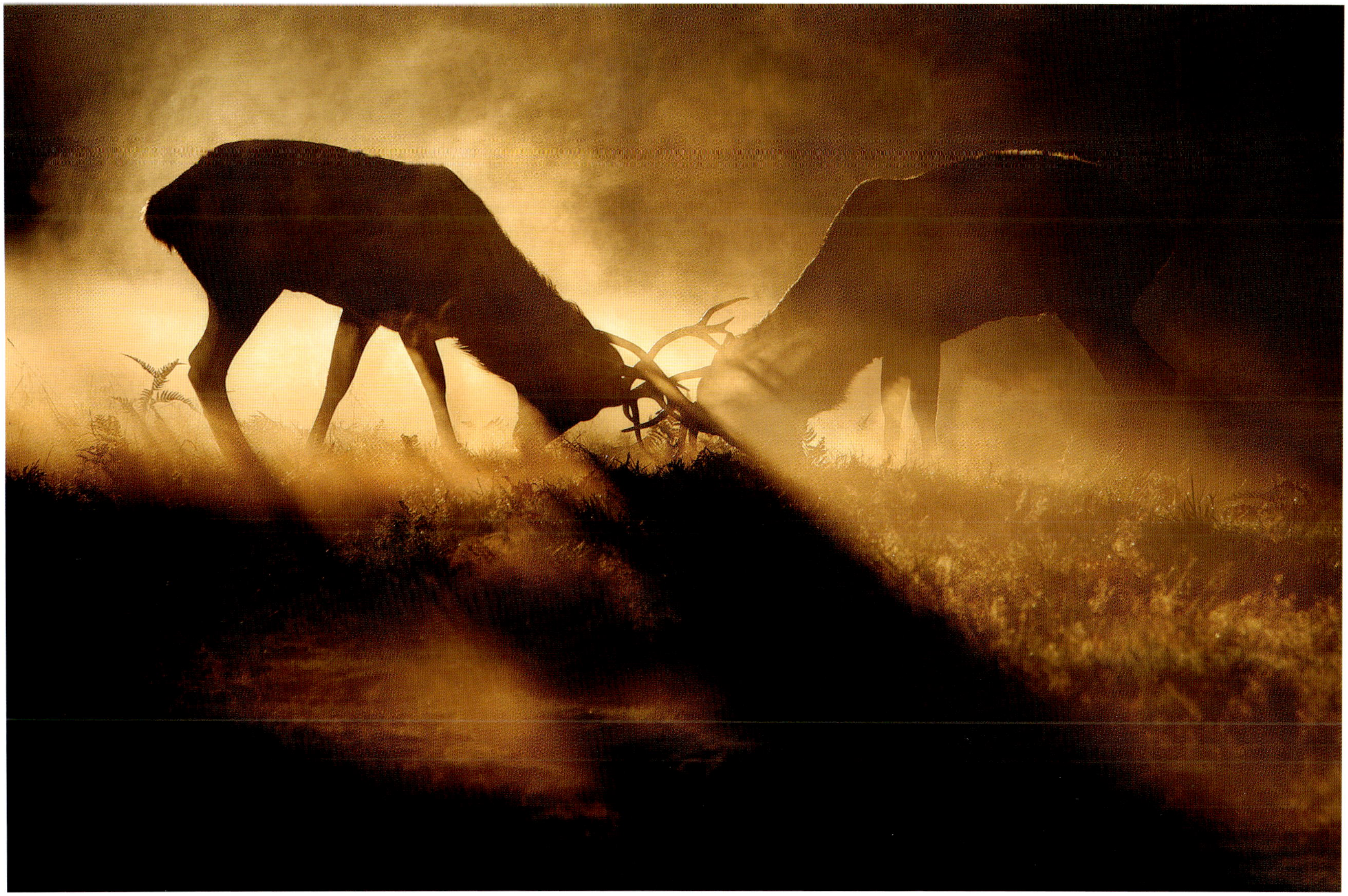

ALEX TATTERSALL < HIGHLY COMMENDED

Symbiosis

(Bluefire Jellyfish, *Cyanea lamarckii*)

Lundy Island, England

Small fishes seeking refuge in the stinging tentacles of a jellyfish. Taken near the surface on scuba.

RICHARD WATERS ∧ HIGHLY COMMENDED

Deer Rutting at Dawn

(Red Deer, *Cervus elaphus*)

Richmond Park National Nature Reserve, Surrey, England

After arriving at Richmond Park before dawn I spent two hours photographing deer, with moderate success. In one particular location though, I saw the potential of a backlit shot, with shafts of sunlight cutting through the early mist. I moved downwind and waited for a small group of deer to move into the frame, repositioning myself slightly to isolate two rutting deer for a few precious moments.

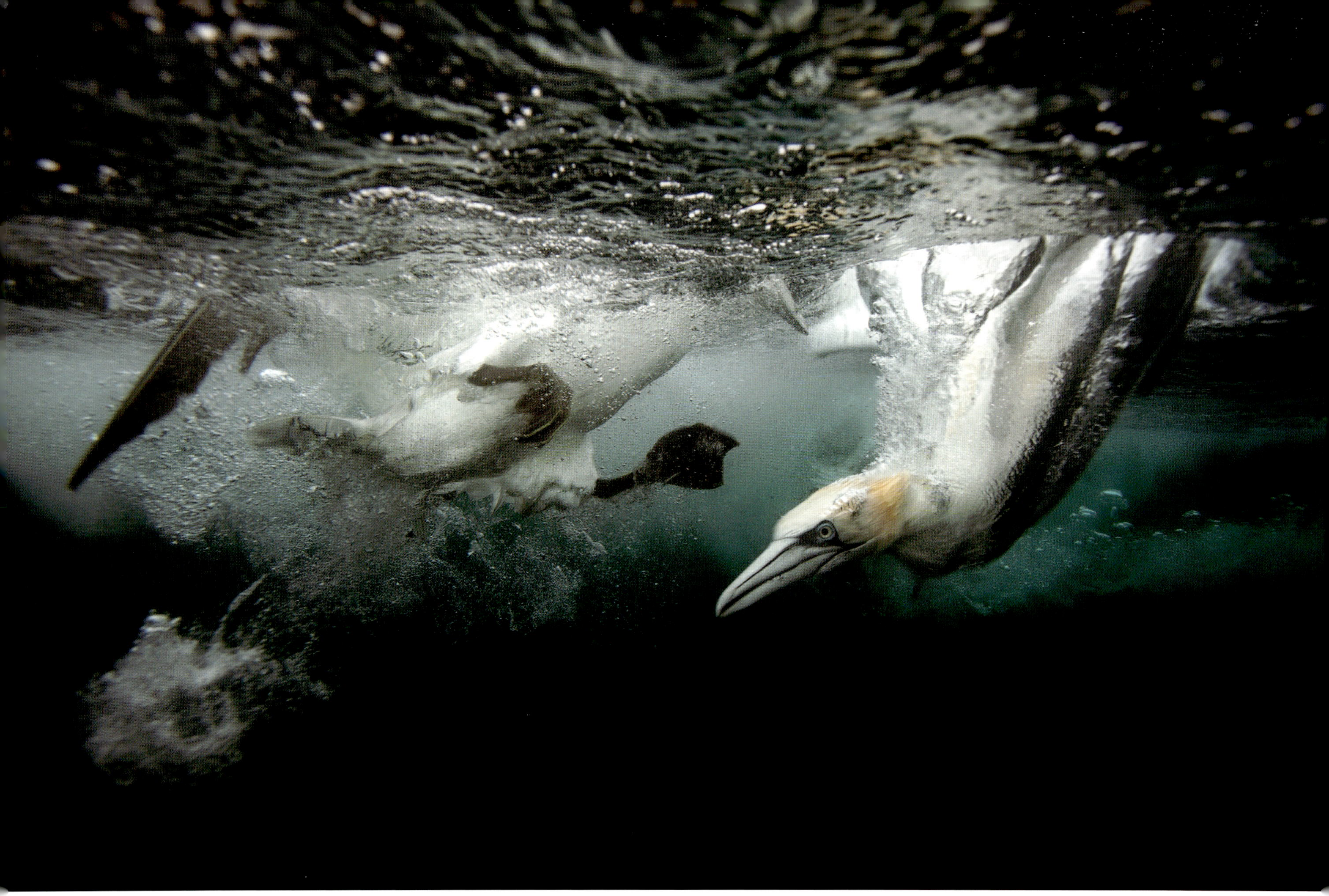

DR MATT DOGGETT HIGHLY COMMENDED

Gannets Half In, Half Out
(Northern Gannet, *Morus bassanus)*
Scotland

Diving gannets are a spectacle to behold and to see them from beneath the waves was a huge privilege. Using dead mackerel as bait I was able to get them diving all around me; they came closer as their confidence increased. It was fast action and often hard to follow, but looking up you could see the birds commencing a dive in the air. Gannets can hit the water at 50–60mph so I was pleased to capture this one arriving just as another left.

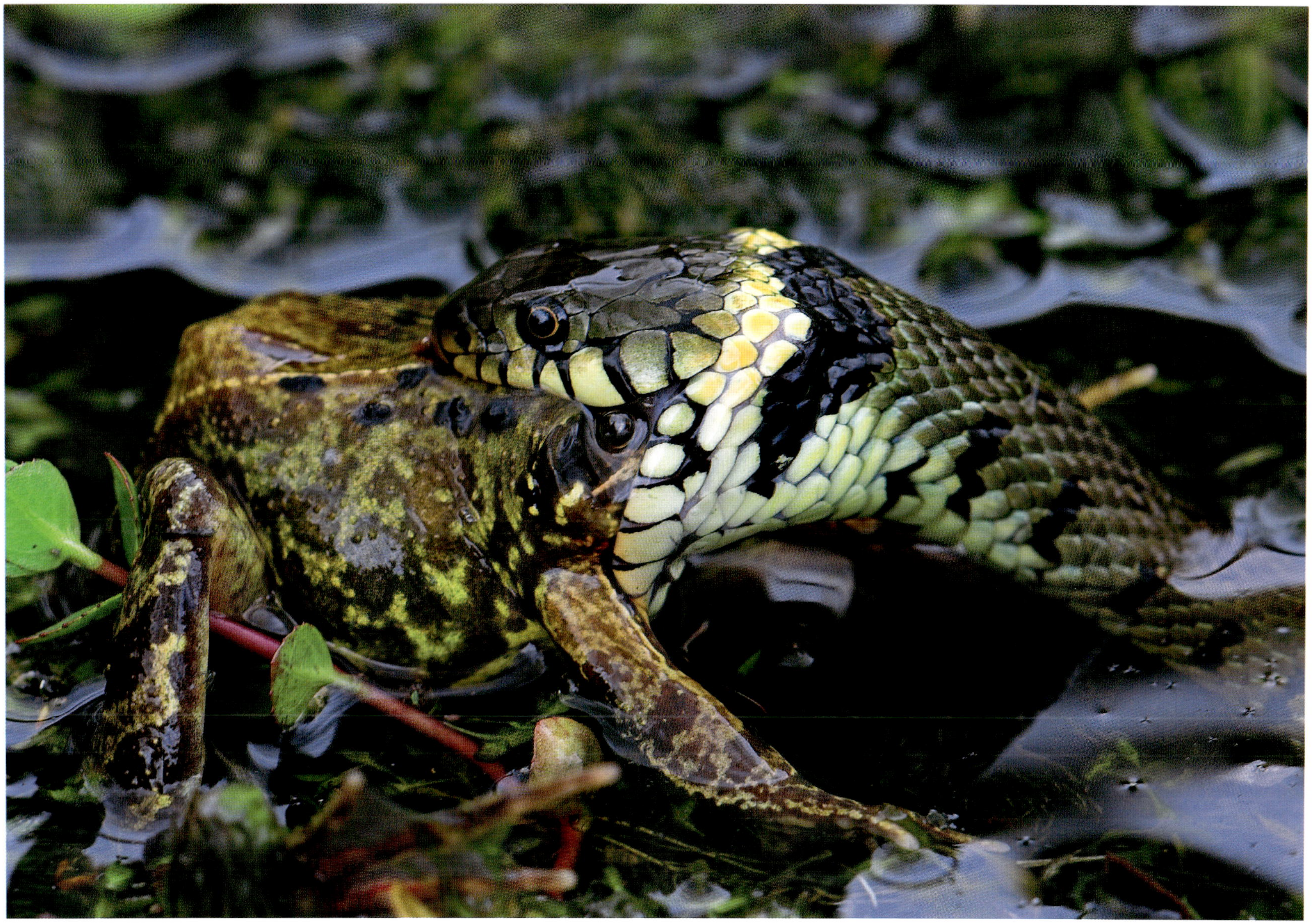

GARY COX

Grass Snake Eating Frog
(Grass Snake, *Natrix natrix*; Common Frog, *Rana temporaria*)
Cirencester, Gloucestershire, England

I walked past my garden pond and saw a grass snake catch a frog. I quickly rushed to fetch my camera and photographed the battle, which lasted for several minutes. Eventually the frog, who had puffed itself up in size, proved just too big for the grass snake and was released.

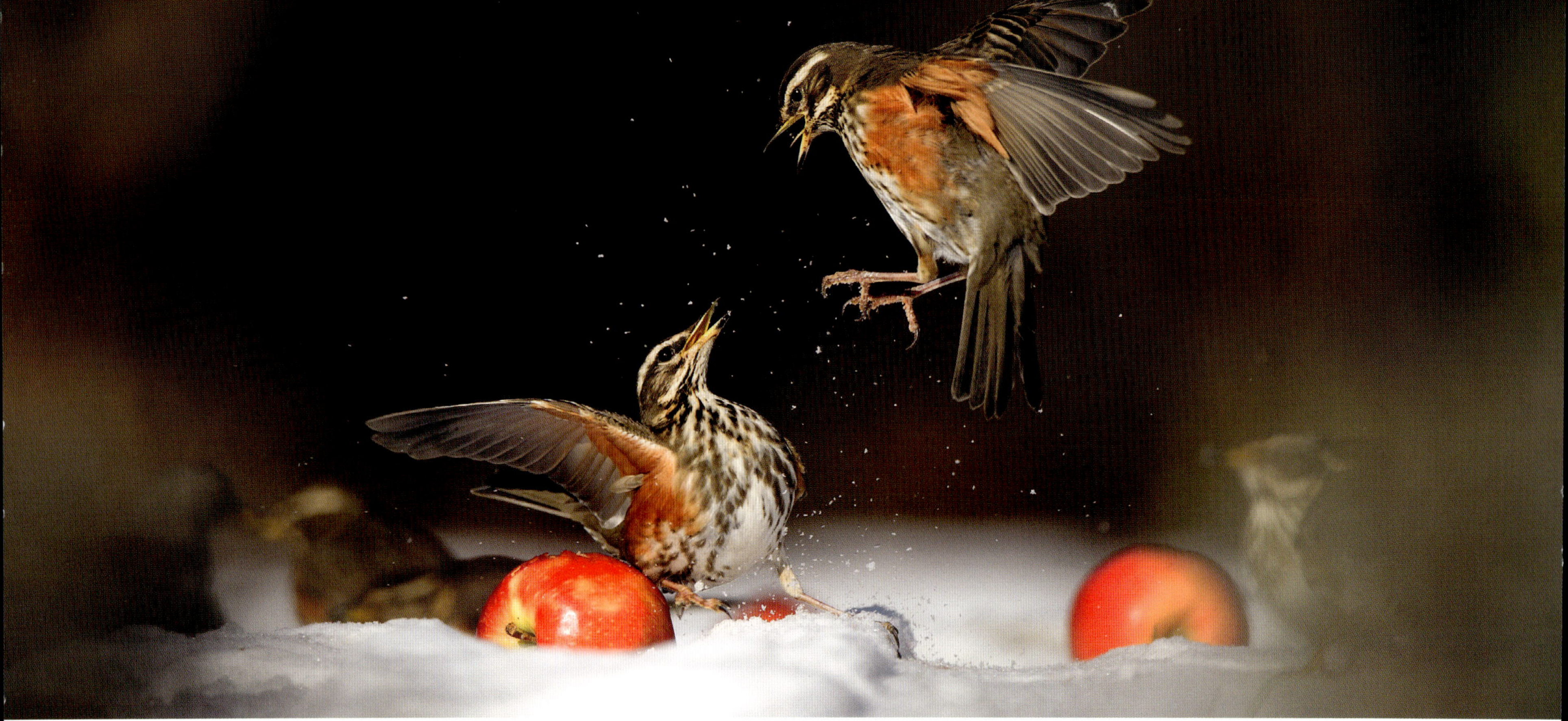

ANDREW PARKINSON ∧

A Pair of Redwings Come to Blows in a Dispute over some Windfall Apples
(Eurasian Redwing, *Turdus iliacus)*
Derbyshire, England

I had sat in my hide from dawn until 3.30pm before I noticed any territorial aggression between the redwings that had gathered. The last remnants of snow lay on the lawn and it was bitingly cold but I felt confident that there were some good images to be made. As far as I know these are the only images in existence that show this explosive behaviour.

NEIL McINTYRE >

Red Squirrel Peering Around Pine Tree
(Red Squirrel, *Sciurus vulgaris)*
Aviemore, Highland, Scotland

Red squirrels have endless charm and appeal and are always great fun to photograph. This image was taken in an area of pine forest near my home at a feeding station I have set up there. I tend to go looking for a specific picture in mind and on this occasion I wanted to show the inquisitive character of the squirrel as it peered round the pine tree.

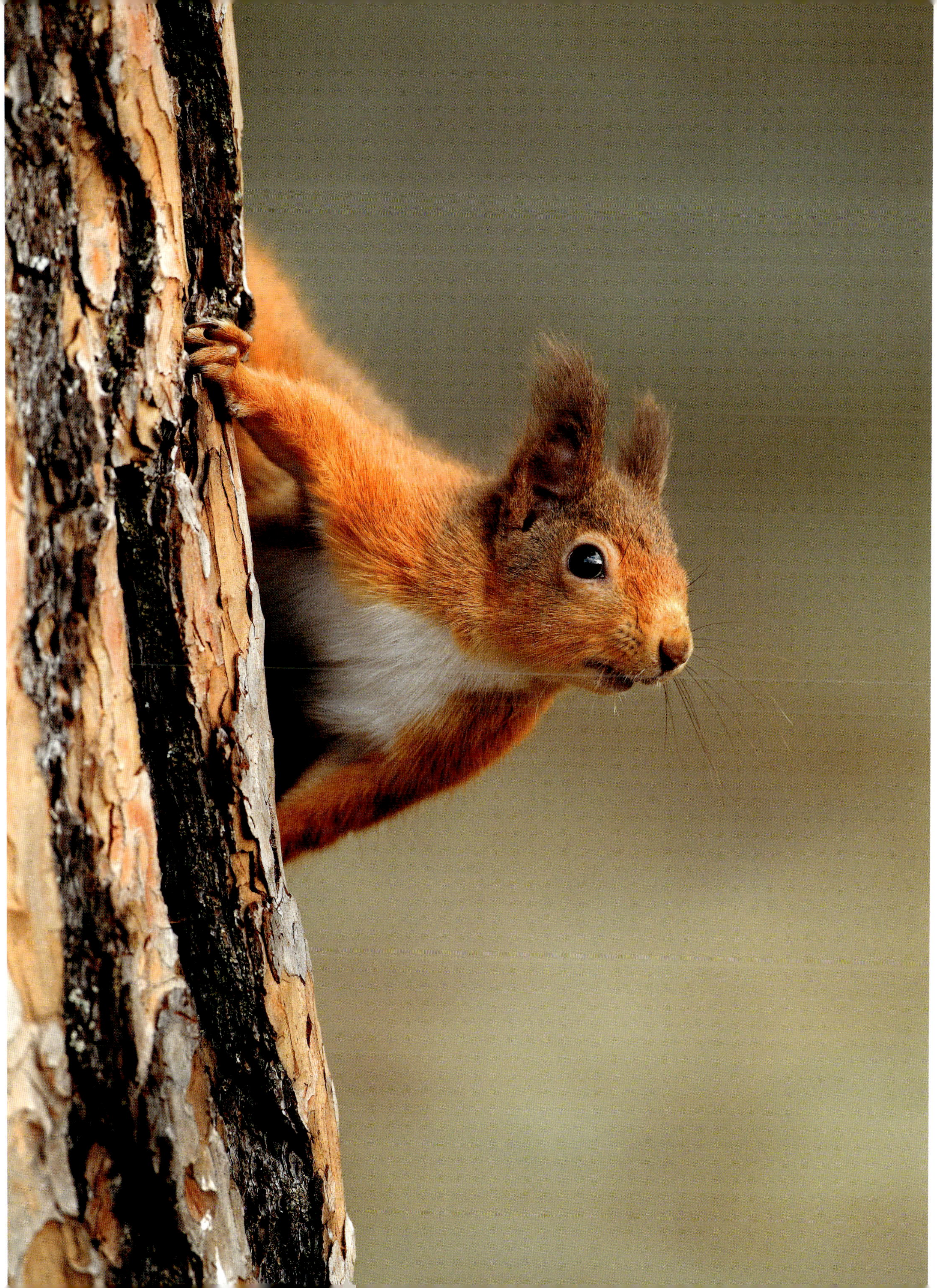

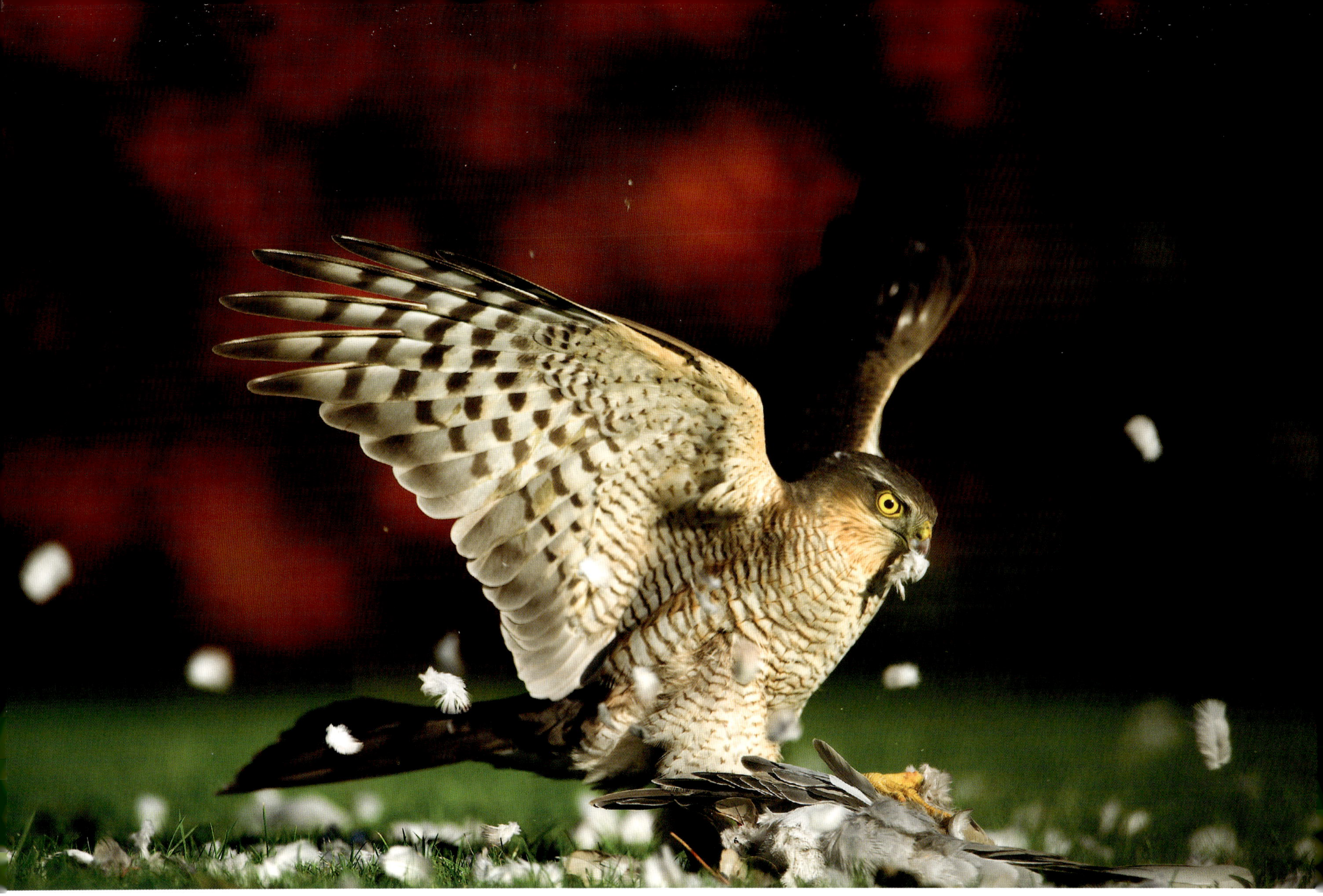

ANDREW PARKINSON

Sparrowhawk Subduing a Collared Dove that it Brought Down in an Urban Garden

(Sparrowhawk, *Accipiter nisus*)
Crich, Derbyshire, England

I had watched a sparrowhawk bring down a collared dove in my neighbour's garden and, knowing how trusting sparrowhawks can be when preoccupied with feeding, I was able to stalk it to within just a few metres. Every now and again it would pause and suddenly open its wings as it repositioned itself on the dead or dying bird.

URBAN WILDLIFE

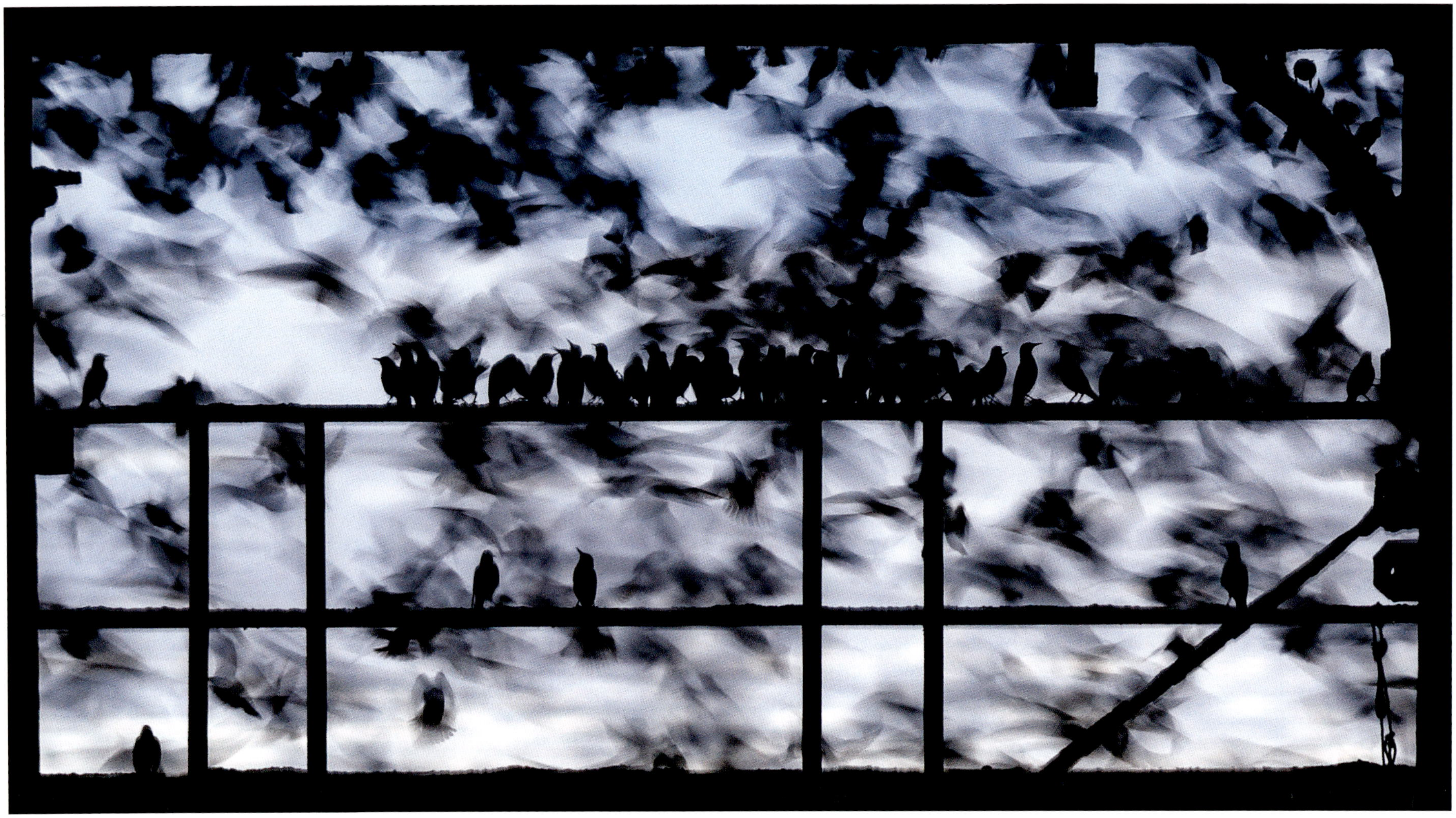

URBAN WILDLIFE WINNER

PHIL JONES

Starlings Watching the Starlings
(Common Starling, *Sturnus vulgarus)*
Aberystwyth, Ceredigion, Wales

Starlings roost under the pier every autumn and winter and put on a great aerial display most nights. I wanted something a little different so concentrated on the perched birds looking out to sea with the movement of the flock beyond them. Given the poor weather it took several visits to get the photograph I had in mind.

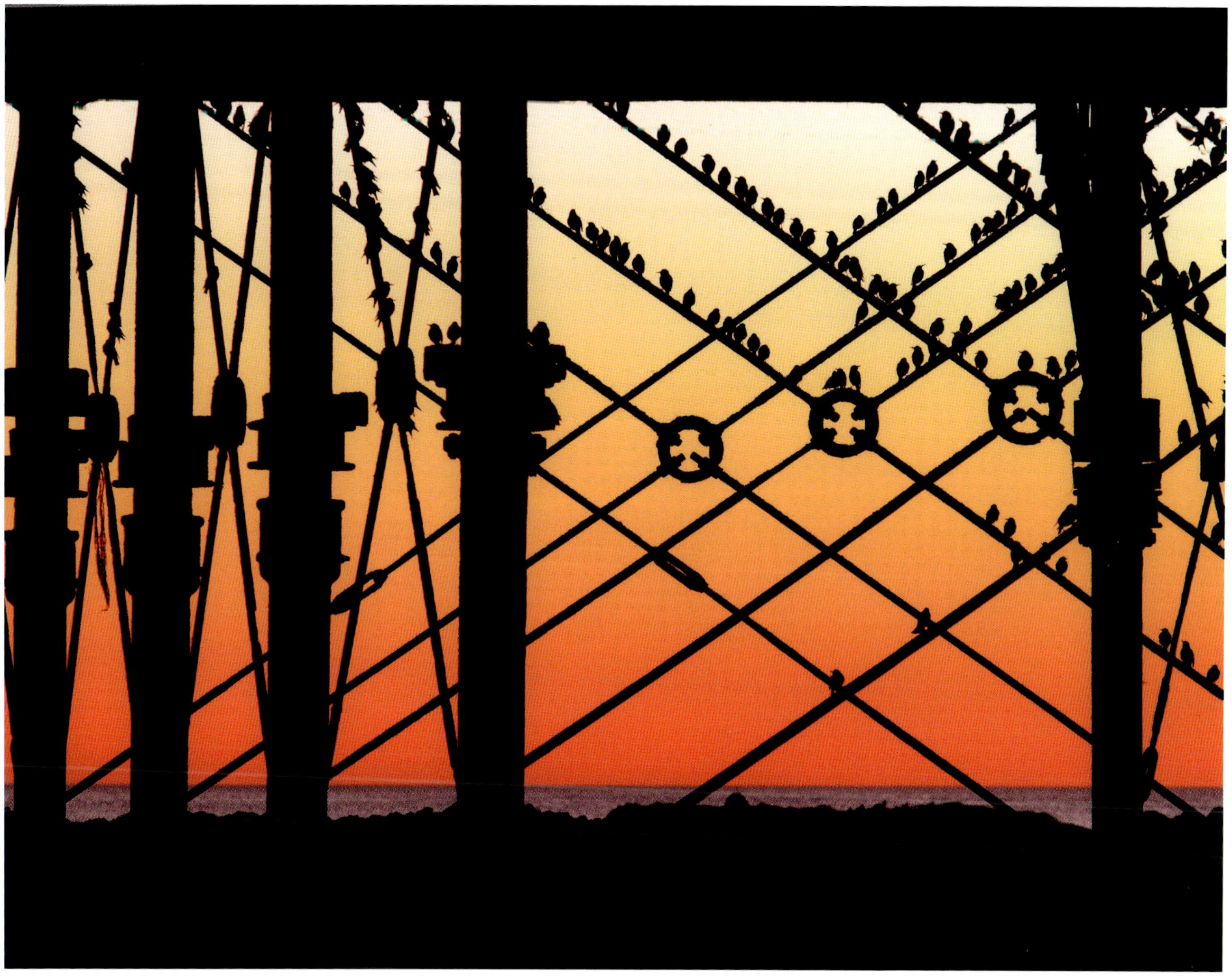

JEREMY MOORE HIGHLY COMMENDED

Starlings under Aberystwyth Pier
(Common Starling, *Sturnus vulgarus)*
Aberystwyth, Ceredigion, Wales

Starlings roost under the pier at Aberystwyth from October until March. I photographed them as they flew in, and then again as they settled on the metal pier supports after sunset.

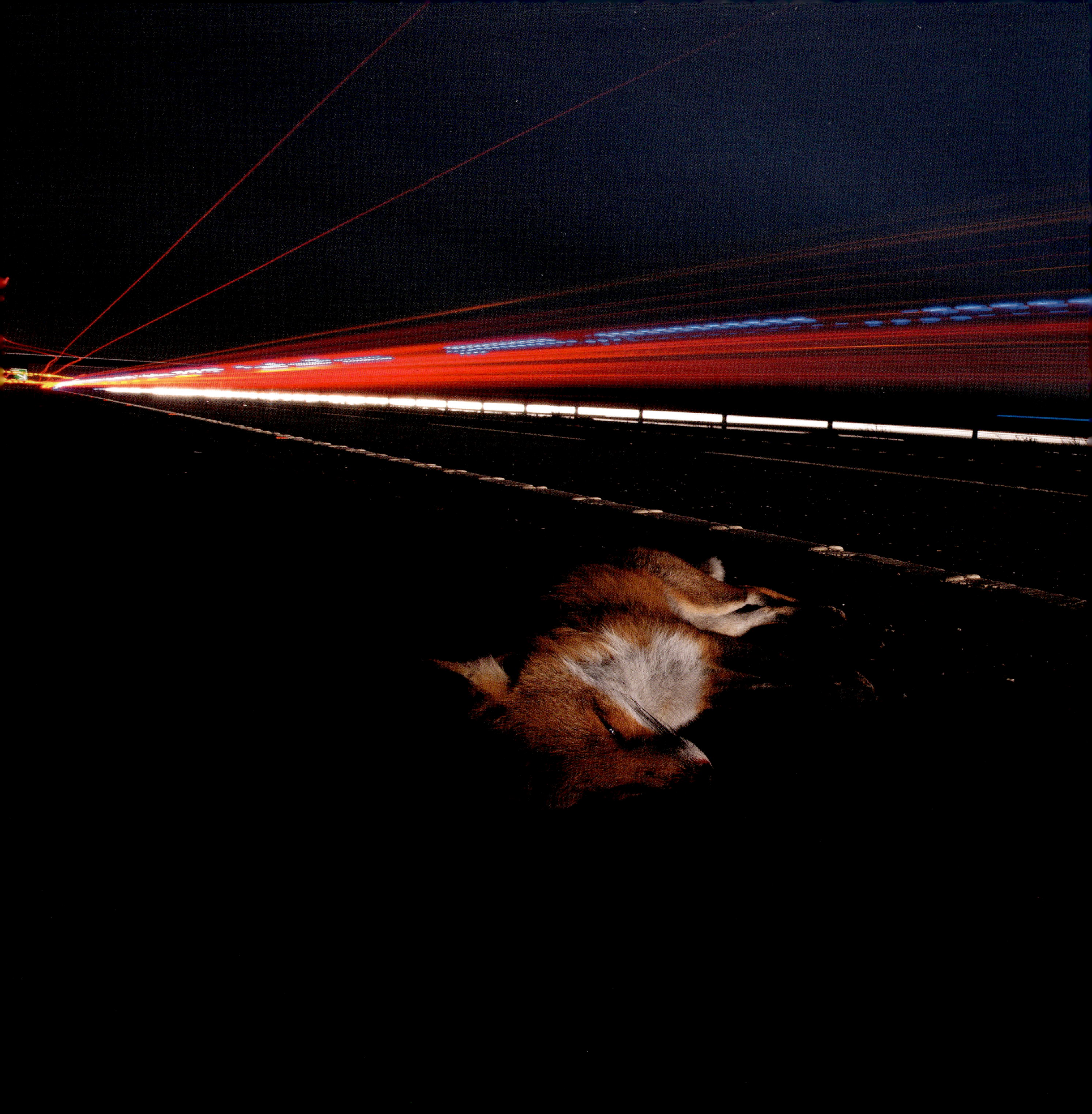

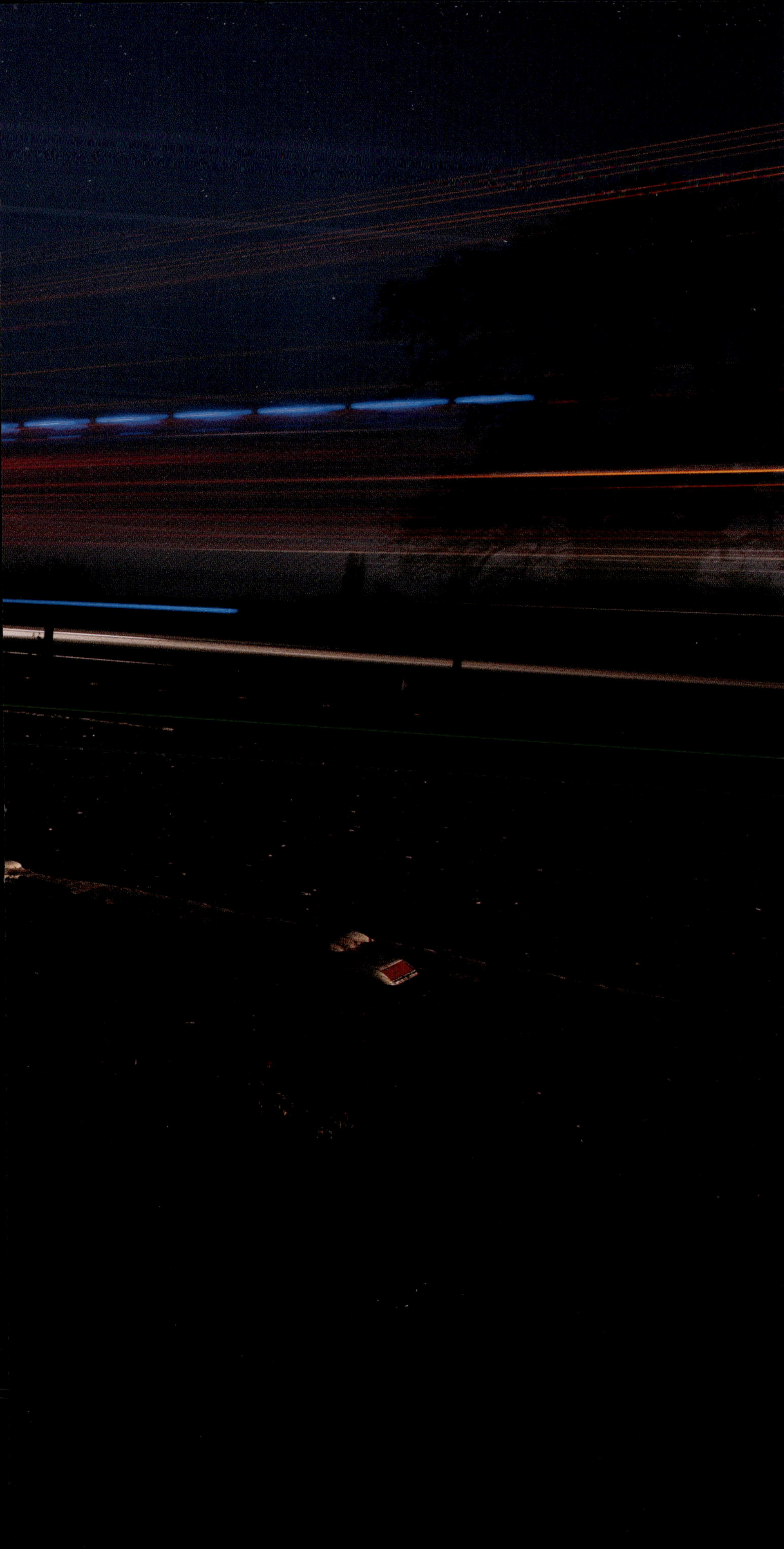

JAMIE HALL HIGHLY COMMENDED

The World Just Passes By Without Any Care...
(Red Fox, *Vulpes vulpes)*
Norwich, Norfolk, England

Whilst driving home from a night out trying to photograph Norwich's urban fox population I noticed a dead vixen lying beside the bypass. I pulled over and just sat next to her for a few minutes and watched how the world just passed by without a care. I decided to try and portray this in the image as I feel this is a side of wildlife photography that is so often neglected and just passed by, even by other photographers...

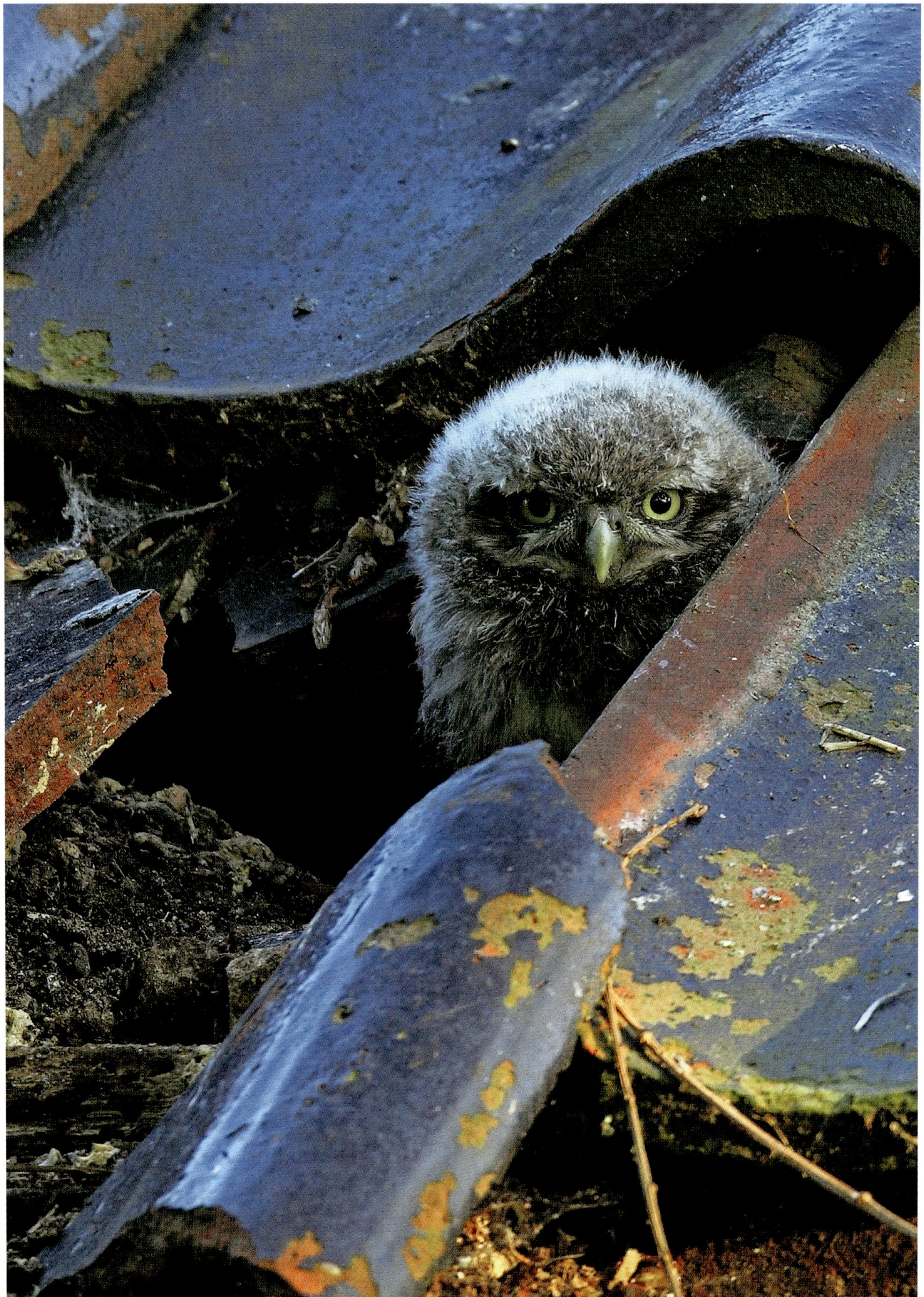

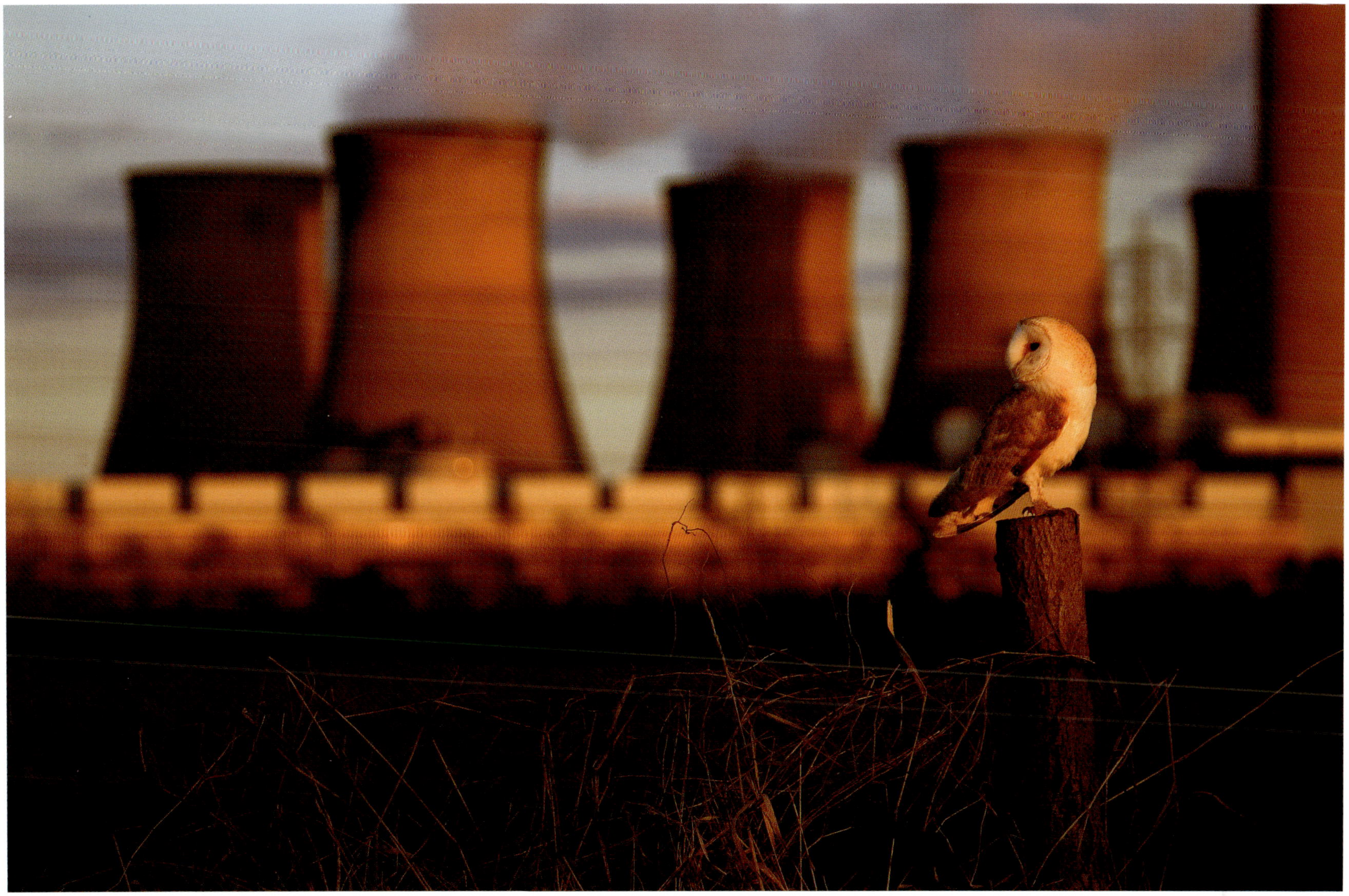

DAVID HERMON < HIGHLY COMMENDED

Little Owl Chick in Stable Roof
(Little Owl, *Athene noctua)*
Bawdsey, Suffolk, England

After weeks of trying to track down where the adult owl was going with all the prey it was catching, a young chick gave the location of the nest away by peering out from beneath the tiles. A further few days of sitting silently in a nearby bush for several hours in the evening rewarded me with this shot of a young owl looking out from beneath the tiles.

LEWIS PHILLIPS Λ HIGHLY COMMENDED

Barn Owl in Suburbia
(Barn Owl, *Tyto alba)*
Rampton, Nottingham, England

I was alongside the River Trent, which is a good feeding area for barn owls during the spring as it is surrounded by flood plains. These birds follow the same areas most days for hunting and Rampton power station is situated just behind this area of the Trent. This bird was on this perch when something spooked it and it quickly turned its head just in time for me to compose and capture.

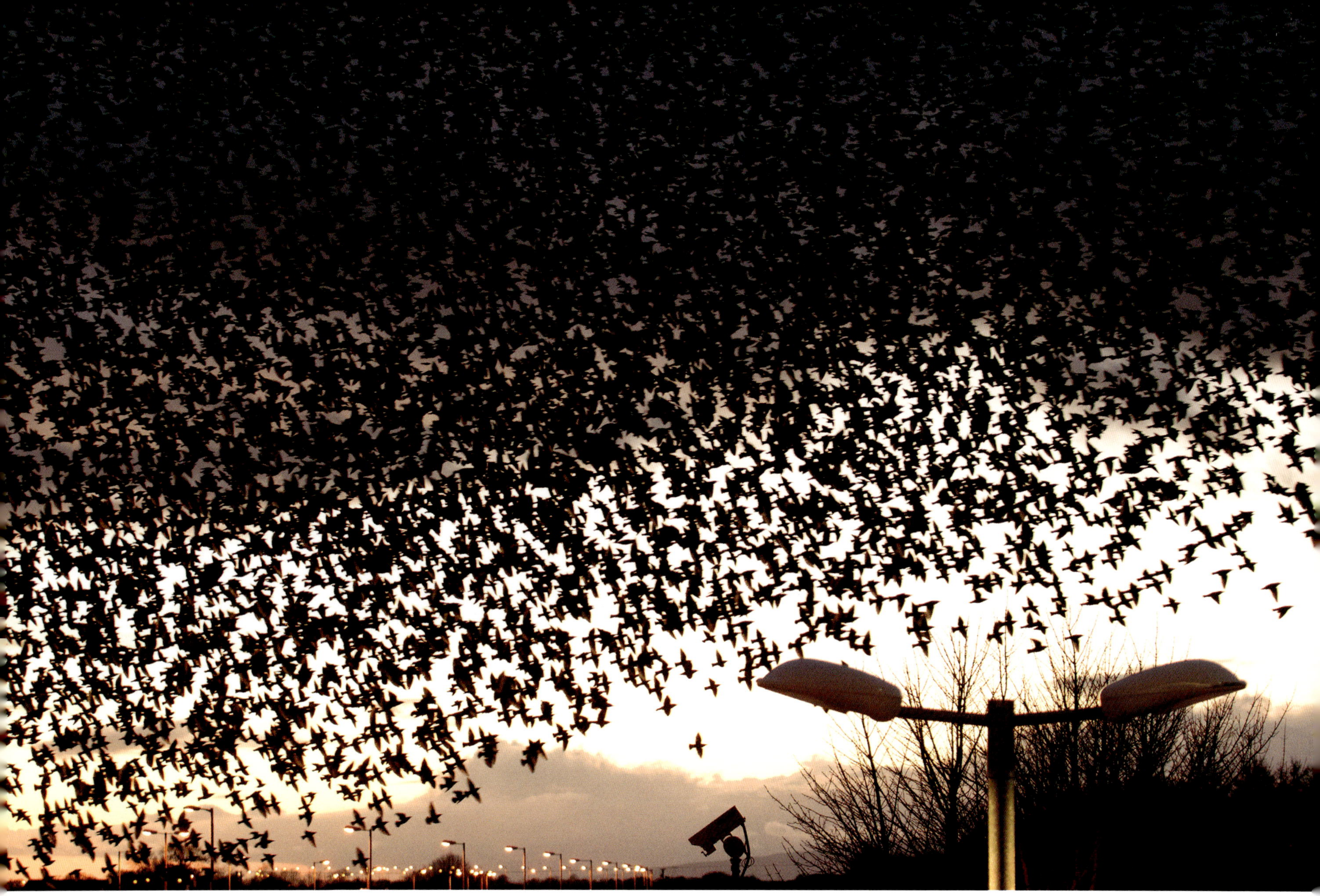

PAUL HOBSON

Starlings Flocking over a Main Road at Dusk
(Common Starling, *Sterna vulgaris)*
Gretna Green station, Scotland

I love urban connections in images so I deliberately waited until the starling flocks swooped over the main road. I was standing on the bridge at the train station. I kept the road and lights at the bottom of the image because I like the way they seem to be leaving the space around the bottom right street light.

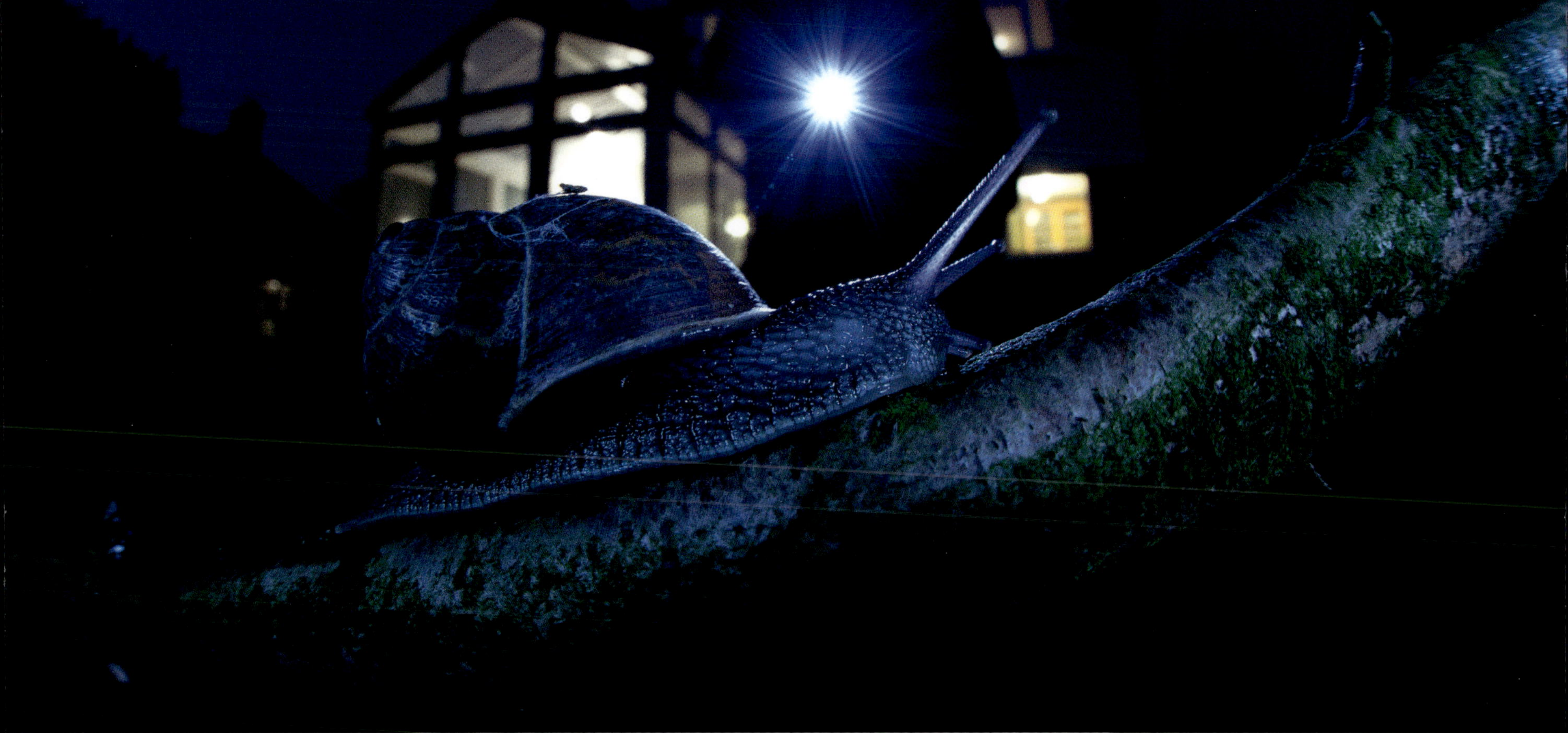

LIAM MARSH

Searching for Snails
(Garden Snail, *Helix aspersa)*
Hampshire, England

I built up this image in layers. First I turned on all the lights in my house, removing shades to maximise light. I lit the snail with a softened flash then added a half blue gel to maintain the feeling of dusk. The final element was to include my father with a torch, which not only added a rim light to the snail but also reaffirmed the feeling of a man-made environment.

HIDDEN BRITAIN

SPONSORED BY
BUGLIFE

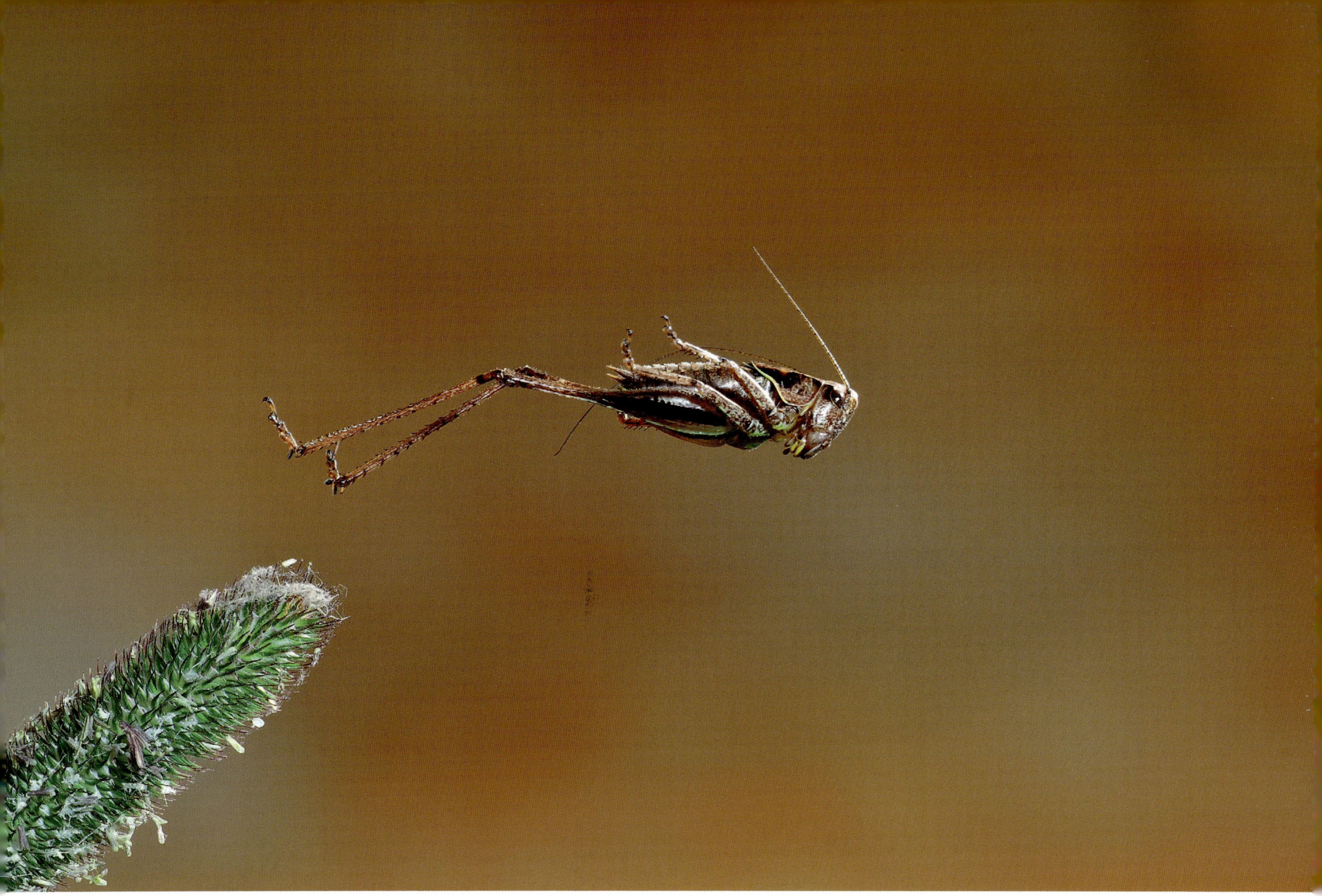

HIDDEN BRITAIN WINNER

DALE SUTTON

Leap for Freedom
(Dark Bush Cricket, *Pholidoptera griseoaptera)*
Battle, East Sussex, England

As I specialise in high-speed photography, I wanted to show people a bush cricket in mid leap. We all see grasshoppers and crickets sitting on grass stems but never after they have just taken off. This had to be done in a controlled environment using a special high-speed flash and an array of equipment. I think I caught the essence of the subject well.

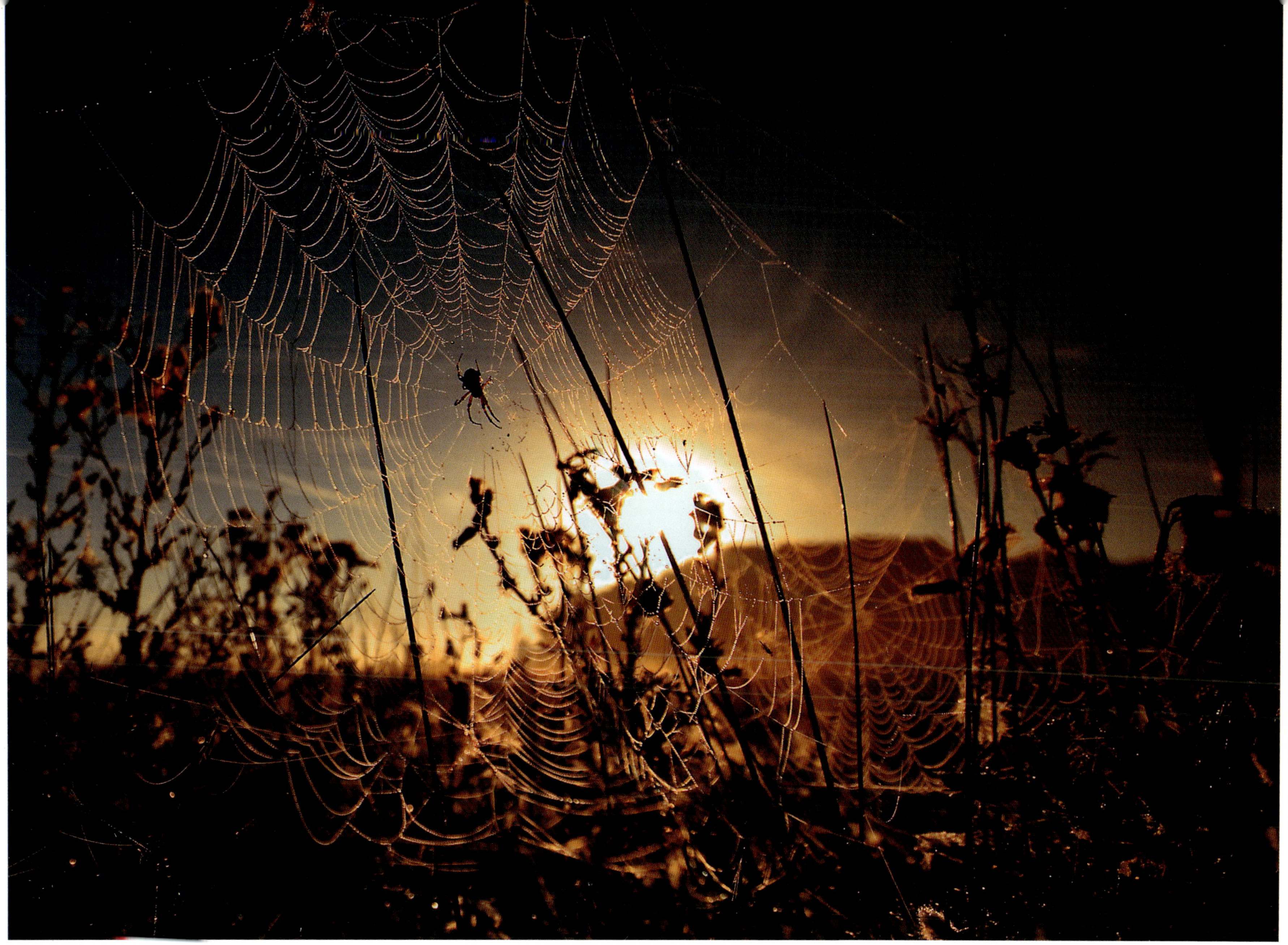

ALEX SABERI HIGHLY COMMENDED

Catcher in the Rise
(Orb-web Spider, *Araneus quadratus)*
Richmond Park National Nature Reserve, Surrey, England

I was busy shooting bigger animals near the pen ponds in Richmond Park when I spotted no less of a subject. I thought that having the spider's web at dawn with its creator would look great.

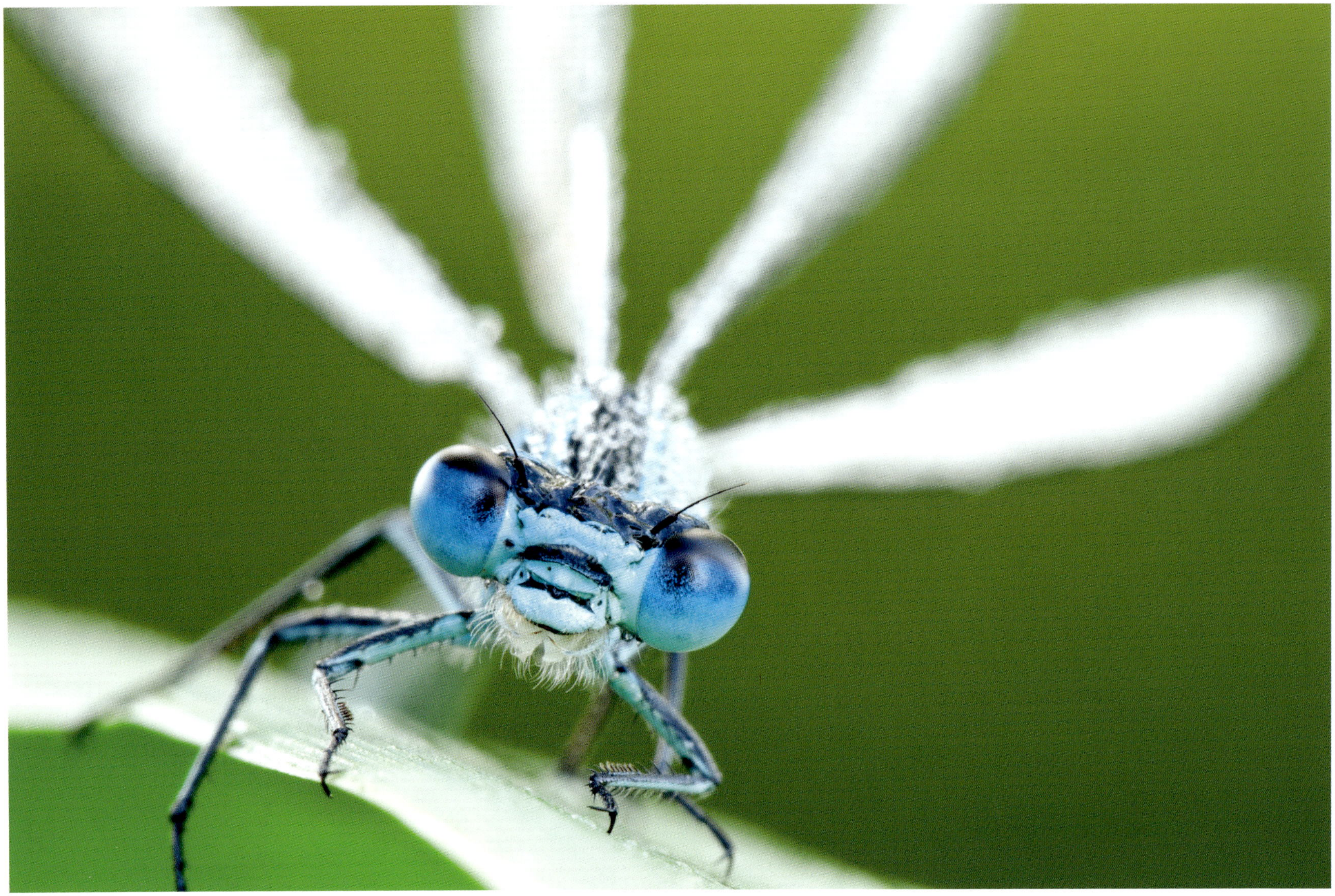

ROSS HODDINOTT/2020VISION ∧ HIGHLY COMMENDED

Damselfly Portrait
(Common Blue Damselfly, *Enallagma cyathigerum)*
Tamar Lakes, Cornwall, England

An early start on a still summer's morning was rewarded by this encounter with a common blue damselfly. The insect was smothered in dew after a still, cool night, but unusually its wings were partly open rather than closed. I opted for a head-on angle to emphasise its large, disproportionate eyes.

ALEX SABERI > HIGHLY COMMENDED

Electric Blue
(Common Blue Damselfly, *Enallagma cyathigerum)*
Richmond Park National Nature Reserve, Surrey, England

Amongst the yellow meadows of Richmond Park one summer I spotted this flash of blue darting around. I loved the contrast of the two colours. And the bokeh was just perfect for me.

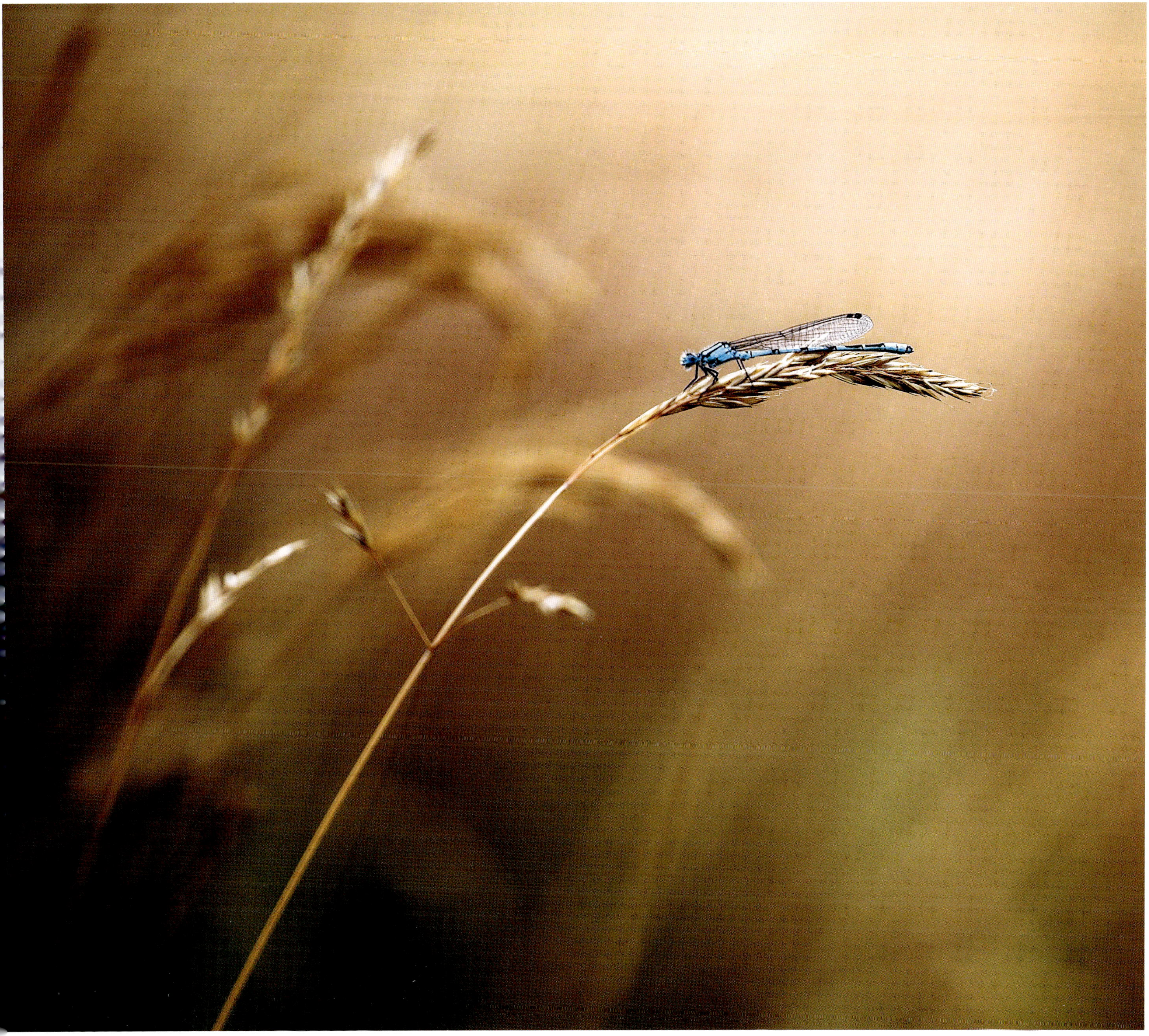

SIMON WHITE HIGHLY COMMENDED

Spider

(Spider, *Mettalina species)*

Clavering, Saffron Walden, Essex, England

Whilst looking for garden tools in my shed one day I noticed this spider which was illuminated by natural light from the skylight above. I underexposed the image by one stop to darken the background.

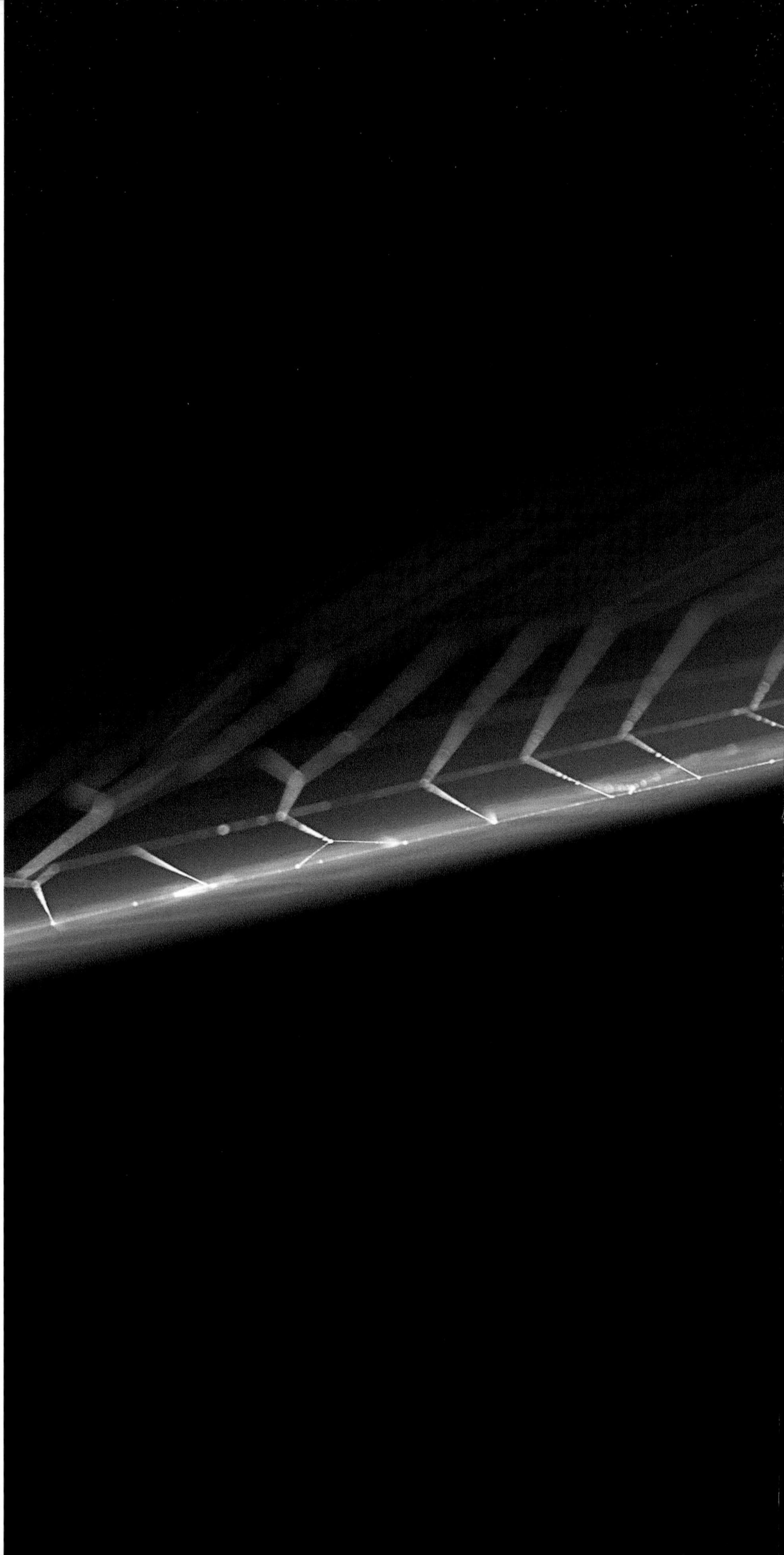

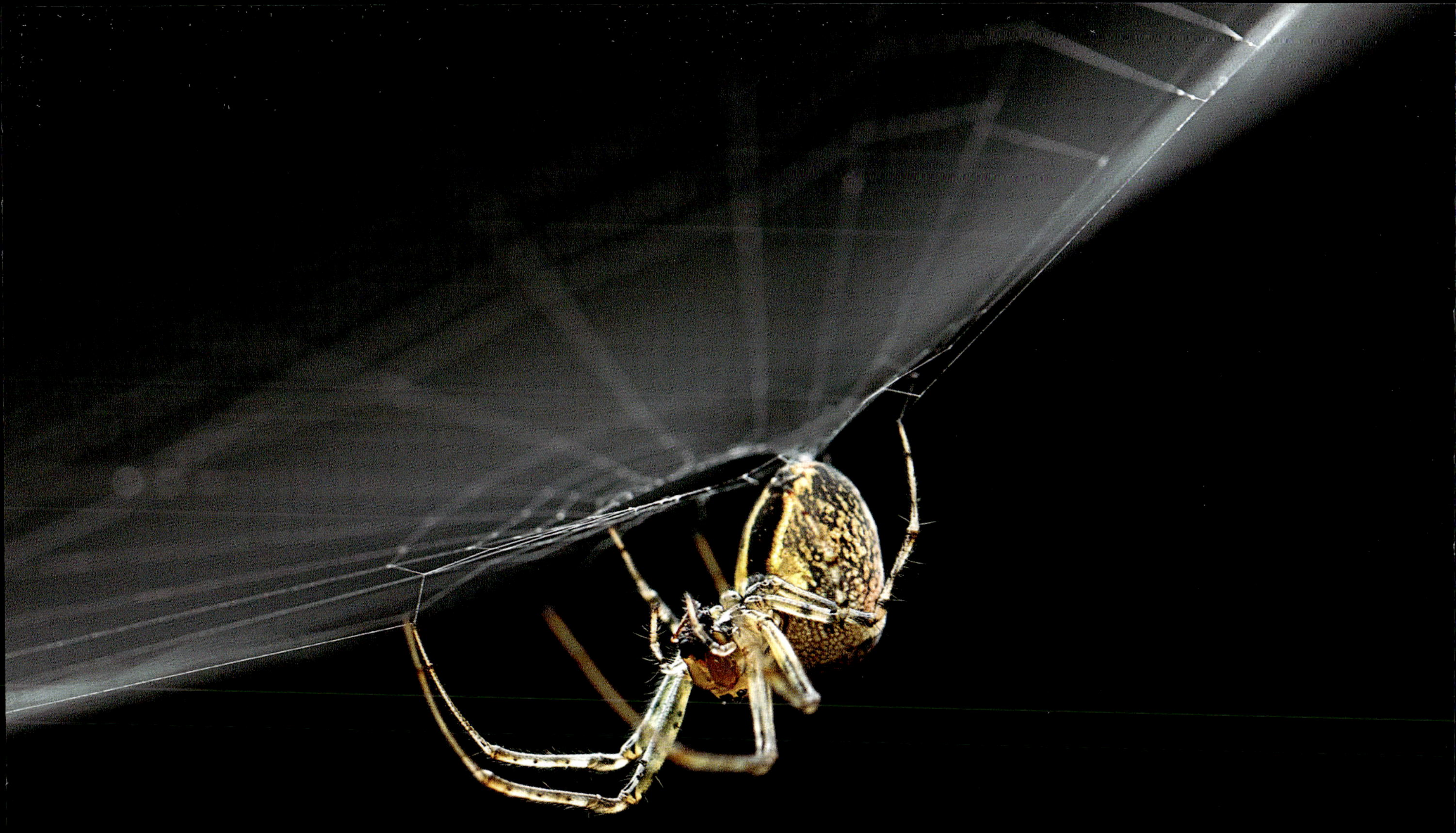

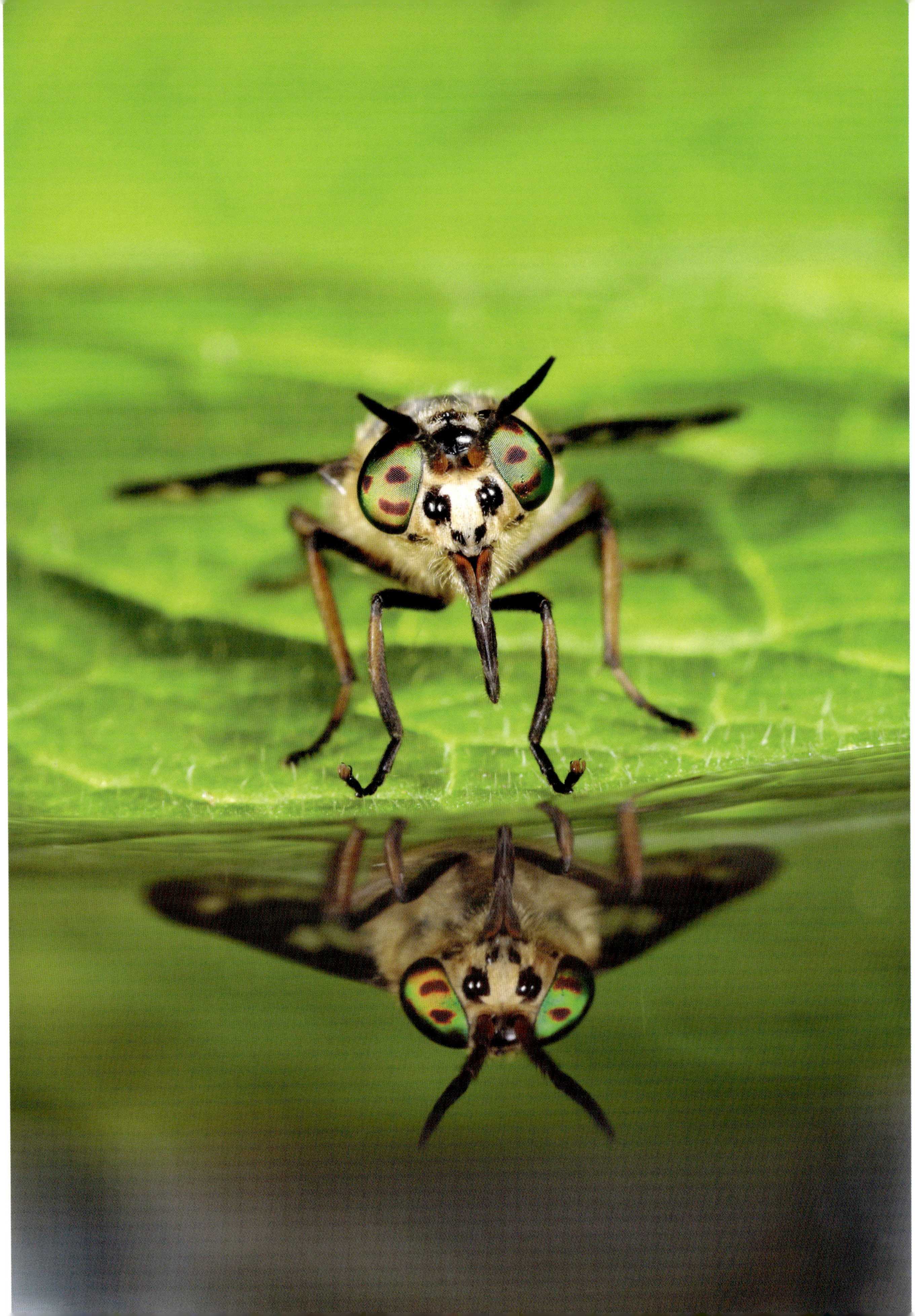

LES GIBBON < HIGHLY COMMENDED

Deer Fly Reflection
(Deer Fly, *Chrysops relictus)*
Arglam Farm, East Yorkshire, England

This photograph may look simple but it has taken me years to capture. I've had this image in my head for a long time. These insects are often found near water, and easily spooked when they are not attacking you. After many trips into the field, I finally came across this individual. Late in the evening it was resting on a leaf that held sufficient water to get the reflection.

JOHN H BRACKENBURY >

Orb-web Spider on Silk Line at Sunset
(Orb-web Spider, *Araneus quadratus)*
Mare's Way, Cambridgeshire, England

Large specimens are common in early autumn after feeding throughout the summer months. This shot was taken looking directly up from ground level, with the subject framed against the sky.

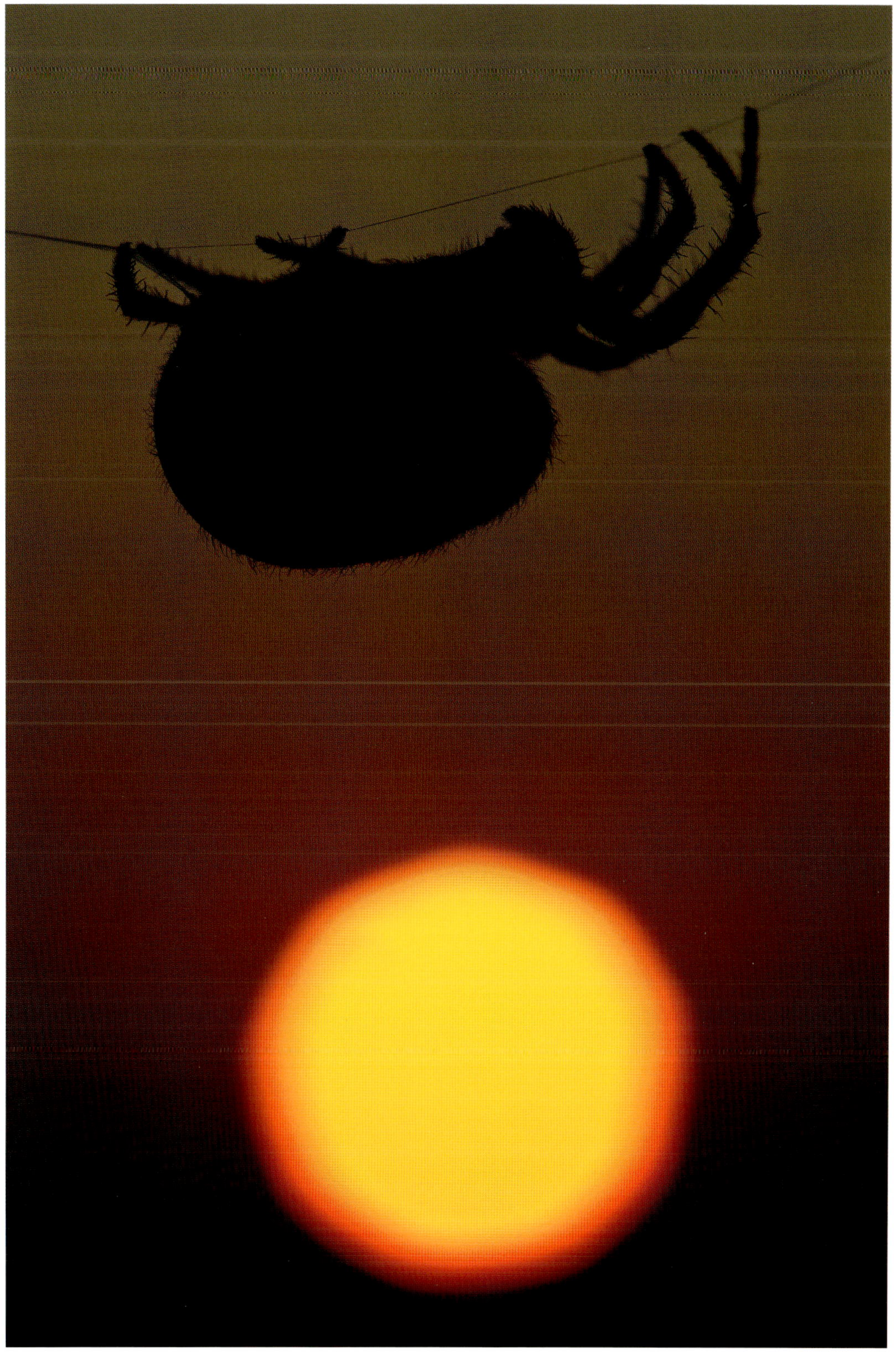

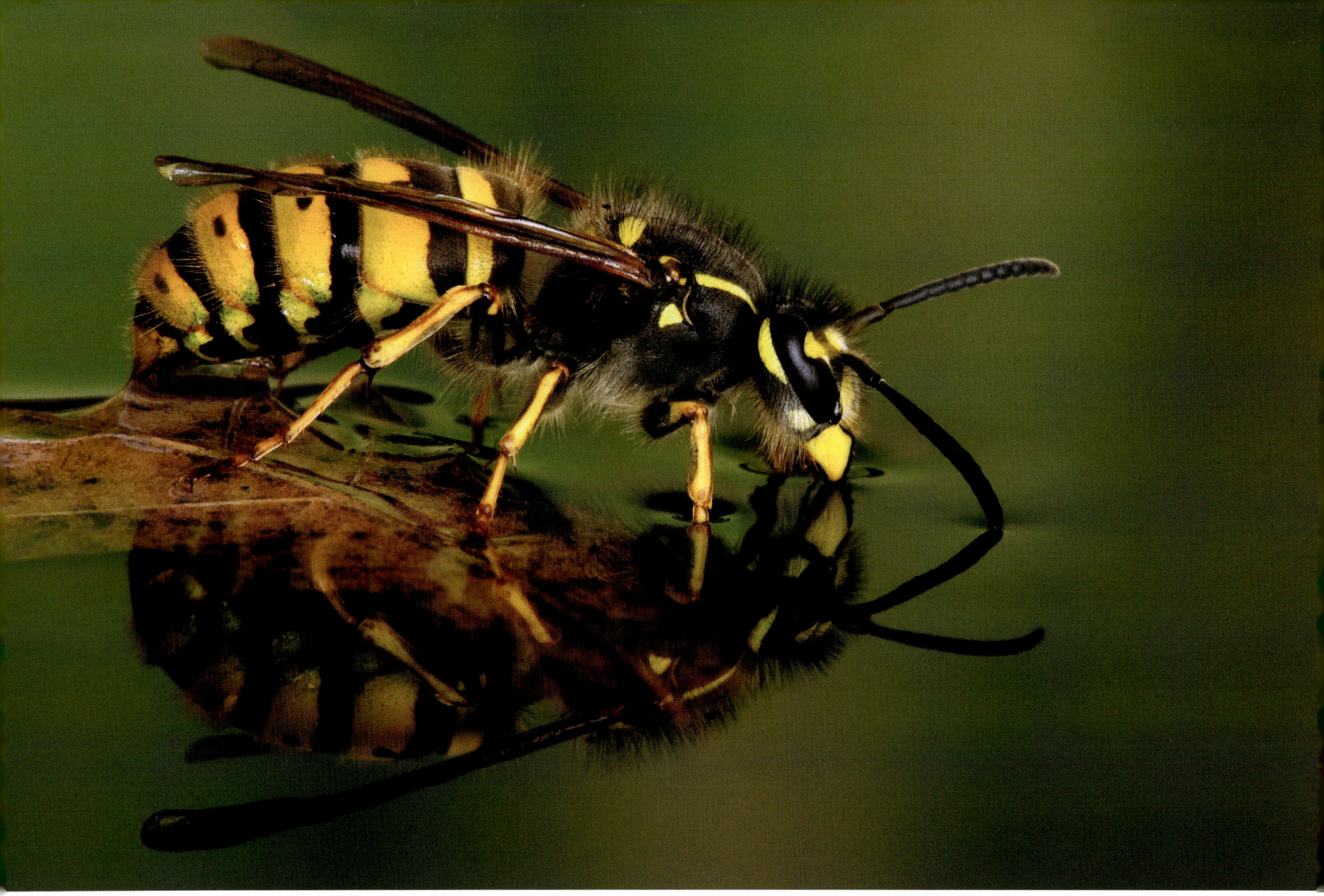

ANDY SANDS

Common Wasp Drinking
(Common Wasp, *Vespula vulgaris*)
Hertfordshire, England

I wanted to get a great shot of a wasp drinking, so after many wild attempts I decided to try in the studio. Several hours later that evening I got the picture I was after and the next morning the wasp was released back into my garden.

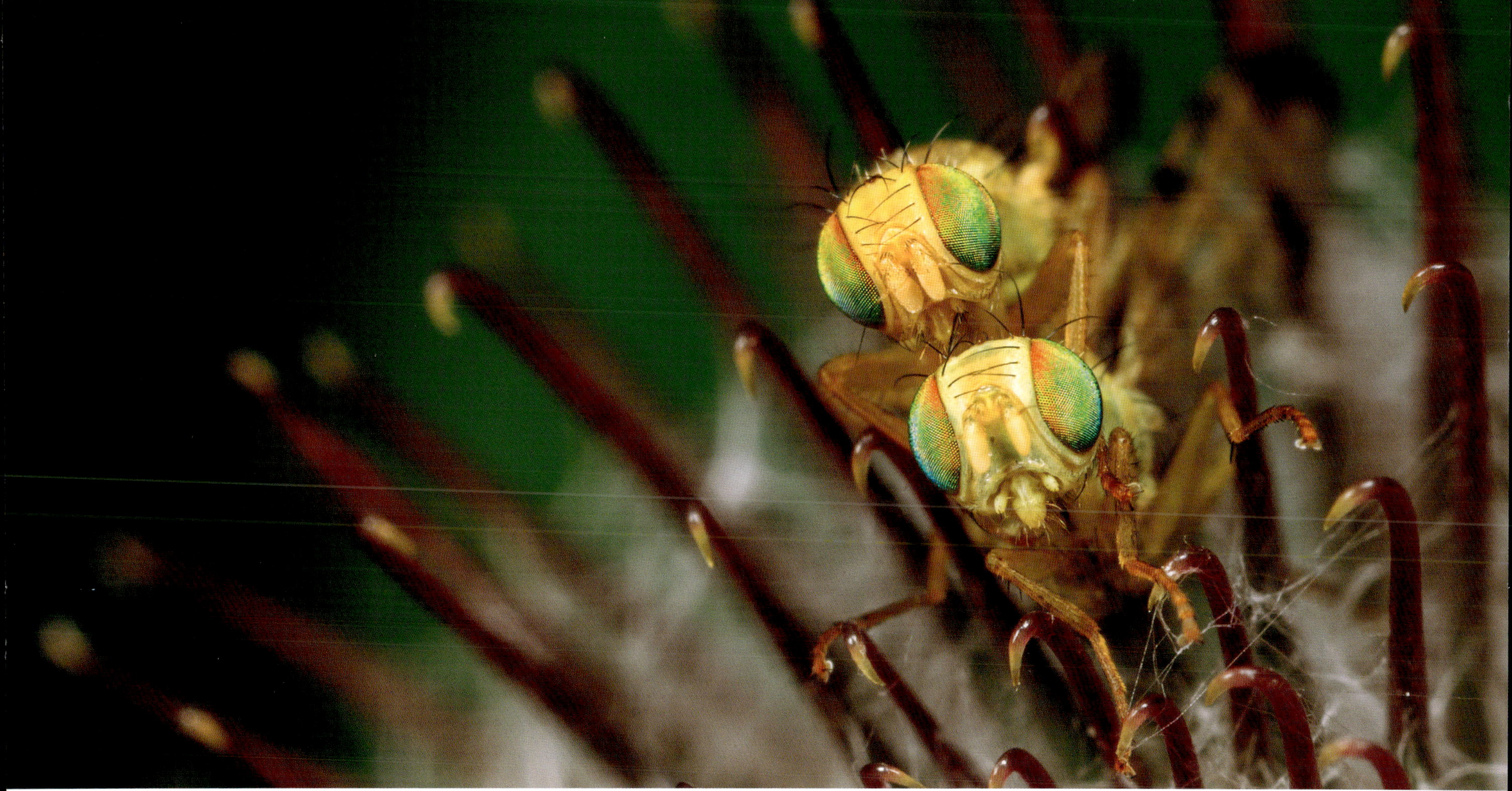

OLIVER CHARLES WRIGHT

Mating Flies on Burdock in Yorkshire
(Fly, *Terellia tussilaginis*)
Fairburn Ings, Yorkshire, England

This was taken handheld at high magnification so it was really important to get the right part of the image in focus. With the minute depth of field, both sets of eyes needed to be in focus.

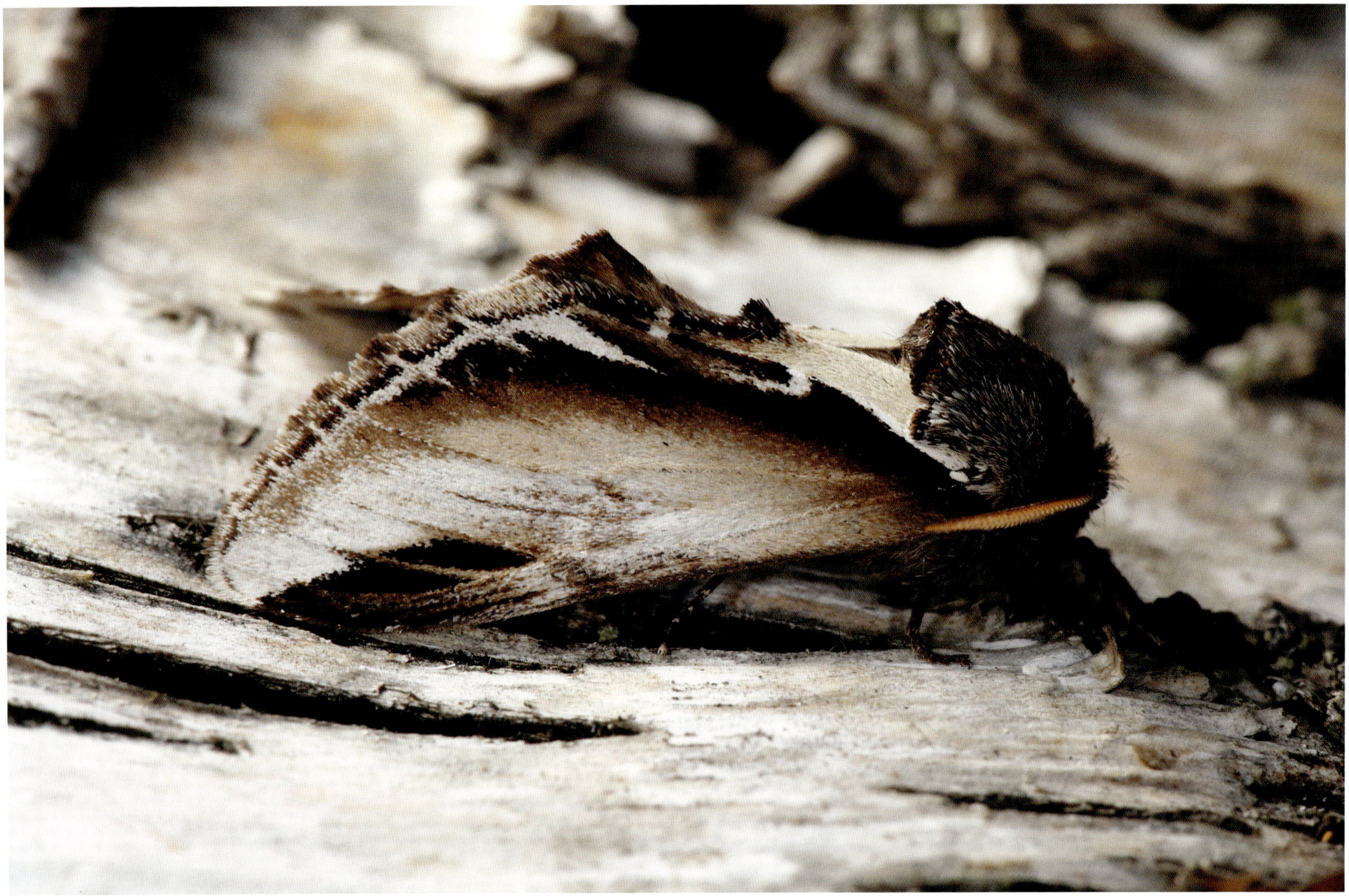

LES GIBBON

Nature's Camouflage
(Lesser Swallow Prominent, *Pheosia gnoma)*
Crowle Moor Reserve, Lincolnshire, England

I have always been fascinated by nature's ability to evolve and problem solve. In this case the moth has evolved its appearance to mimic the colour and pattern of the silver birch bark to such an extent that the two appear to merge into one, creating perfect camouflage. This allows the moth to avoid predation when at rest.

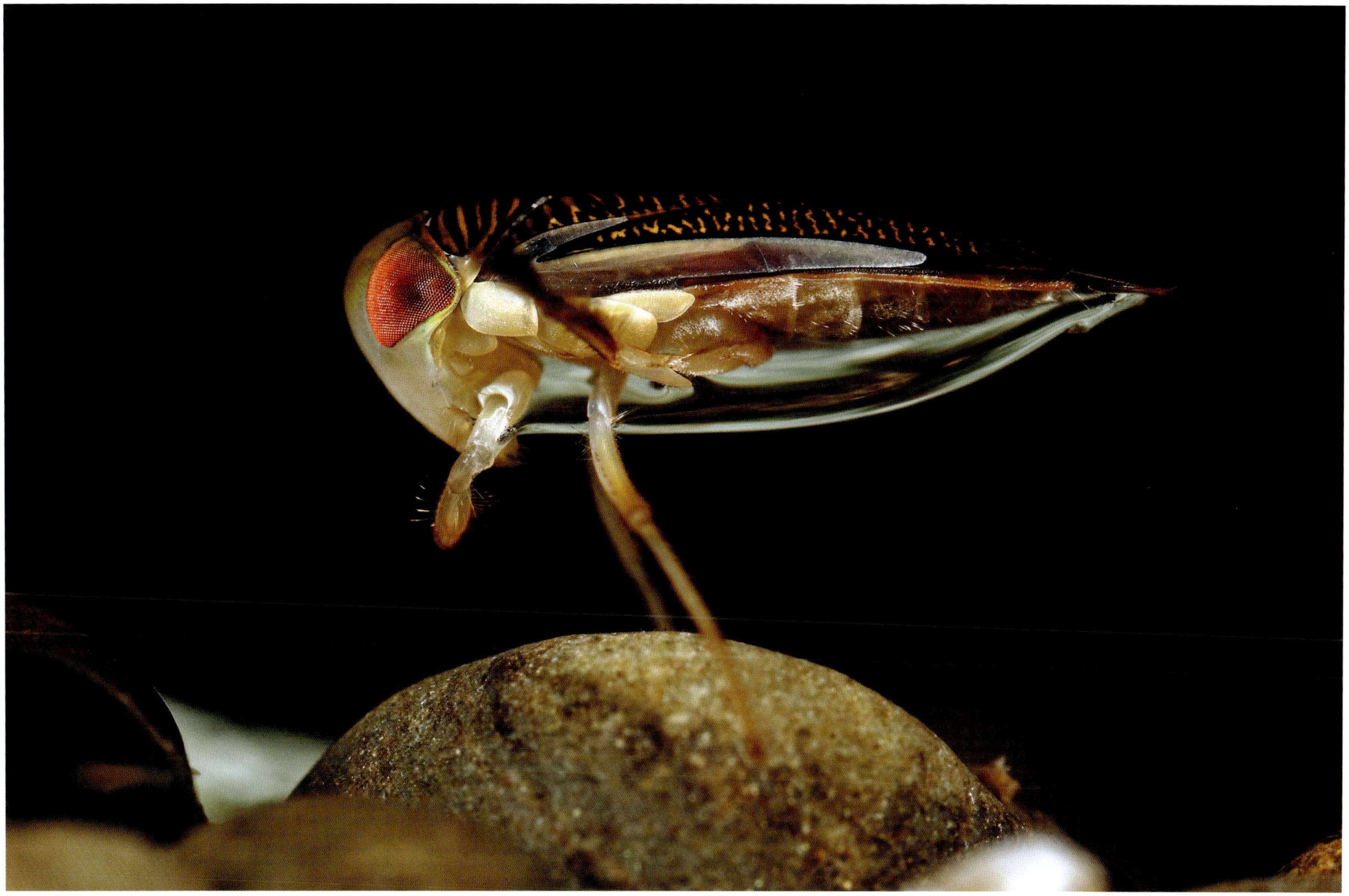

DUNCAN COOKE

Lesser Water Boatman
(Lesser Water Boatman, *Corixa punctata)*
Barrow-in-Furness, Cumbria, England

I set up a freshwater tank and stocked it with various pond dwellers, including this lesser water boatman. I wanted to capture some of the fine detail, especially the compound eye and the air sack held beneath its body. These are very fast movers, but every now and then they settle for a time, as this one did, enabling me to capture this streamlined side view, which I particularly liked.

LES GIBBON

A Cluster of Emerald Damselflies
(Emerald Damselfly, *Lestes sponsa)*
Cumbria, England

Sometimes you just get lucky. This picture is one example of that. I often visit this pond when I'm in the area and on this occasion these emerald damselflies were ovipositing *en masse*. It was late in the day and the low sunlight helped to bring out the beautiful colours on the insects.

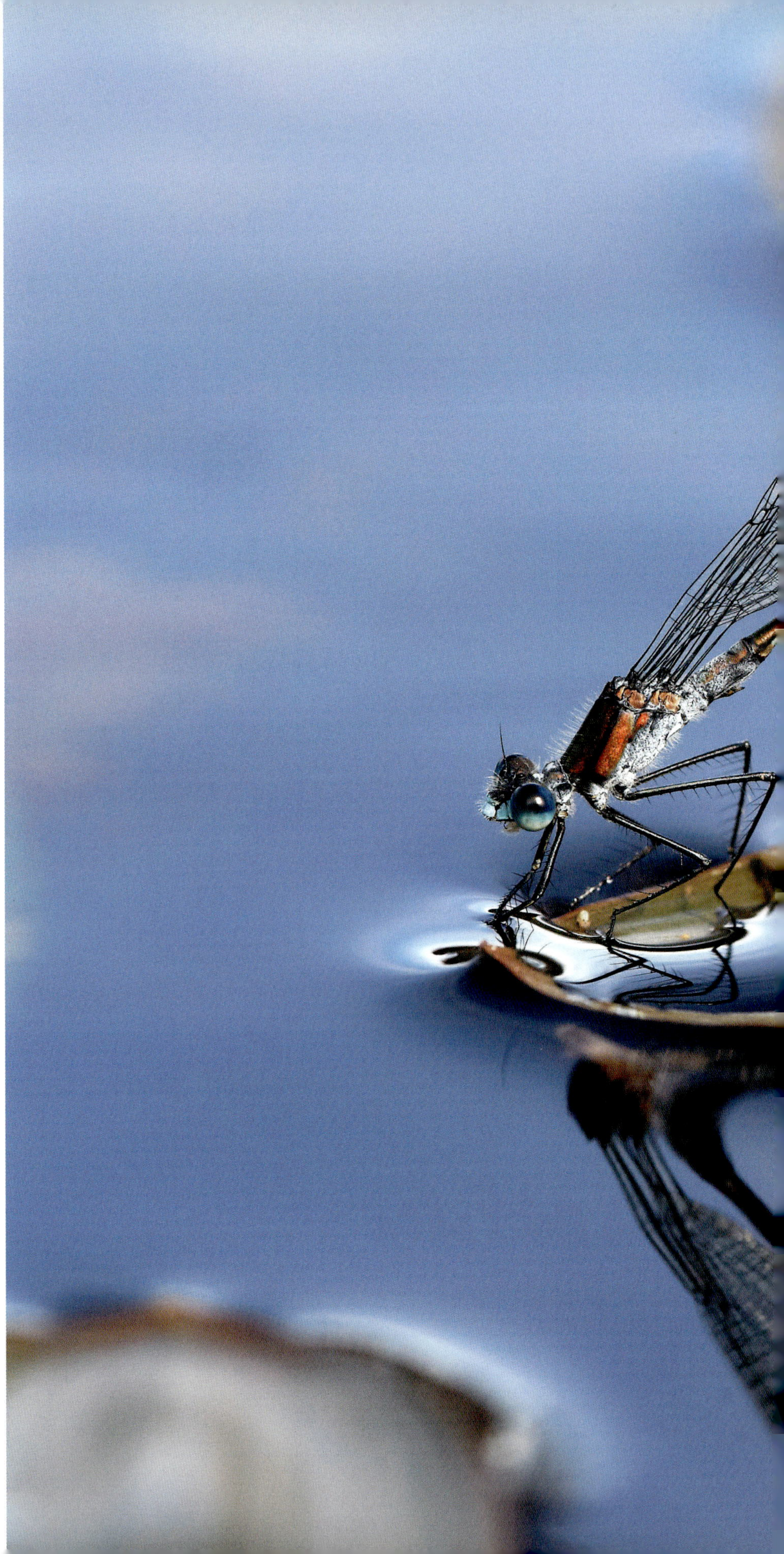

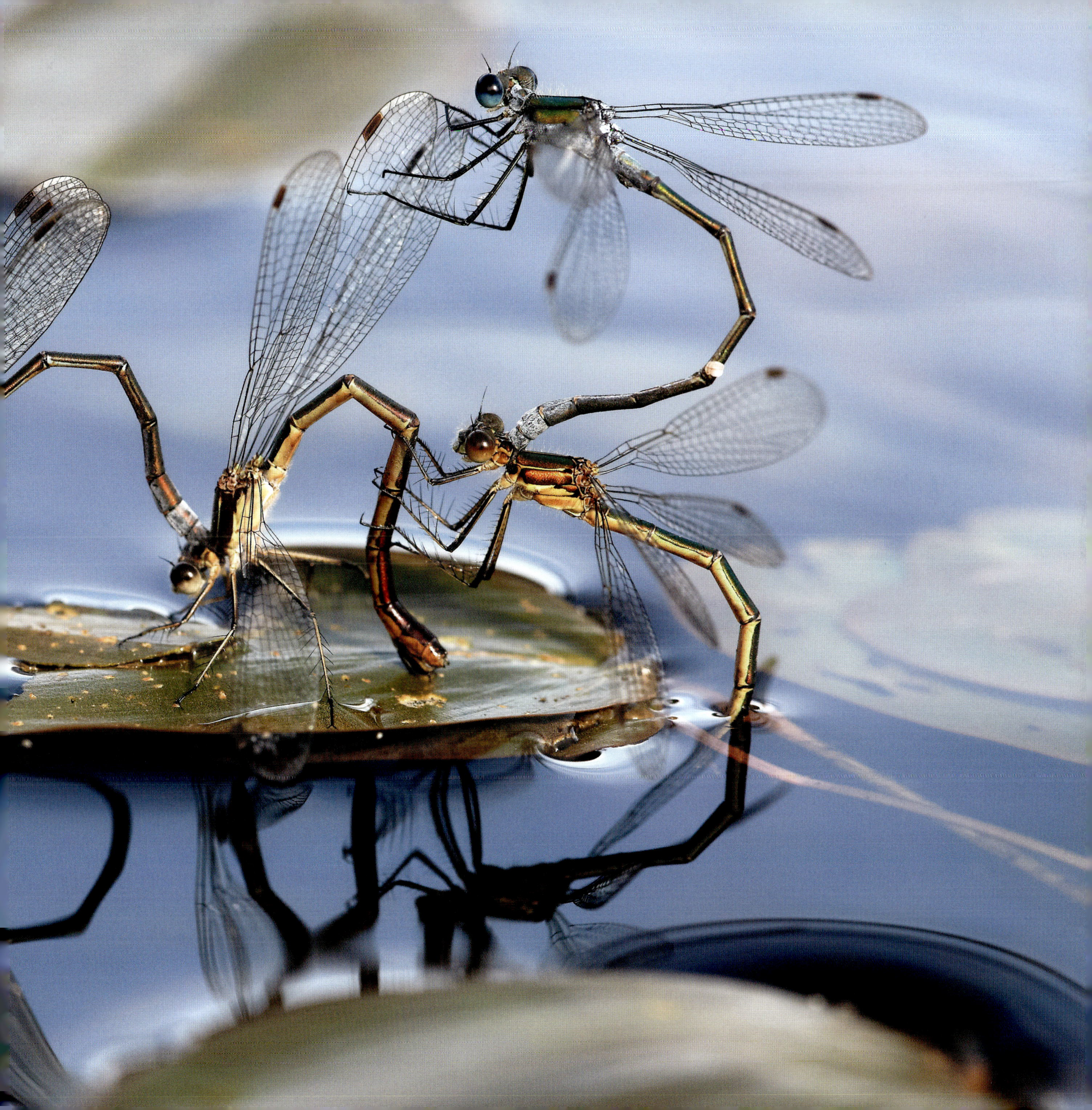

COAST AND MARINE

SPONSORED BY

WWF

COAST AND MARINE WINNER

BWPA 2012 OVERALL WINNER

DR MATT DOGGETT

Gannet Jacuzzi
(Northern Gannet, *Morus bassanus)*
Scotland

See caption on page 20

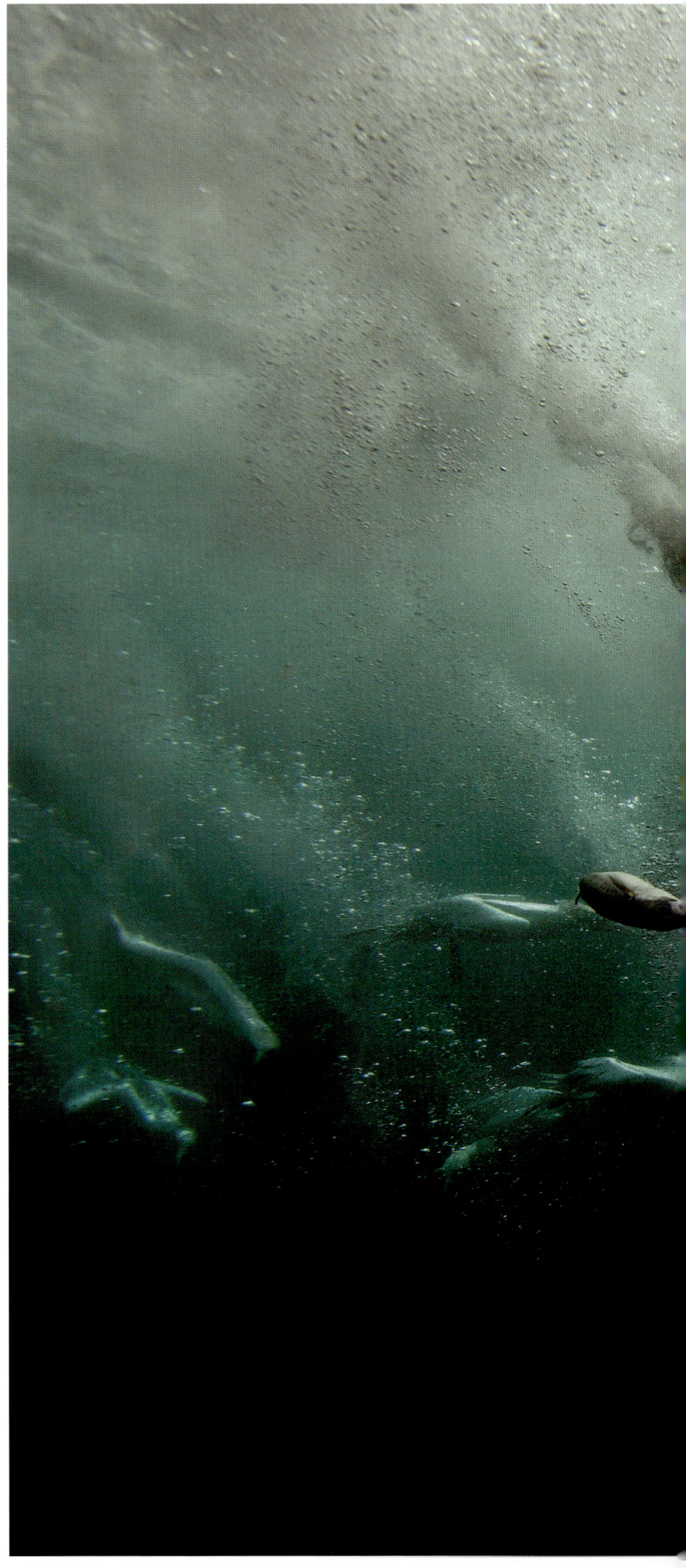

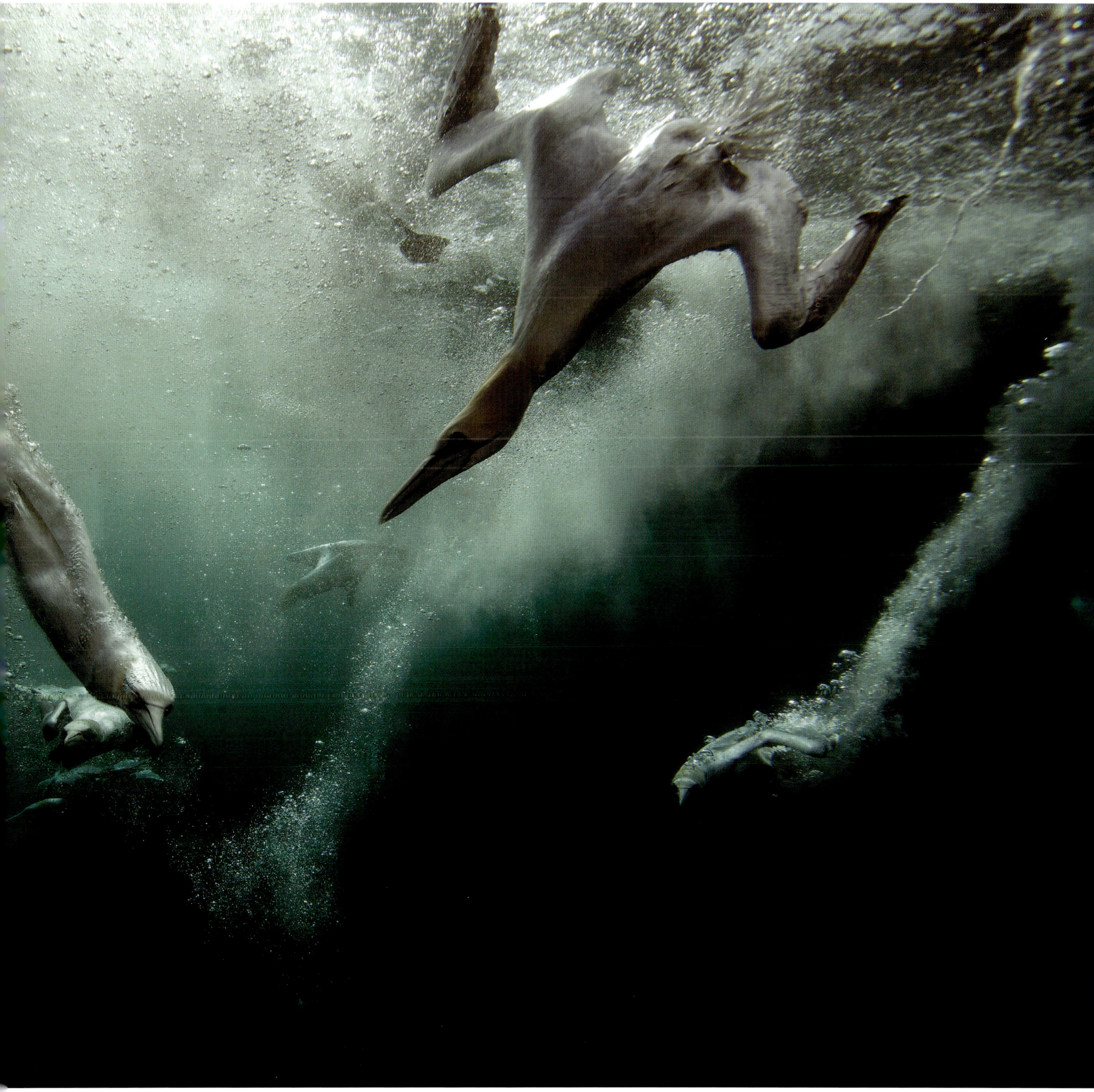

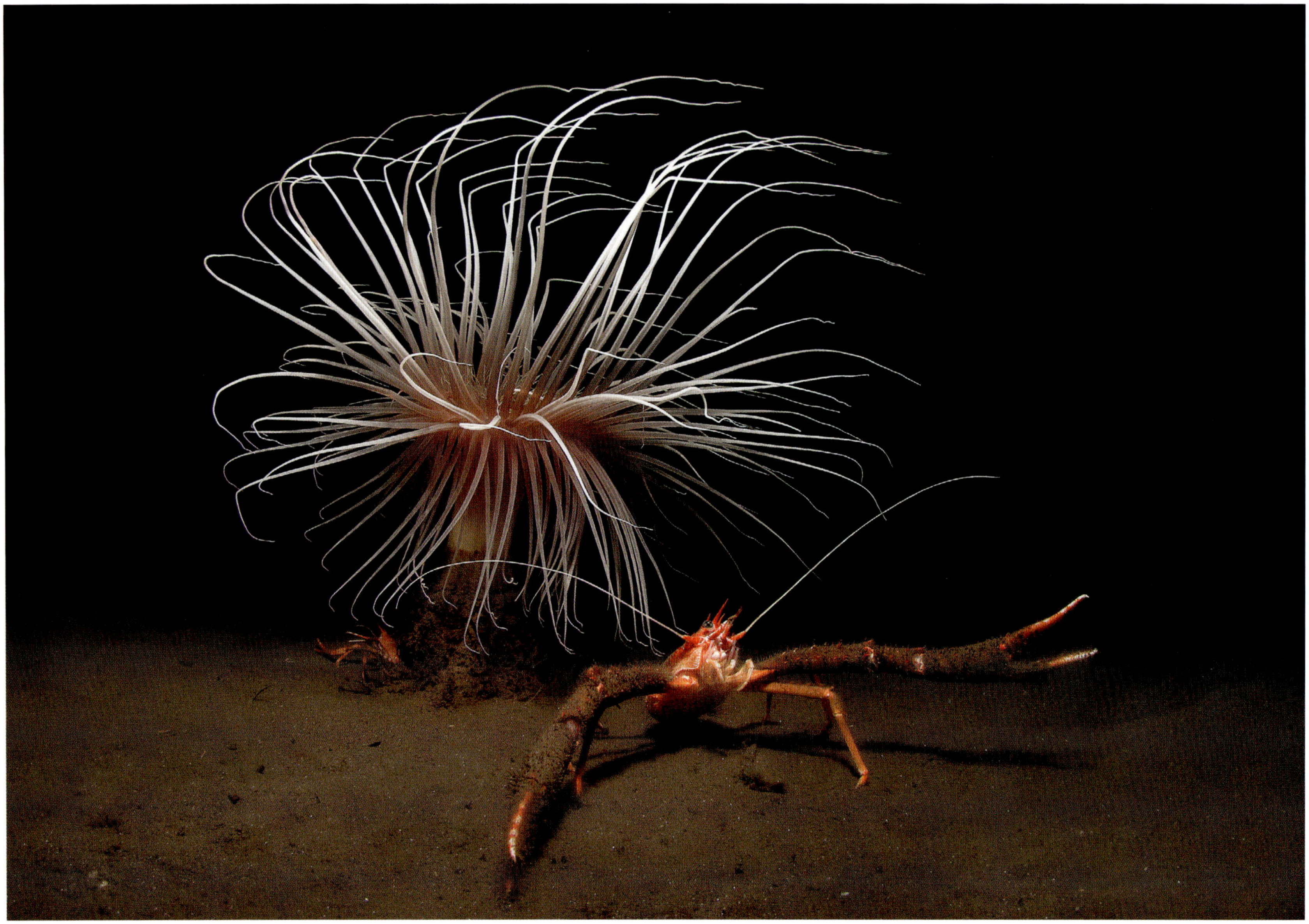

TREVOR REES HIGHLY COMMENDED

Fireworks Anemone and Squat Lobster
(Fireworks Anemone, *Pachycerianthus multiplicatus*;
Long-clawed Squat Lobster, *Munida rugosa*)
Loch Goil, Argyll, Scotland

These beautiful giant anemones (up to 30cm across and the largest in the UK) thrive in the deep, silty environments of certain Scottish sea lochs. The long-clawed squat lobster emerging from under the anemone helps give scale and insight into the deep, dark waters just off the shore. Great care was needed not to stir up the silt when photographing in this environment.

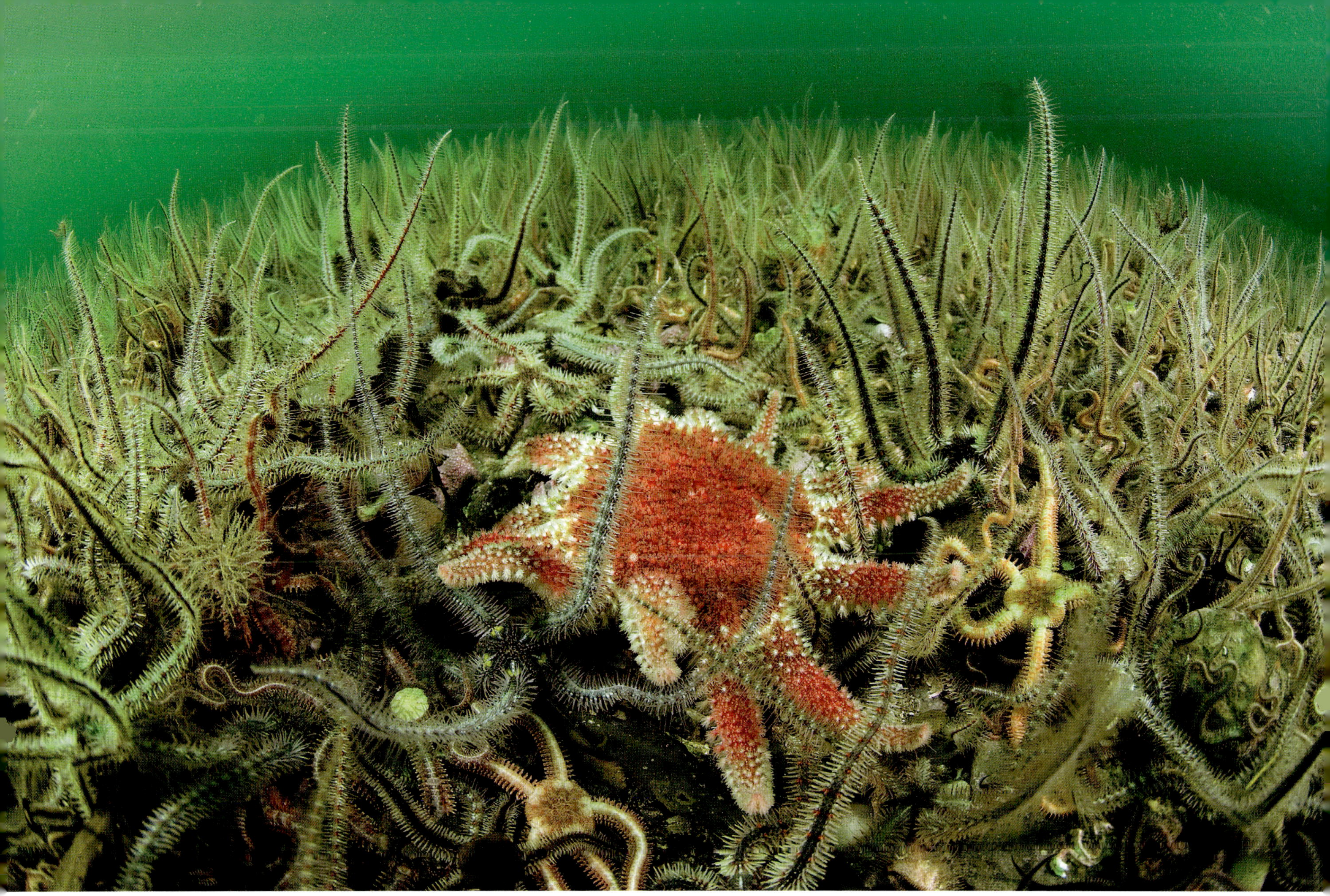

PAUL NAYLOR HIGHLY COMMENDED

Sunstar in Brittlestar Bed
(Common Sunstar, *Crossaster papposus*;
Common Brittlestar, *Ophiothrix fragilis*)
Loch Carron, Highland, Scotland

The movement of the common brittlestars desperately trying to avoid this predatory sunstar caught my eye. Brittlestars carpet much of the seabed in this beautiful sea loch.

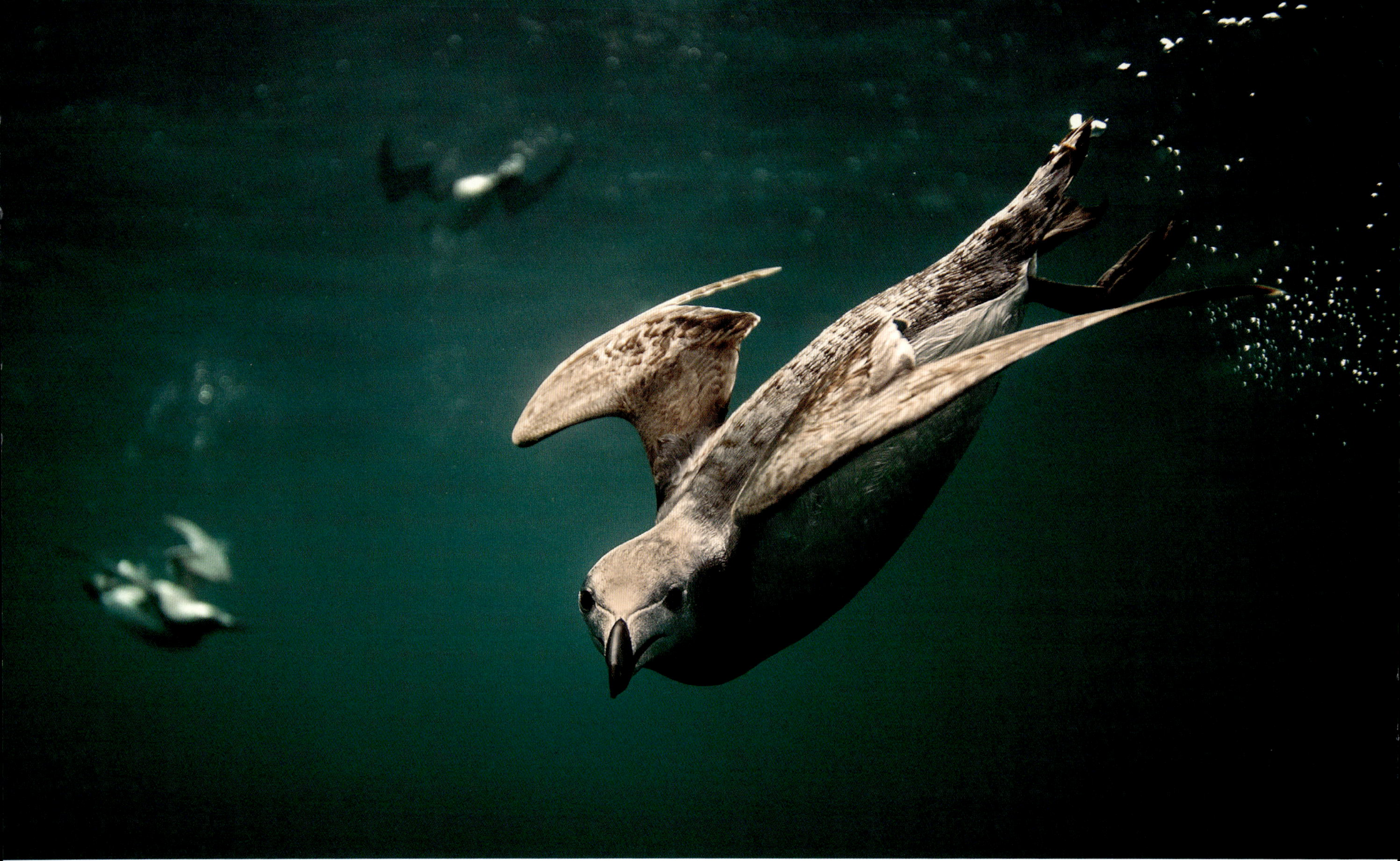

GEORGE KARBUS HIGHLY COMMENDED

Flying in Liquid Space
(Razorbill, *Alca torda)*
Rathlin Island, Northern Ireland

Snorkelling off Rathlin Island in summer is amazing, especially when you are surrounded by guillemots and razorbills which are constantly diving under water. It's very hard to photograph those agile birds. I spent several hours with them, then I came across some curious razorbills which started flying straight in to the camera. This incredible moment lasted for about two minutes and during that time I was able to get a few sequences.

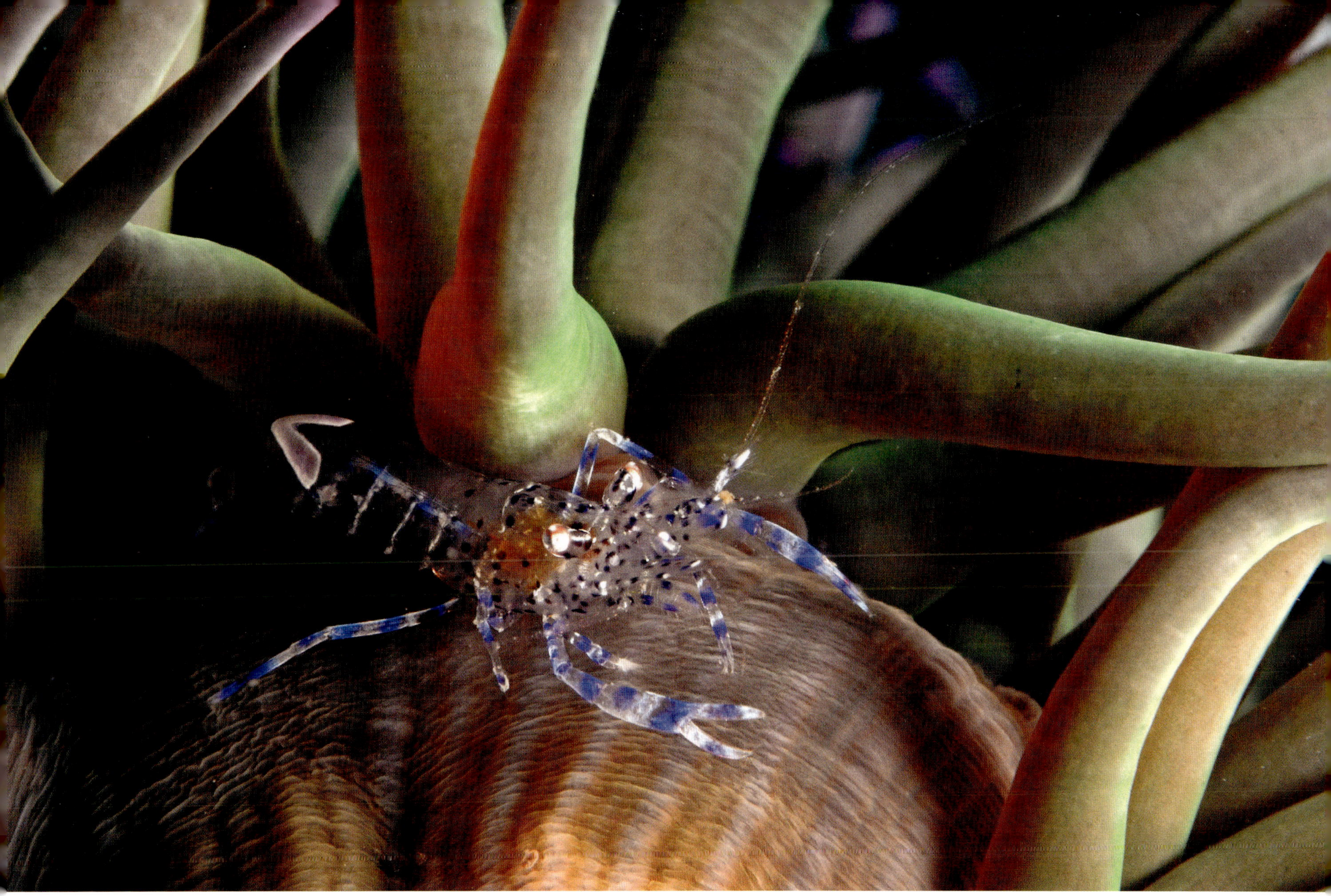

ALEX TATTERSALL HIGHLY COMMENDED

Periclimenes Sagittifer on Snakeslocks Anemone
(Anemone Shrimp, *Periclimenes sagittifer*;
Snakelocks Anemone, *Anemonia viridis)*
Swanage, Dorset, England

Every September for four years this colourful but tiny shrimp has been found living in the many snakelocks anemones under Swanage pier.

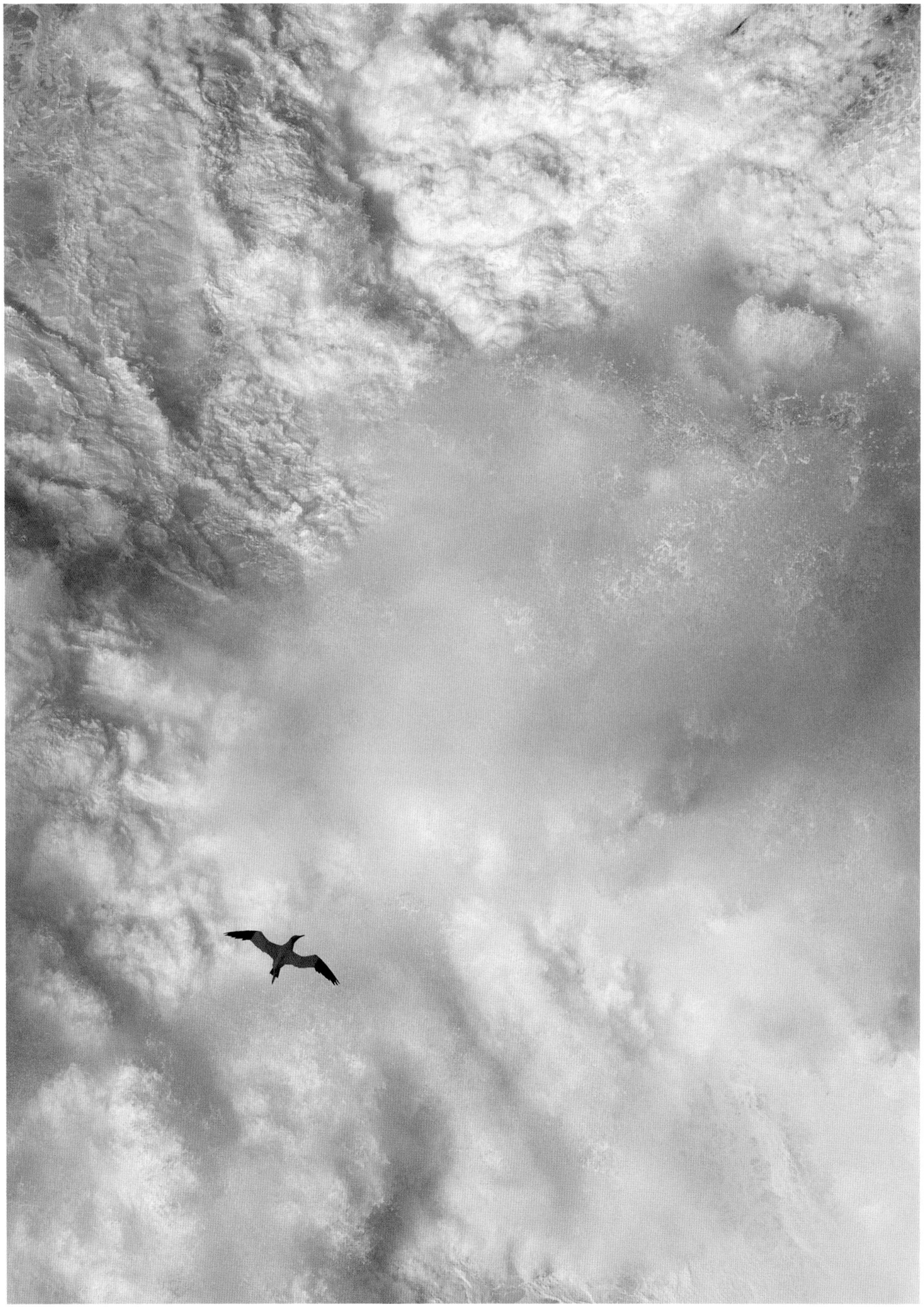

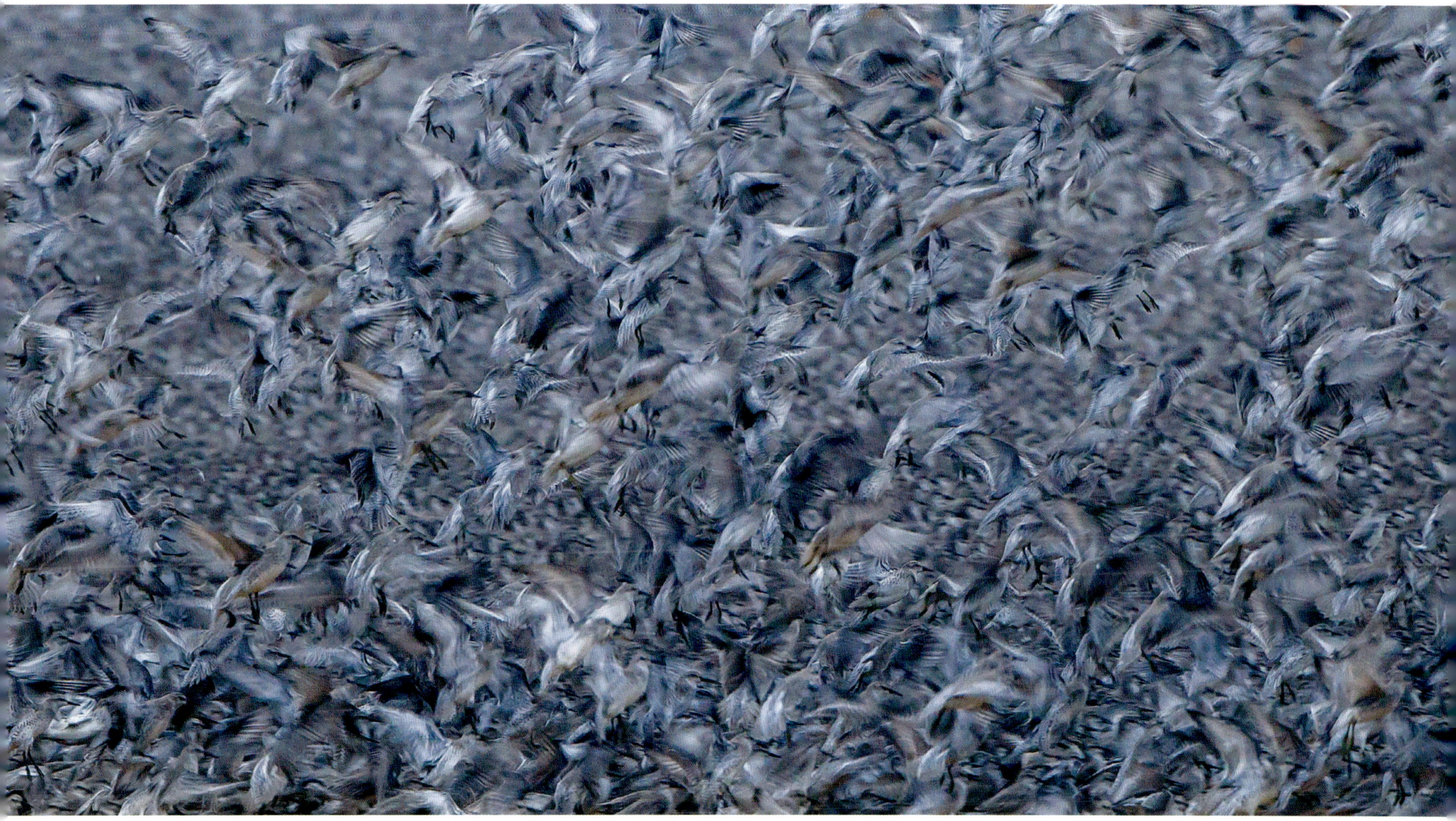

ANDREW PARKINSON < HIGHLY COMMENDED

A Gannet in Flight over Raging Seas
(Northern Gannet, *Morus bassanus)*
Hermaness National Nature Reserve, Shetland Isles, Scotland

I was perched on the top of Hermaness' treacherous cliffs as a bitingly cold Force 8 Northerly battered the cliffs and churned the sea. Whilst I battled the numbing cold the gannets appeared to thrive, riding the gusts with consummate ease as they soared over the tumult below. I journeyed to Hermaness specifically to show the relationship between bird and ocean and to document the occasional ferocity of their natural habitat.

MARK SMITH Λ HIGHLY COMMENDED

Snettisham Stereograms
(Red Knot, *Calidris canutus)*
Snettisham, Norfolk, England

The autumnal equinox normally brings with it the highest tides of the year and this year was no exception, driving a huge flock of knots into the gravel pits on the north Norfolk coast. After setting off at 3:30 in the morning and waiting in a hide for two hours I had about 10 seconds to get my shot through the fog before the flock thinned out. Worth the effort? Absolutely!

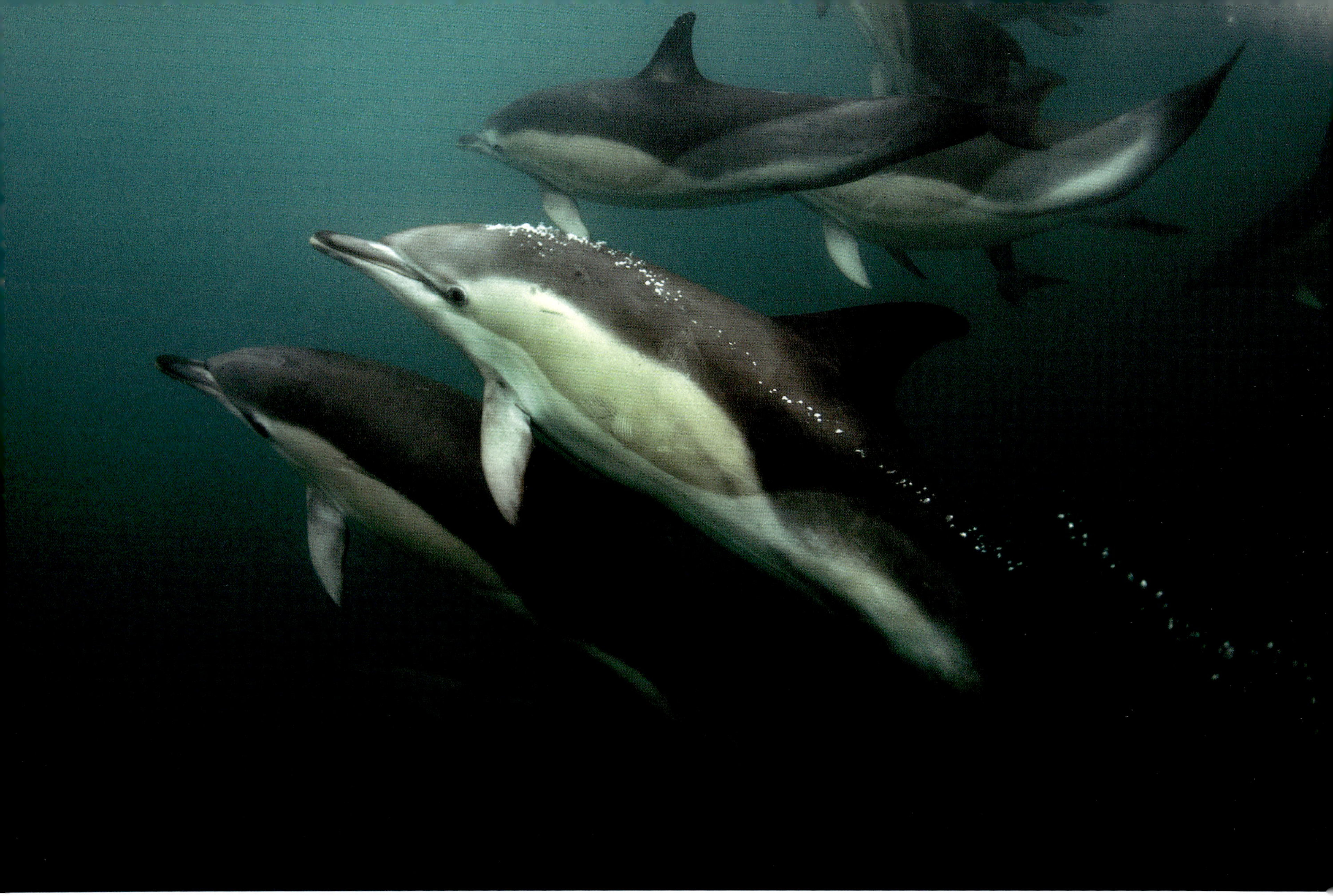

GEORGE KARBUS HIGHLY COMMENDED

Magic Souls in Emerald World
(Common Dolphin, *Delphinus delphis*)
St George's Channel between Wales and Ireland

I spend a lot of time on the sea searching for marine mammals every winter. We met this pod of very chilled common dolphins in St George's Channel. After a few minutes of interacting with the boat I decided to slip into the water and try to get close to them. Once I was in I had a group of 15 dolphins passing very close to me. The light wasn't perfect but this was one of my greatest and luckiest wildlife experiences this winter.

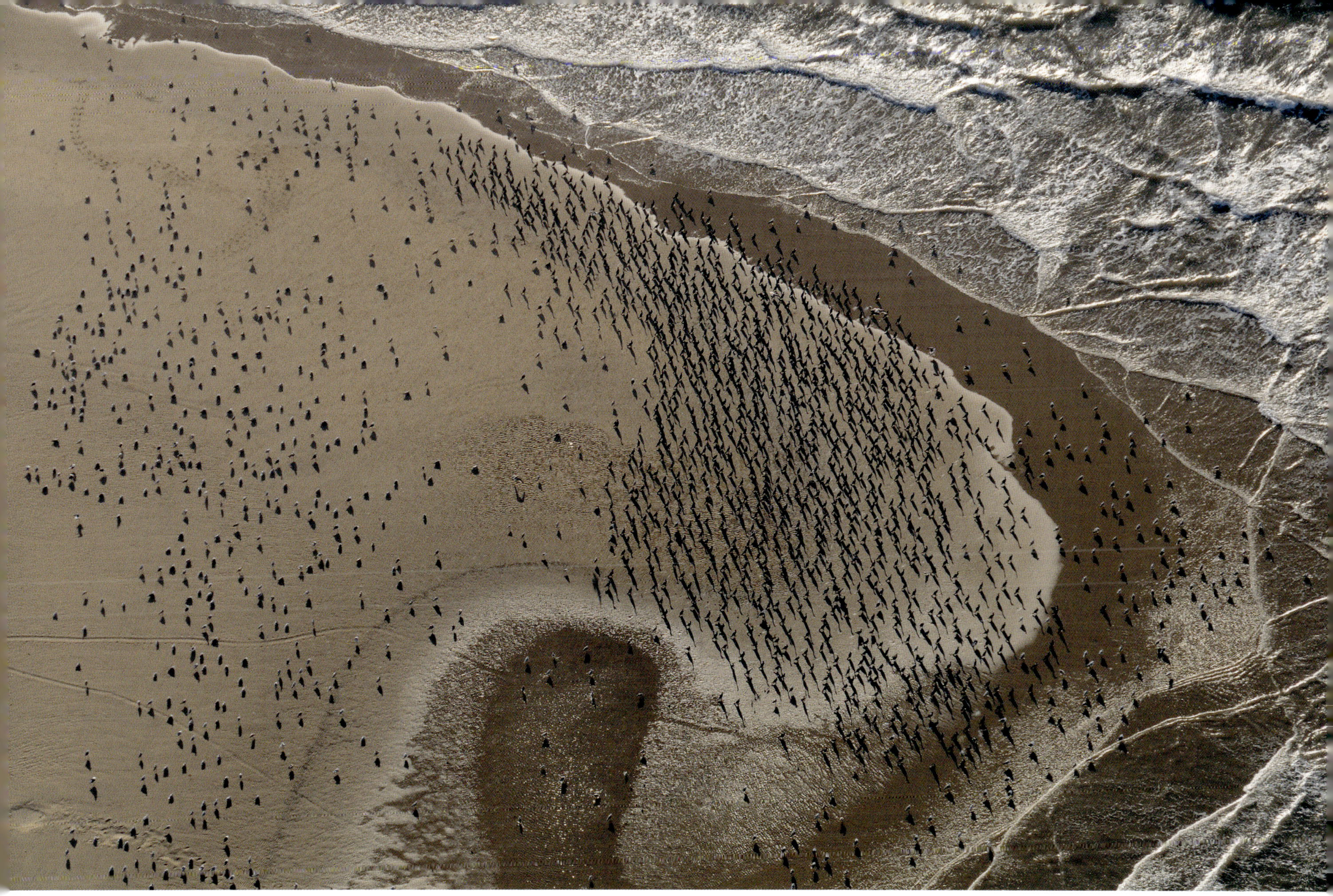

NICK UPTON HIGHLY COMMENDED

Cormorants and Gulls Resting on an Offshore Sandbank
(Great Cormorant, *Phalacrocorax carbo*;
Mixed Gull species, *Laridae)*
Great Yarmouth, Norfolk, England

I was in a helicopter, directing aerial filming of the Norfolk coast on a sunny winter's day we'd waited weeks for. As we travelled from one filming location to another, I spotted this sandbank exposed at low tide and loved the patterns created by the mix of sand, sea and birds casting shadows. The trails left by seals hauling themselves over the sand are also just visible.

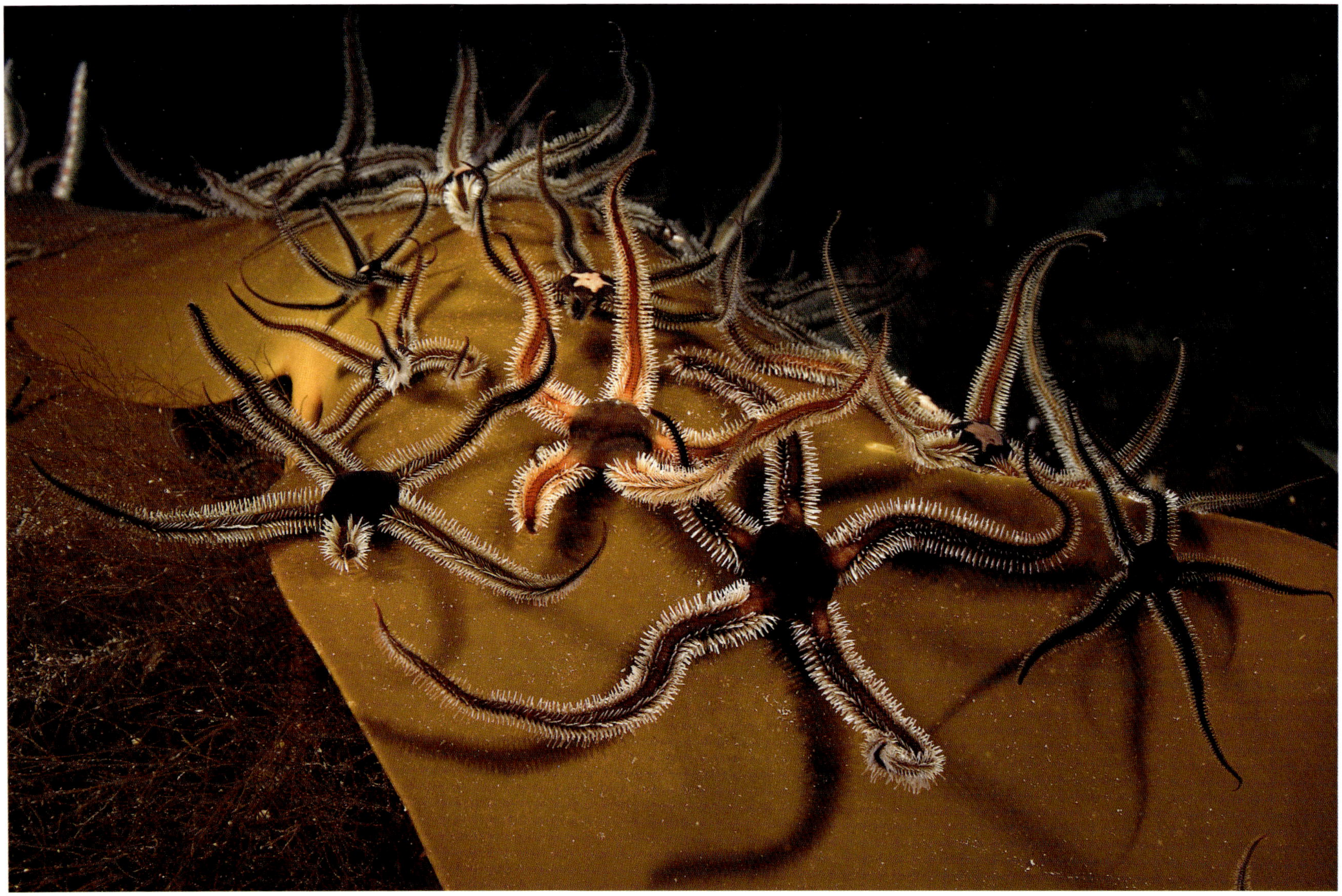

PAUL NAYLOR ∧ HIGHLY COMMENDED

Brittlestars on Kelp
(Black Brittlestar, *Ophiocomina nigra*)
Loch Carron, Highland, Scotland

These black brittlestars were using the elevated vantage point of a kelp frond to catch their planktonic food. I was attracted by the spacing of the brittlestars and their contrast against the background.

ALEXANDER MUSTARD/2020VISION > HIGHLY COMMENDED

Pink Prawn Portrait
(Pink or Northern Prawn, *Pandalus montagui)*
Loch Carron, Highland, Scotland

This is a high-magnification portrait of a pink prawn, photographed living within the 'fingers' of a soft coral known as deadman's fingers. It was a challenge to shoot a moving subject at such high magnification.

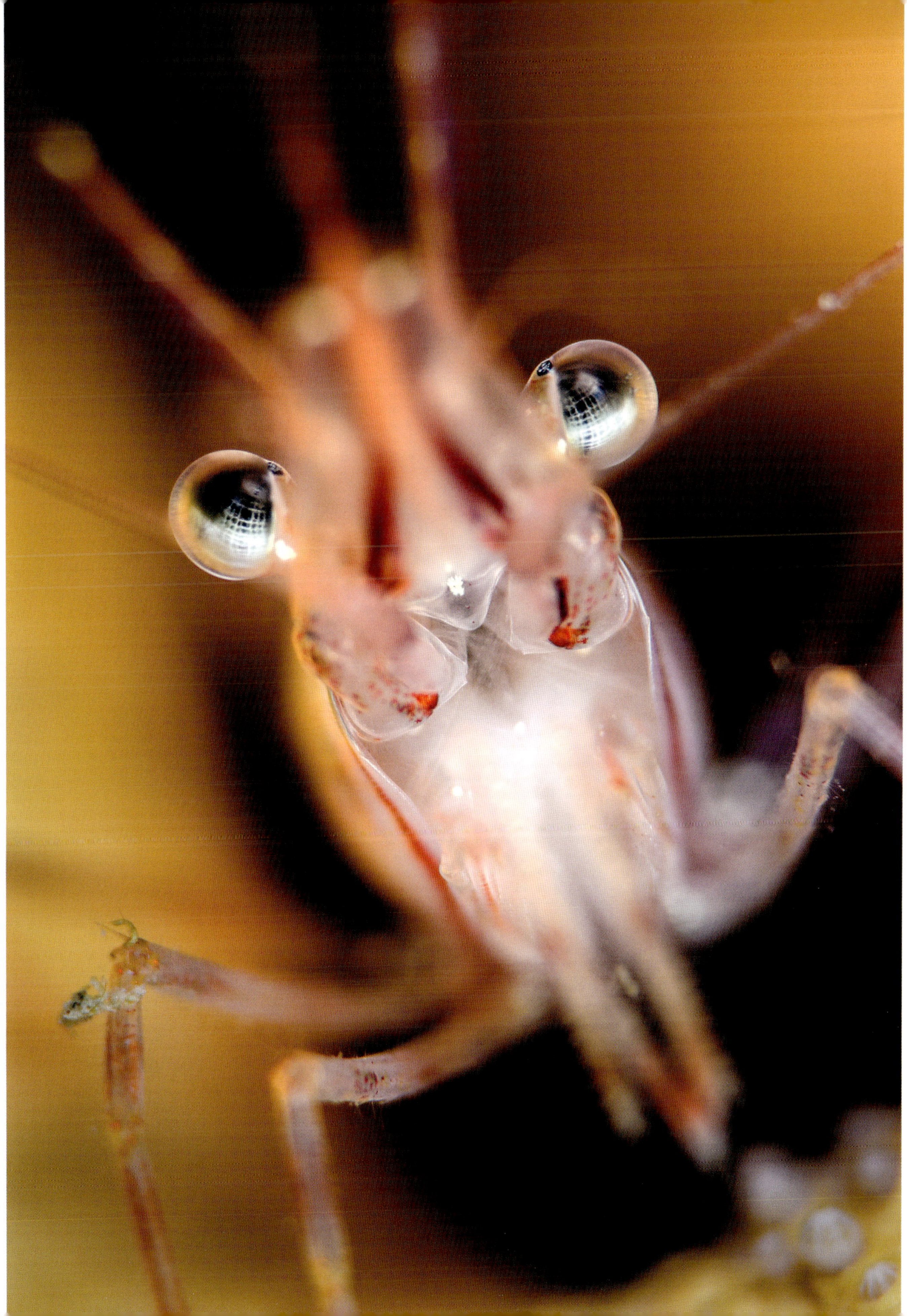

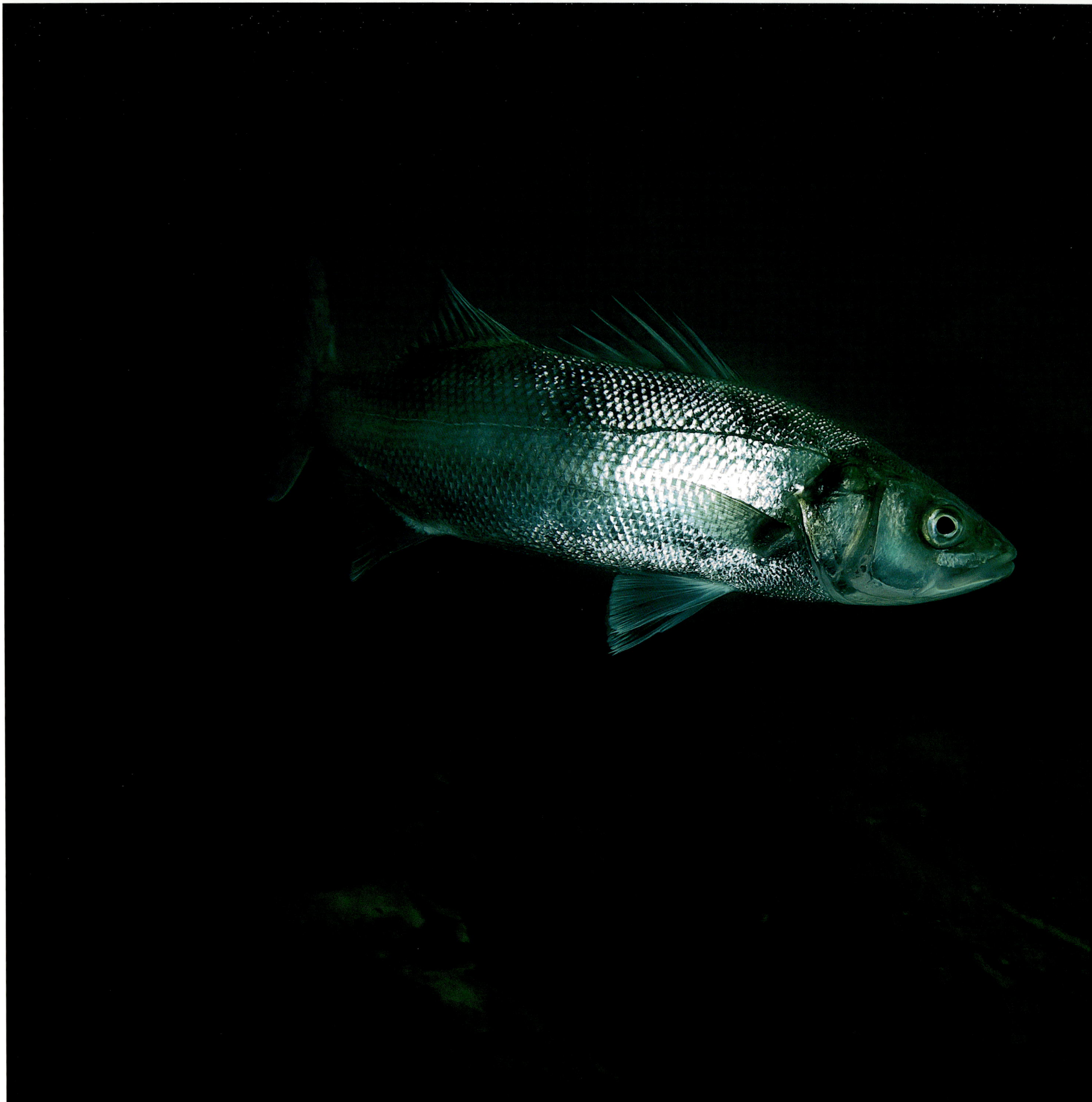

GRAHAM EATON HIGHLY COMMENDED

Bass Hunting over Kelp
(Bass, *Dicentrarchus labrax*)
Abersoch, Llyn, Wales

I had photographed bass in Cardigan Bay many times, and knew how difficult they are to approach and photograph. This particular day the water was very clear and each time I exhaled the bubbles frightened the fish. Holding my breath and wishing that I had a rebreather, I persevered. I took many shots of them before it all came together; the lighting, the distance and most of all, the bass with its spiny dorsal fin erect.

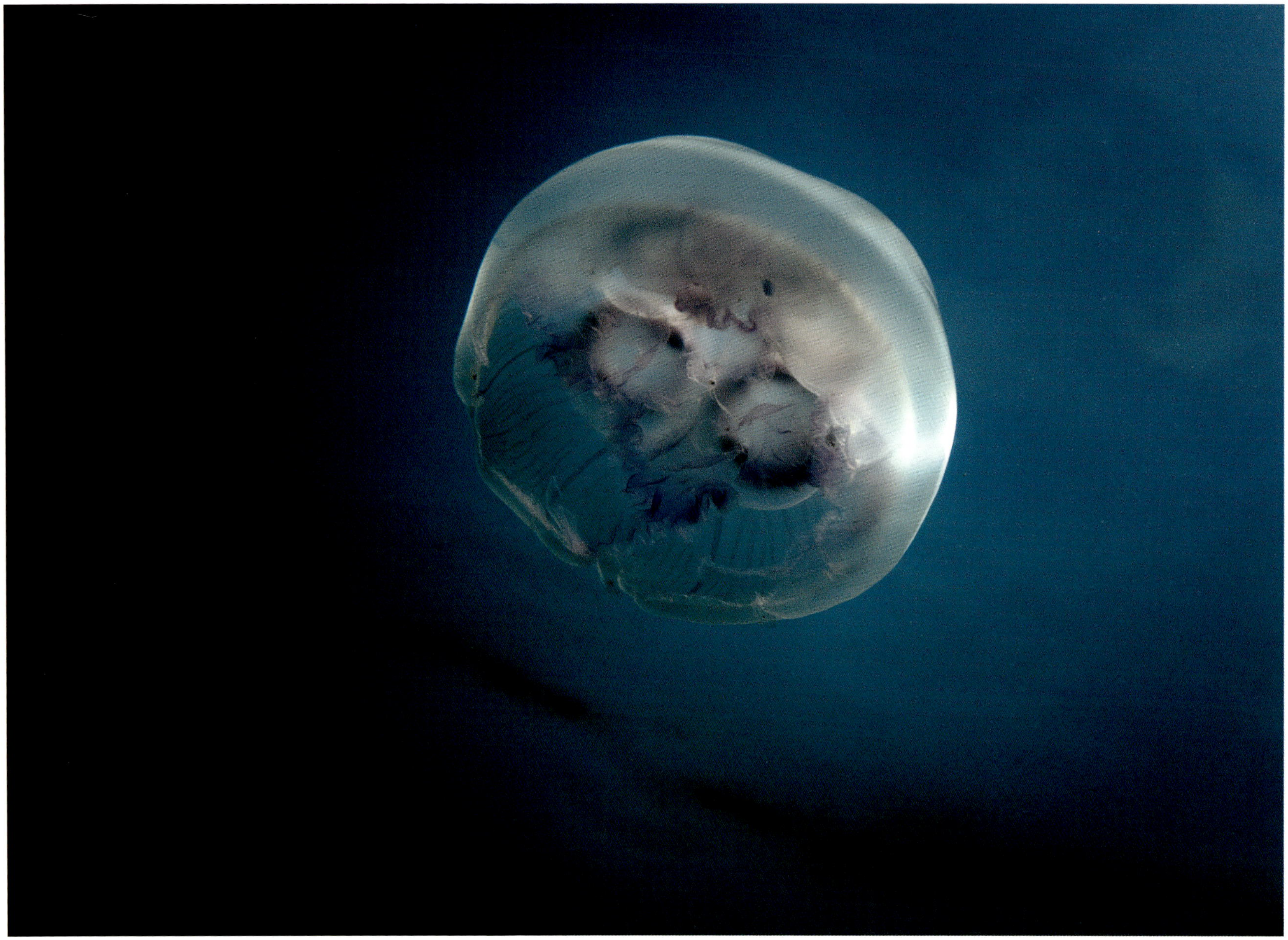

GRAHAM EATON HIGHLY COMMENDED

Common Jellyfish
(Common Jellyfish, *Aurelia aurita)*
Abersoch, Llyn, Wales

I wanted to capture an image of a common jellyfish looking upwards towards the water surface. I was snorkelling to take this image, in unusually clear water.

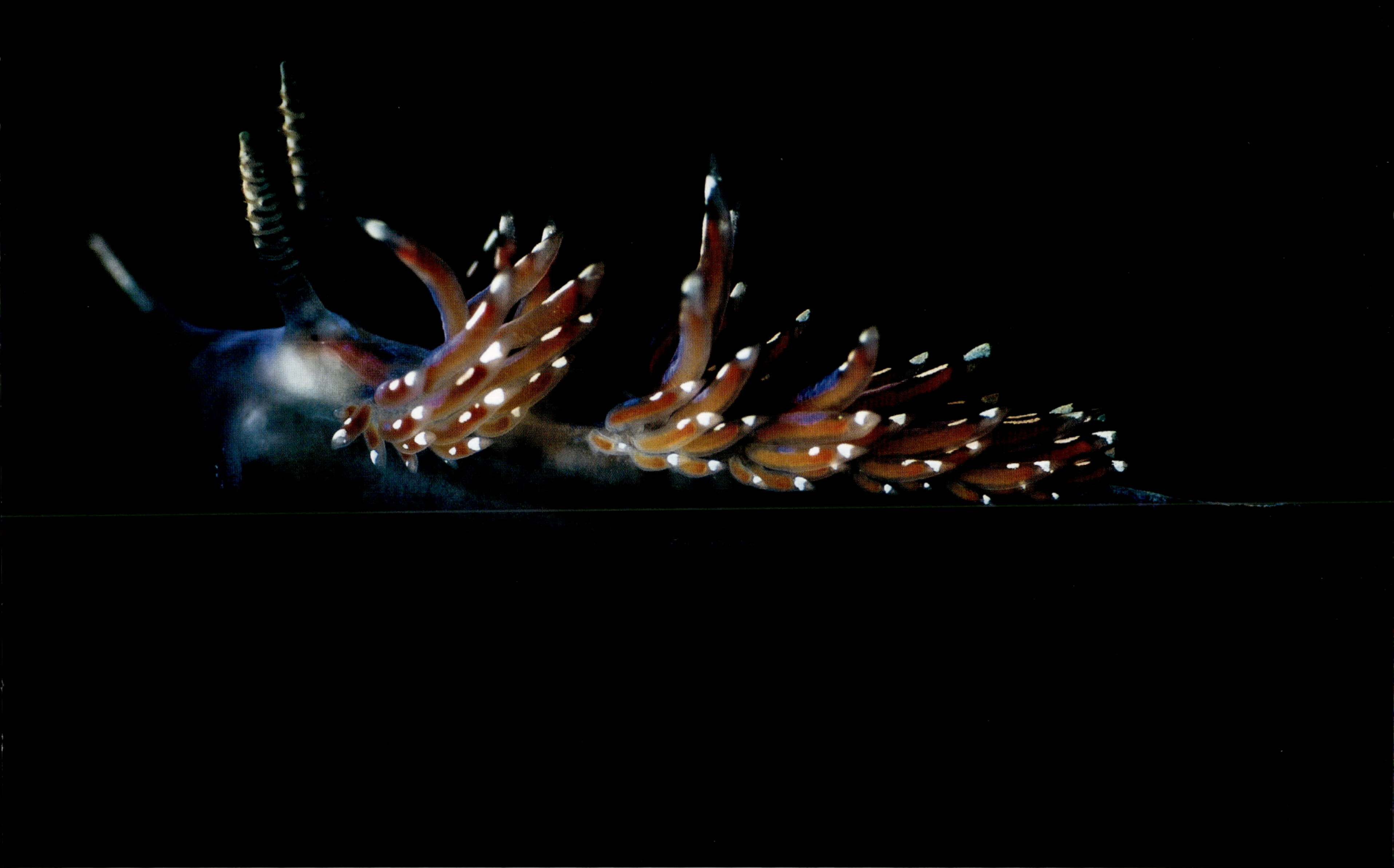

GEORGE STOYLE

Nudibranch at Night
(Nudibranch or Sea Slug, *Facelina auriculata)*
North Rona, Scotland

On a night dive just off the island of North Rona I noticed this tiny nudibranch floating freely in the water column, obviously having become dislodged by the current. Maintaining buoyancy and quickly positioning the focus lights I lit up the miniscule subject so my camera had something to focus on. Focusing on a moving subject underwater with a macro dioptre was tricky, but after some persistence I got the shot I was after.

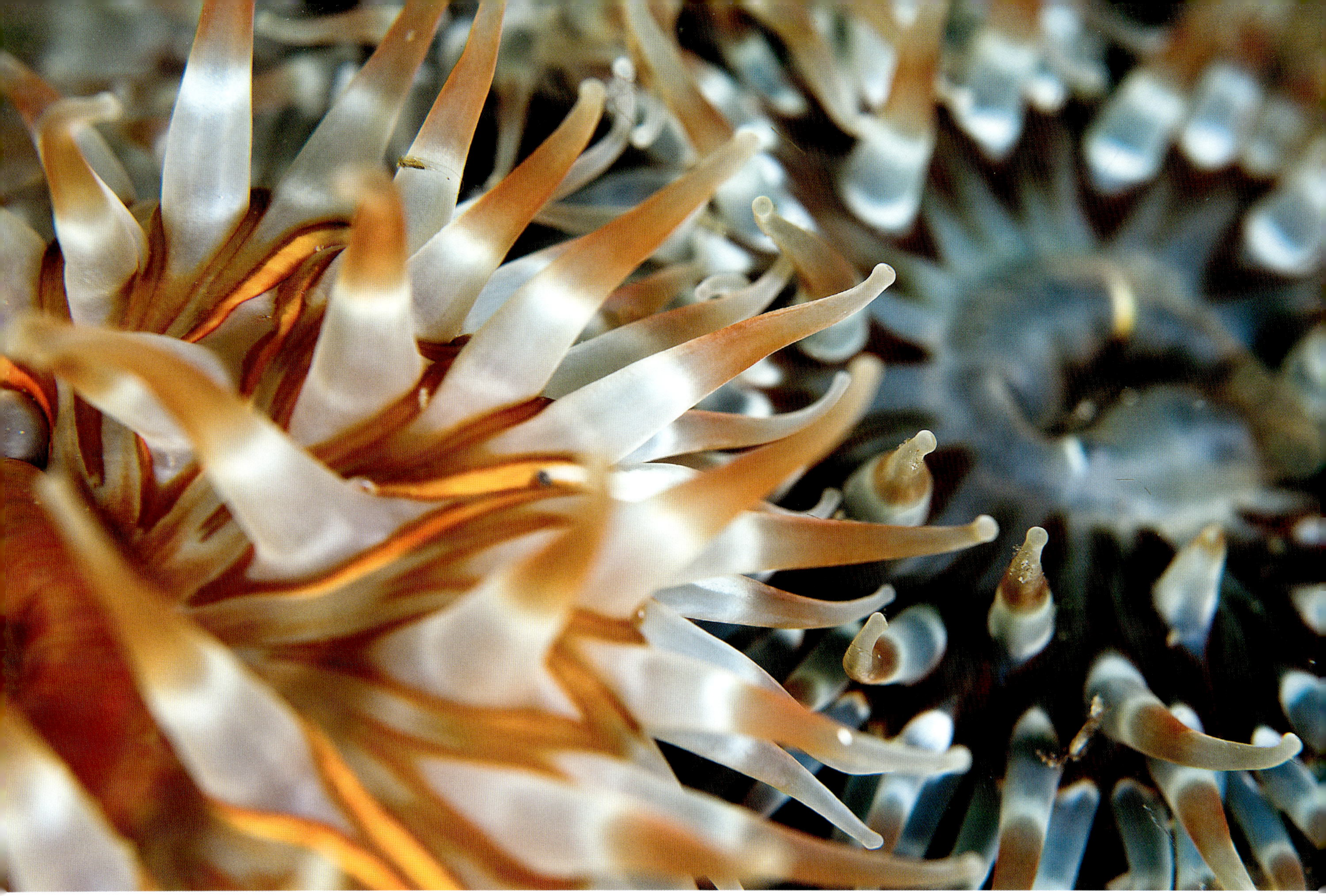

RICHARD SHUCKSMITH

Beautiful but Deadly Tentacles
(Dahlia Anemone, *Urticina felina)*
Shetland Isles, Scotland

The very colourful and beautiful dahlia anemone are always a welcome sight on any dive and look like the flowers of the sea, giving the impression of harmless beauty. However, anything swimming close or getting caught in these deadly stinging tentacles is food for this animal.

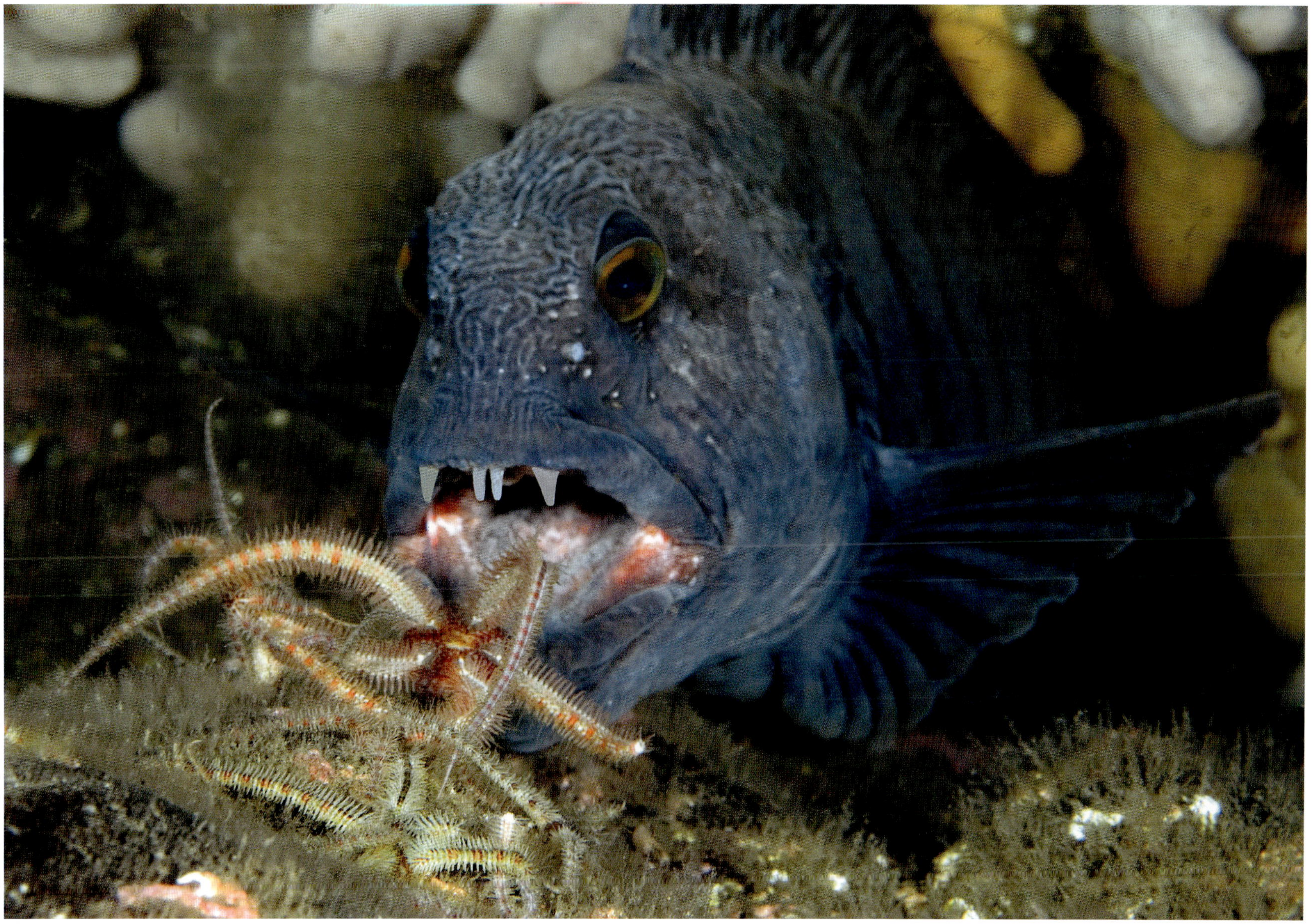

TREVOR REES

Hungry Wolf-fish
(Wolf-fish, *Anarhichas lupus*)
Eyemouth, Berwickshire, Scotland

A hungry wolf-fish emerges briefly from its hiding hole to feed on brittlestars. To find wolf-fish it's a case of diving in the sea off the cooler north-east coast of the UK. The Eyemouth area is a good spot. Wolf-fish are fairly lethargic creatures and tend to look a little more ferocious than the actually are. However, with their impressive teeth they are still fearsome predators. Feeding action occurs pretty rapidly so I needed to press the shutter at just the right time to capture the peak of the action.

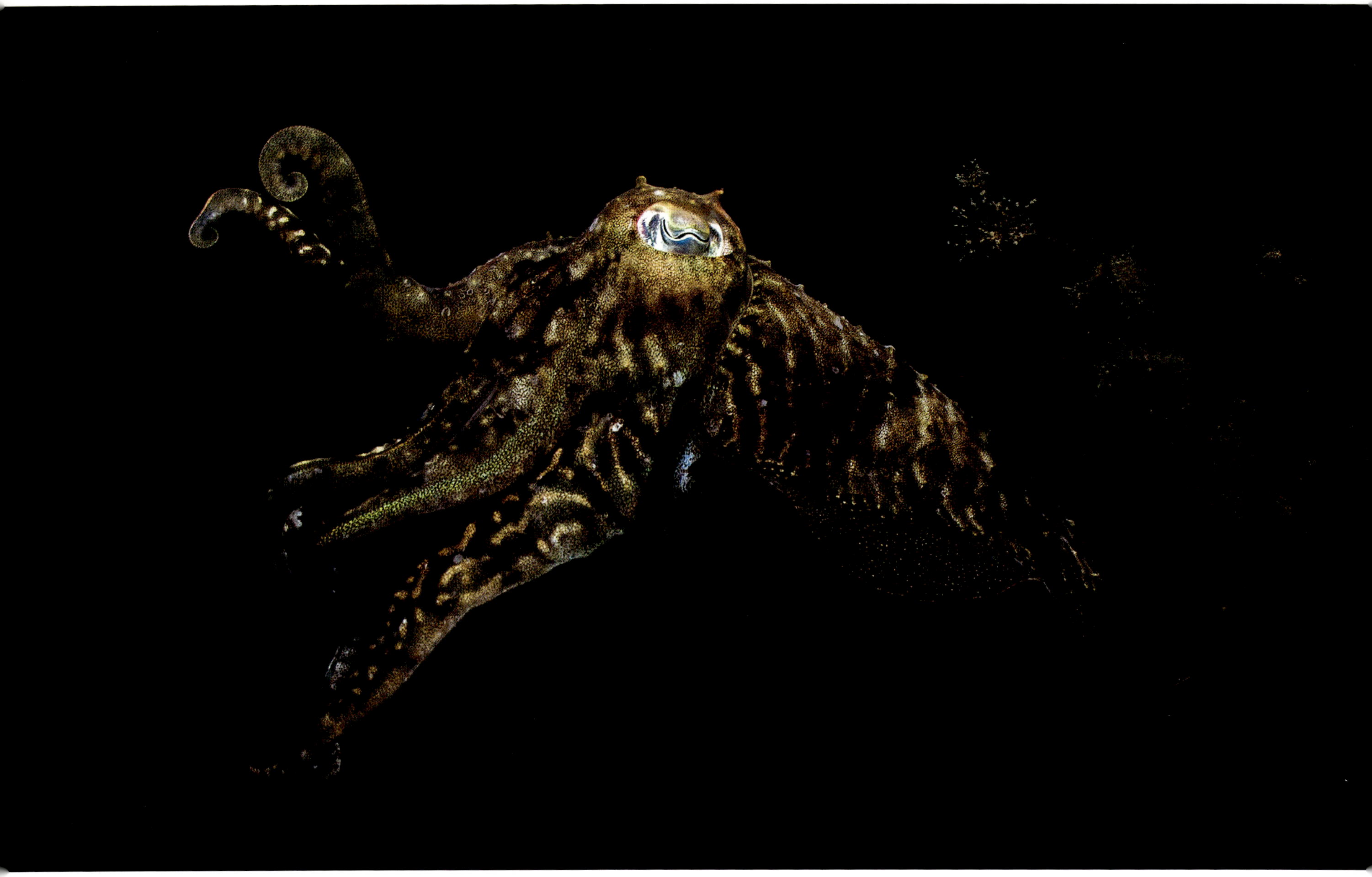

DANY ROS

Cuttlefish at Babbacombe Beach
(Cuttlefish, *Sepia officinalis)*
Babbacombe, Torquay, Devon, England

It was an early morning dive at the beginning of the cuttlefish season and the fishermen hadn't yet cordoned the bay with pots. I came across this cuttlefish that seemed to be guarding something in the different seaweeds. I approached as slowly as I could, stretching my arms so the camera was as close as possible. The cuttlefish came out of its hideout spreading its tentacles. This surprised me as cuttlefish are usually timid.

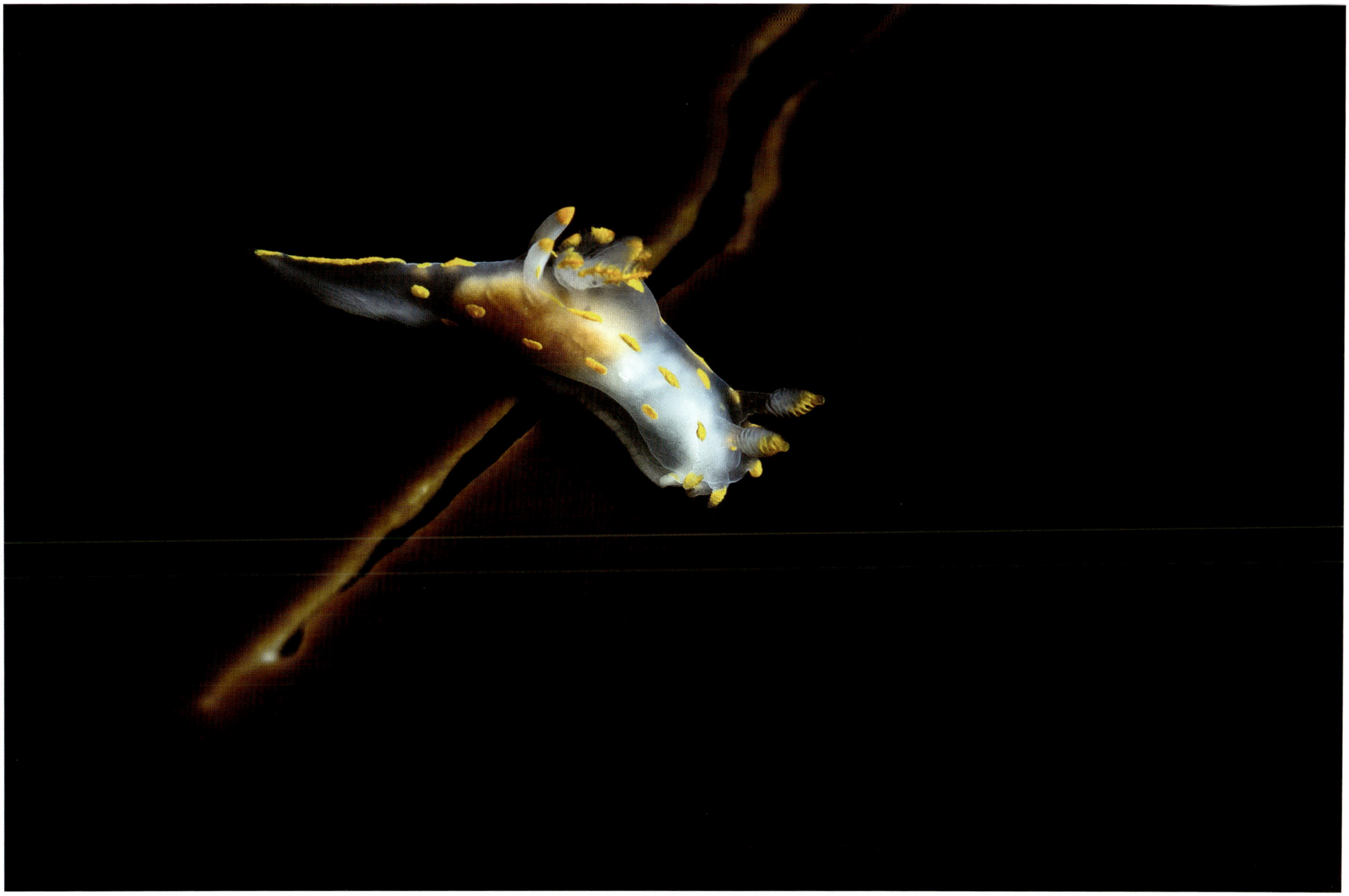

GEORGE STOYLE

Nudibranch on Kelp at Night
(Nudibranch or Sea Slug, *Polycera quadrilineata)*
North Rona, Scotland

In the turbulent waters surrounding the uninhabited island of North Rona these tiny nudibranchs can be tricky subjects to photograph. Although the kelp fronds were covered in these tiny creatures, this particular one caught my eye as it was perfectly positioned over a divide, which looked like a lightning bolt. This charismatic little creature waved gently in the tide as I waited patiently for the right moment when both my buoyancy and subject were stable.

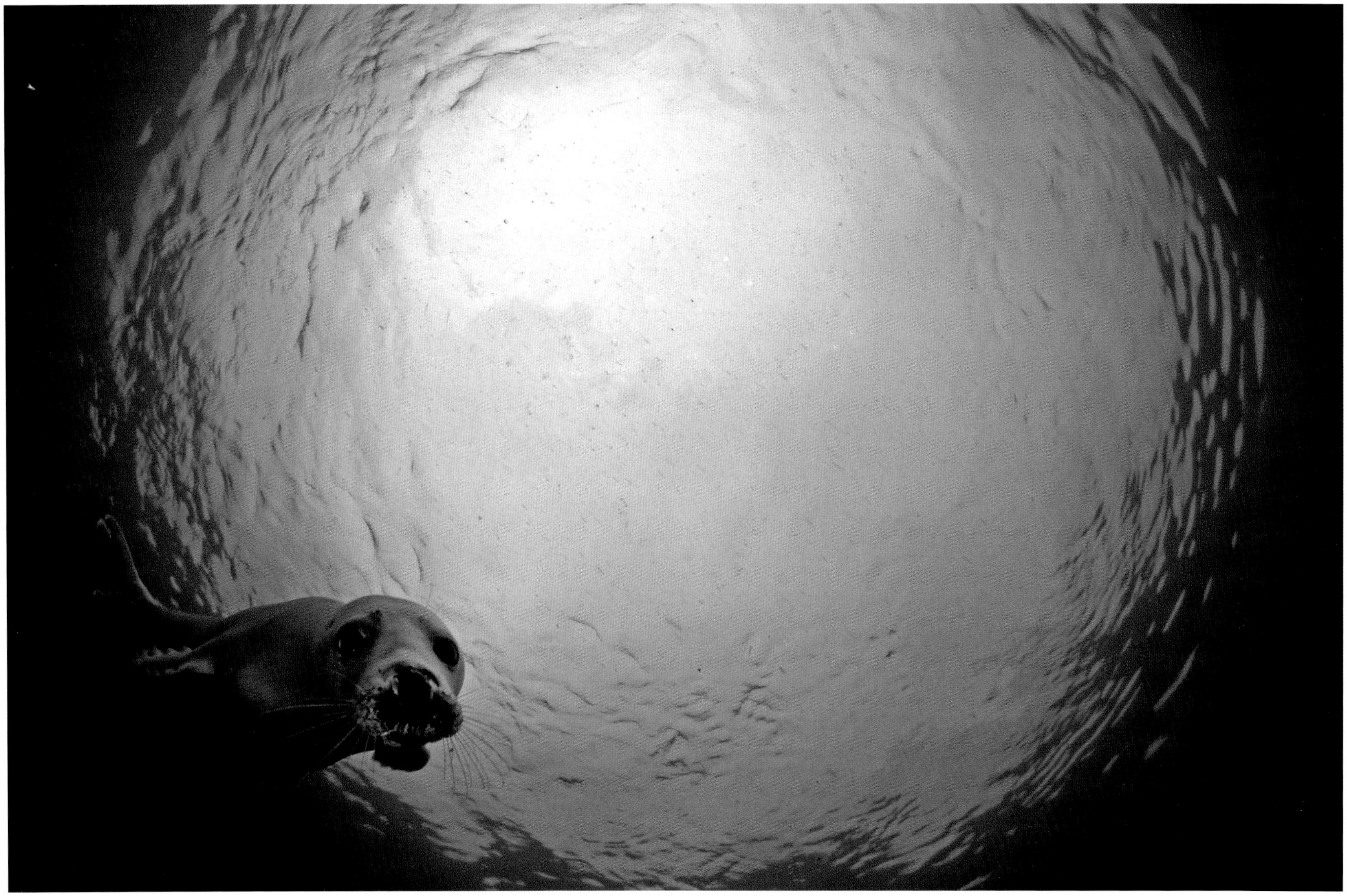

ALEX TATTERSALL

Seal in Snell's Window
(Grey Seal, *Halichoerus grypus*)
Farne Islands, England

Taken lying on my back in five meters of water, waiting for a seal to enter the frame. Wonderful experience.

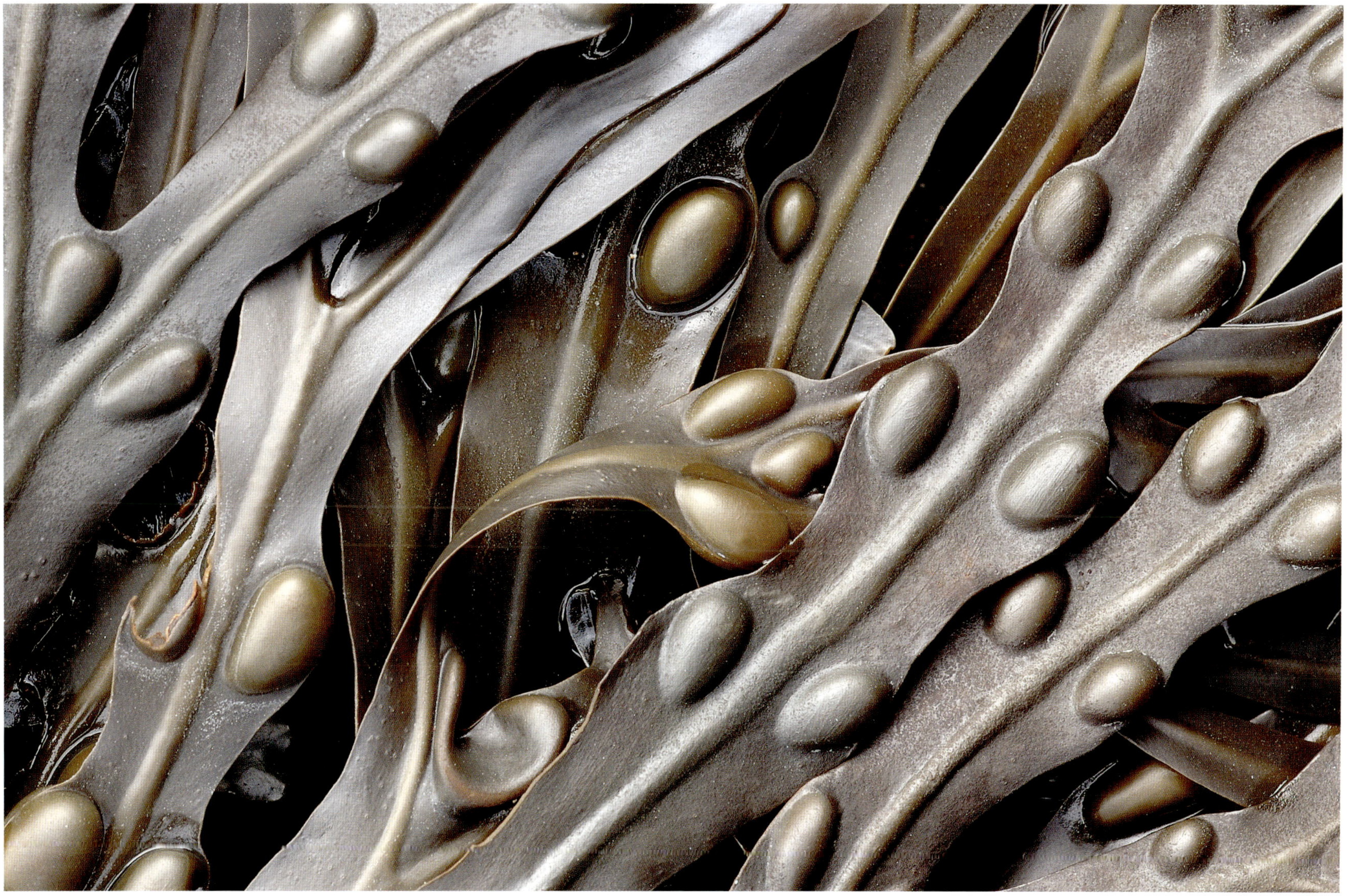

ROSS HODDINOTT

Seaweed
(Bladder Wrack Seaweed, *Fucus vesiculosus)*
Sandymouth Bay, Cornwall, England

Coastal birds and marine wildlife are impressive, so it is easy to overlook less glamorous species like molluscs and seaweed. I love the texture and natural patterns created by rocky outcrops smothered with seaweed. I photographed this bladder wrack using a macro lens, so I could isolate just a small area. Although taken a few years ago, it is still one of my favourite beach close-ups.

WILDLIFE IN MY BACKYARD

SPONSORED BY
COUNTRYSIDE COUNCIL
FOR WALES

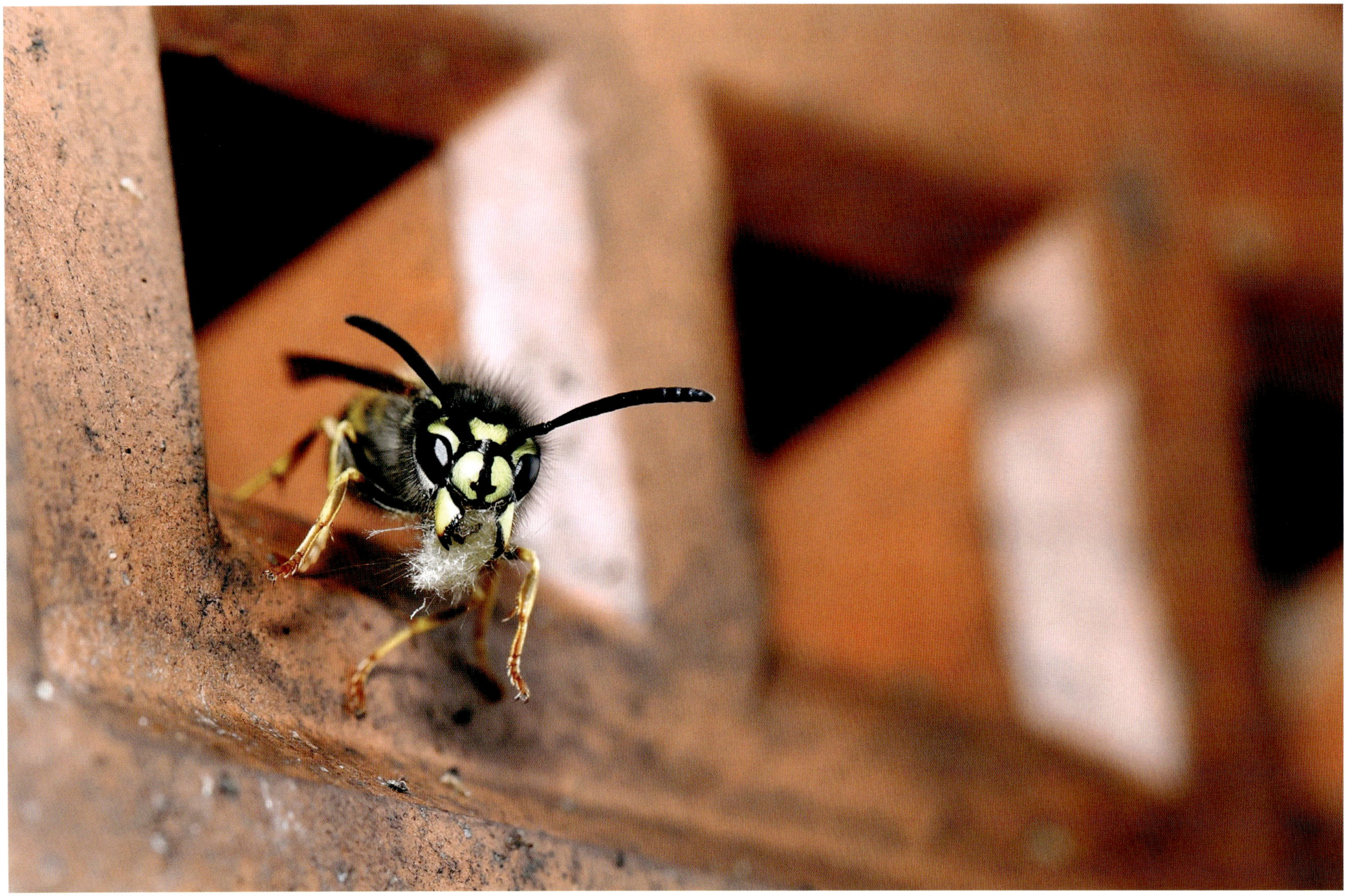

WILDLIFE IN MY BACKYARD WINNER

DAVID THOMAS HANDLEY ∧

Wasp House Cleaning from a Nest in a Cavity Wall
(Wasp, *Vespula vulgaris)*
Tan-y-fron, near Wrexham, North Wales

This shot was taken in my backyard just above the dustbin. As it was the weekend and there was nobody available to come and deal with the wasp's nest, I thought I may as well try and get some photos. This proved challenging as the nest was high on the wall so I had to use a small set of stepladders. My courage lasted about 20 minutes before deserting me.

DENISE BISHOP > HIGHLY COMMENDED

Dragonfly
(Hawker Dragonfly, *Aeshnidae)*
Chelmsford, Essex, England

Over the past few years I have tried to establish a garden that attracts various wildlife species, and on a sunny summer's day last year I was rewarded for those efforts when this beautiful dragonfly flew into my garden and landed on my newly established verbena. The light was perfect and with the dragonfly having landed fairly low, I was able to position myself amongst the undergrowth to take this aerial view.

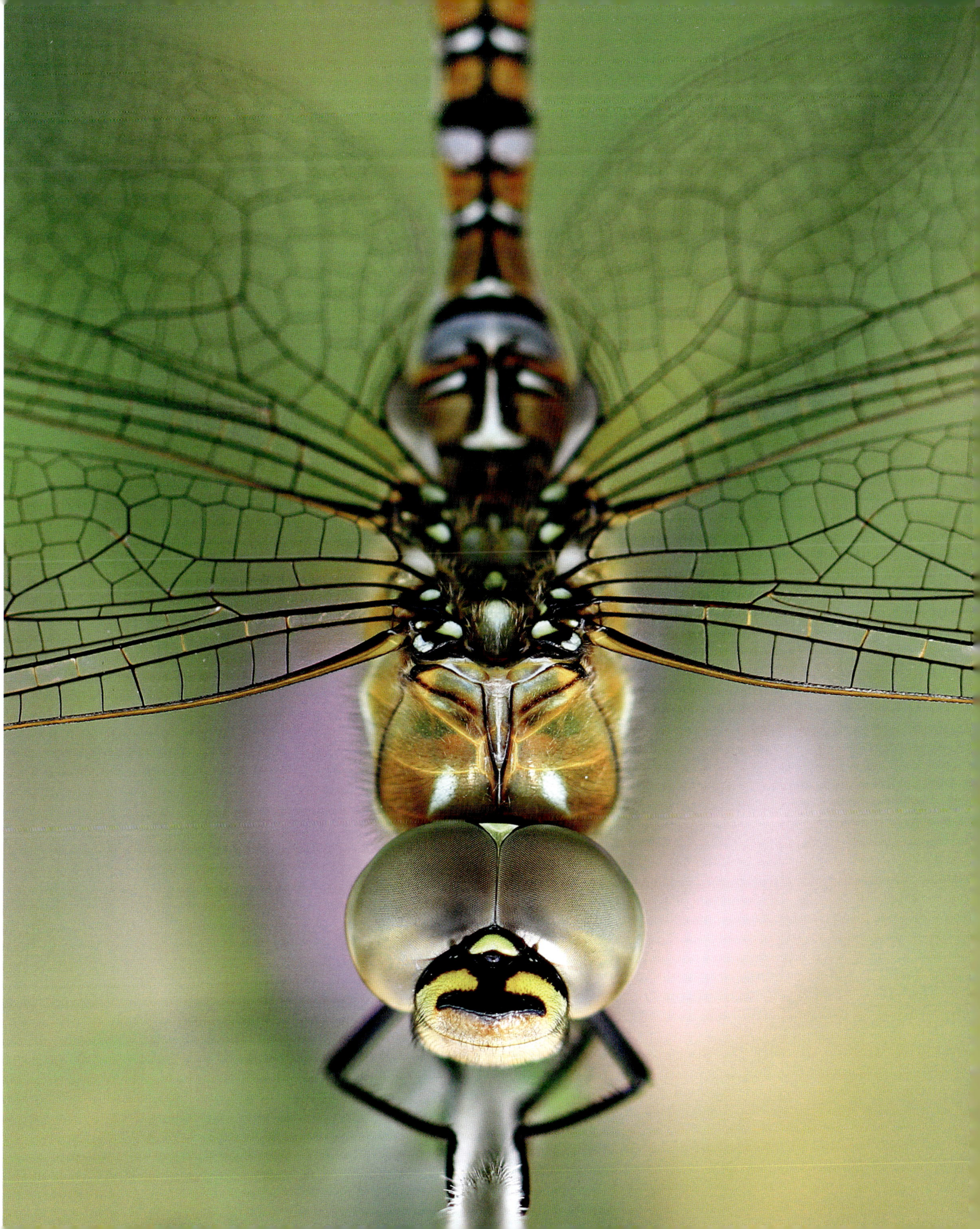

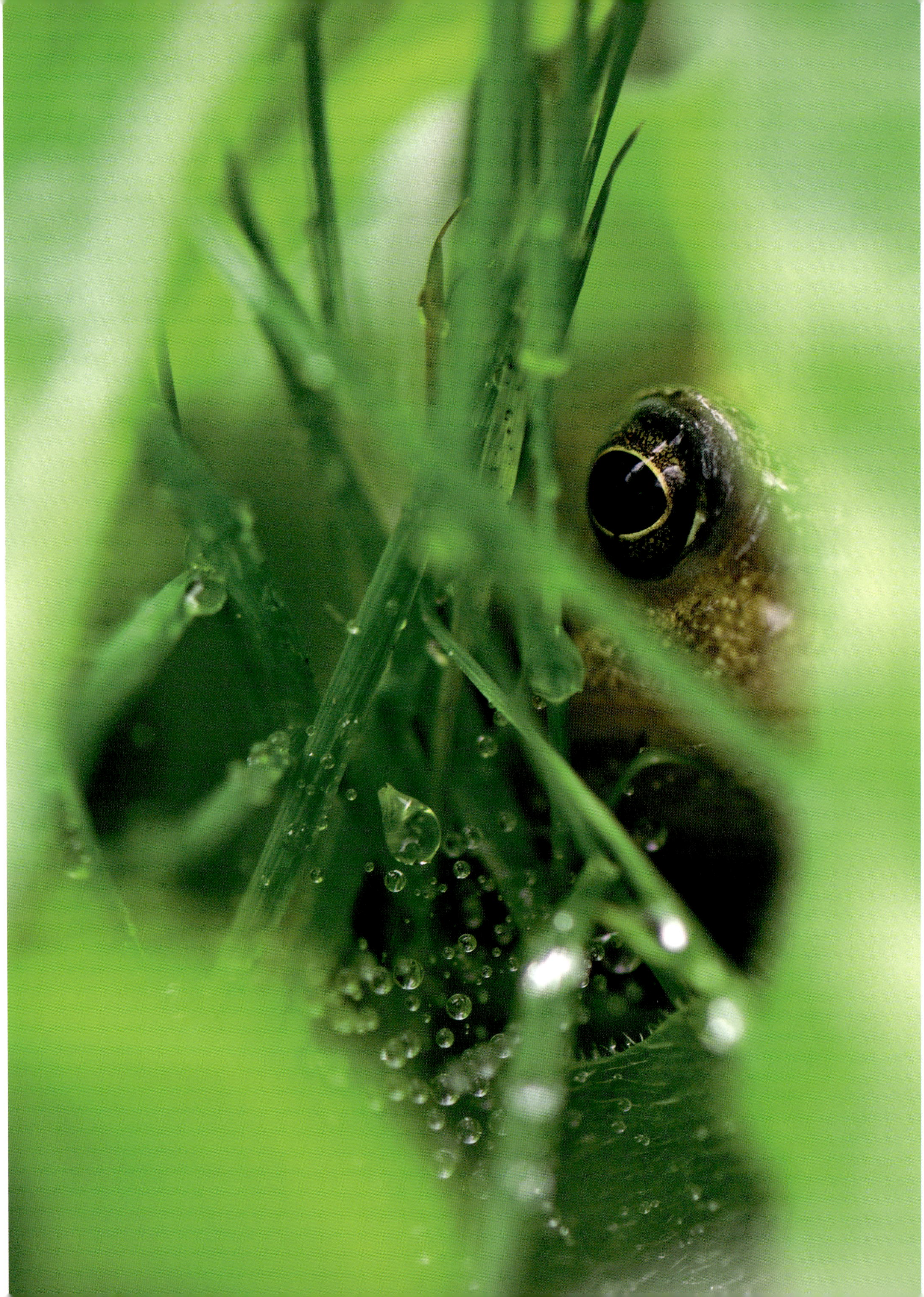

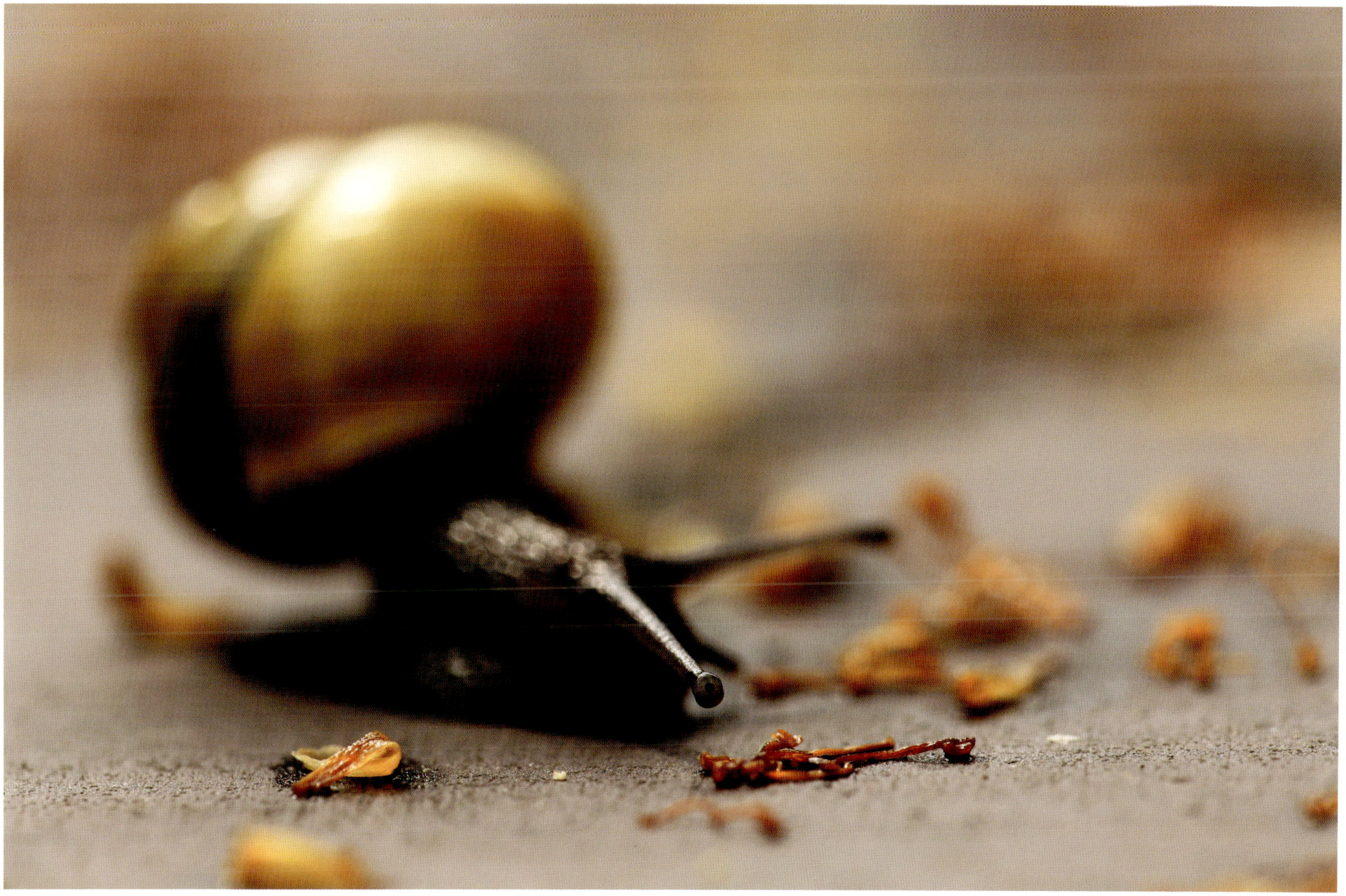

JESS F MASLEN < HIGHLY COMMENDED

Watching Me, Watching You
(Common Frog, *Rana temporaria)*
Tavistock, Devon, England

After a downpour I went out in my front garden and spotted this frog in the grass. I lay down in front of it and, using a short depth of field, focused on the eye, still keeping some of the raindrops in focus… I wanted to infer both the secretive/shy character of the frog and the fact that nature is watching us, even though it's not always obvious. I only managed two shots before the frog was gone.

JAMES YAXLEY Λ

Snail's Eye View
(Grove Snail, *Cepaea nemoralis)*
Norwich, Norfolk, England

I lay on the ground and focused on the pupil of the snail's eye as it looked at some leaf litter. The shallow depth of field and low camera angle emphasise that this is a snail's-eye view of the world; a view that most of us rarely see!

RACHEL PIPER

Frog in My Garden Pond
(Common Frog, *Rana temporaria*)
Hemel Hempstead, Hertfordshire, England

This frog, which was located in my garden pond, seemed unperturbed by my presence and posed happily for the camera. These are fascinating creatures and I never grow tired of watching them and observing their incredible life cycle.

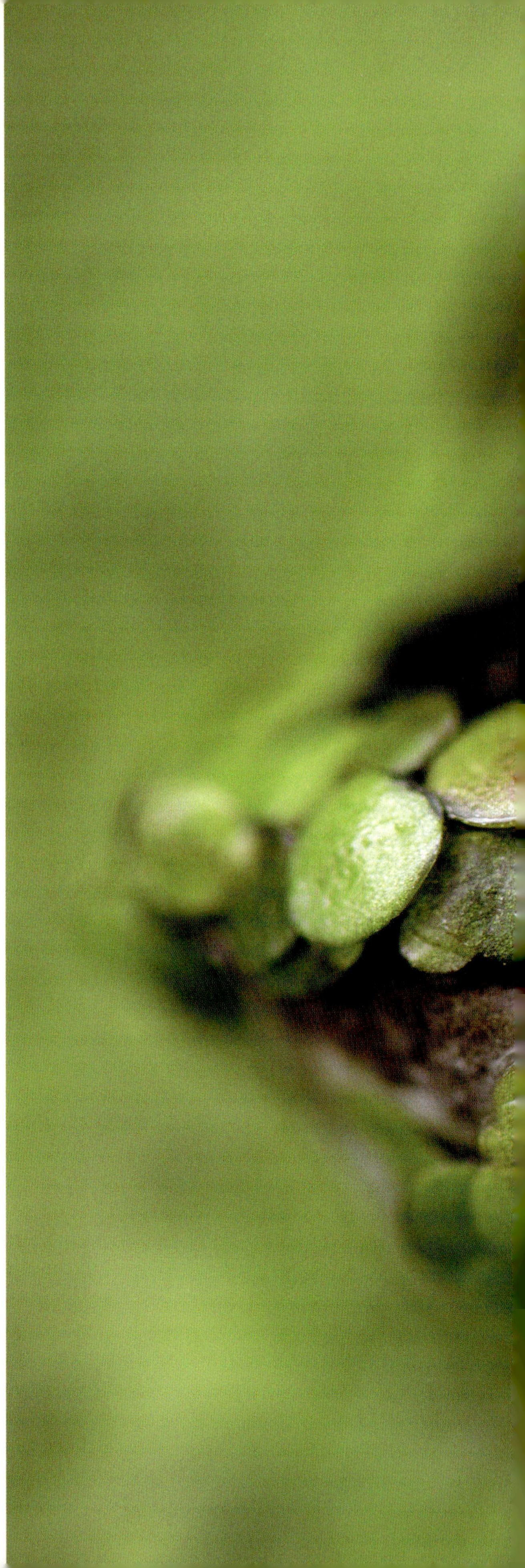

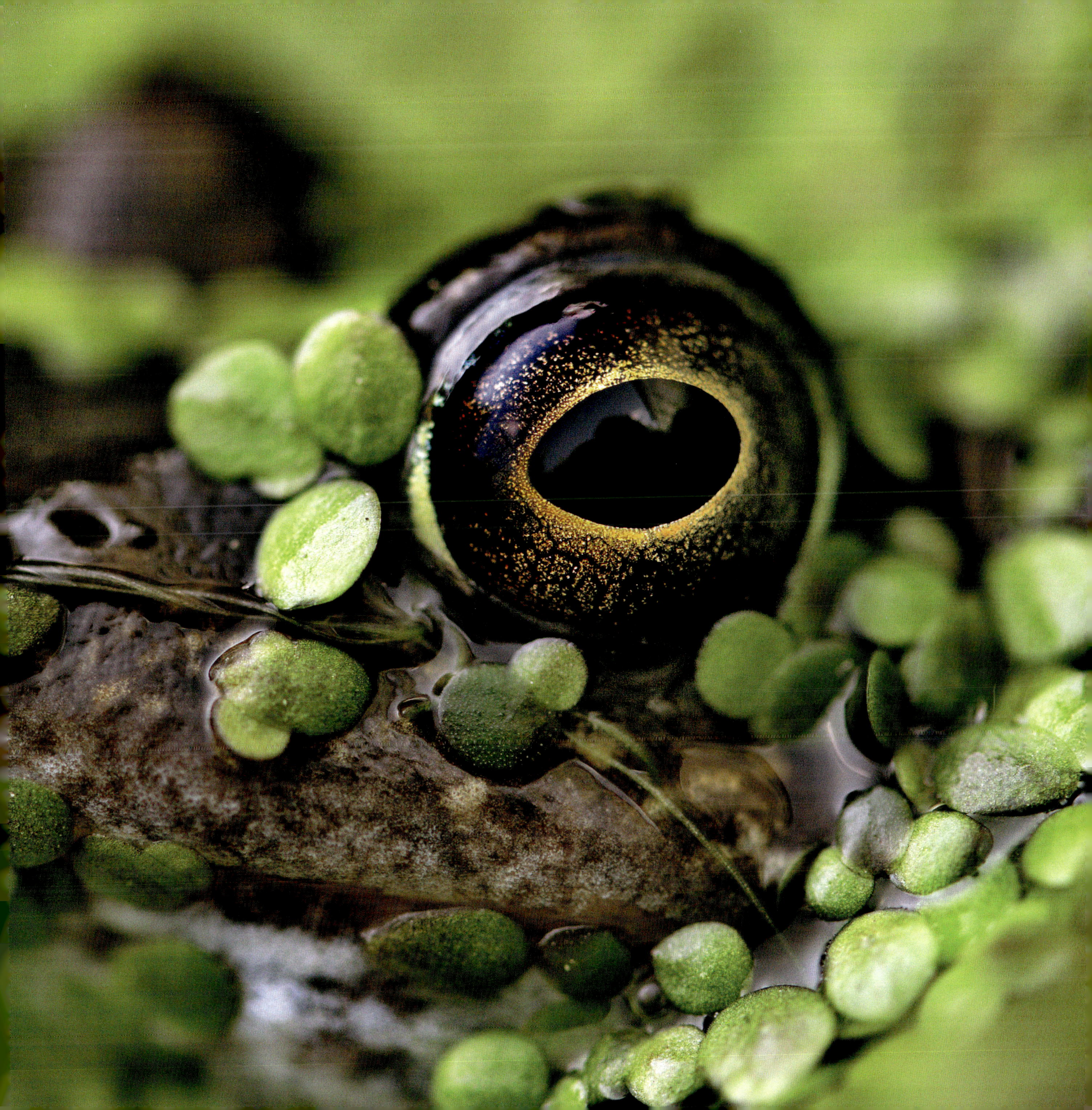

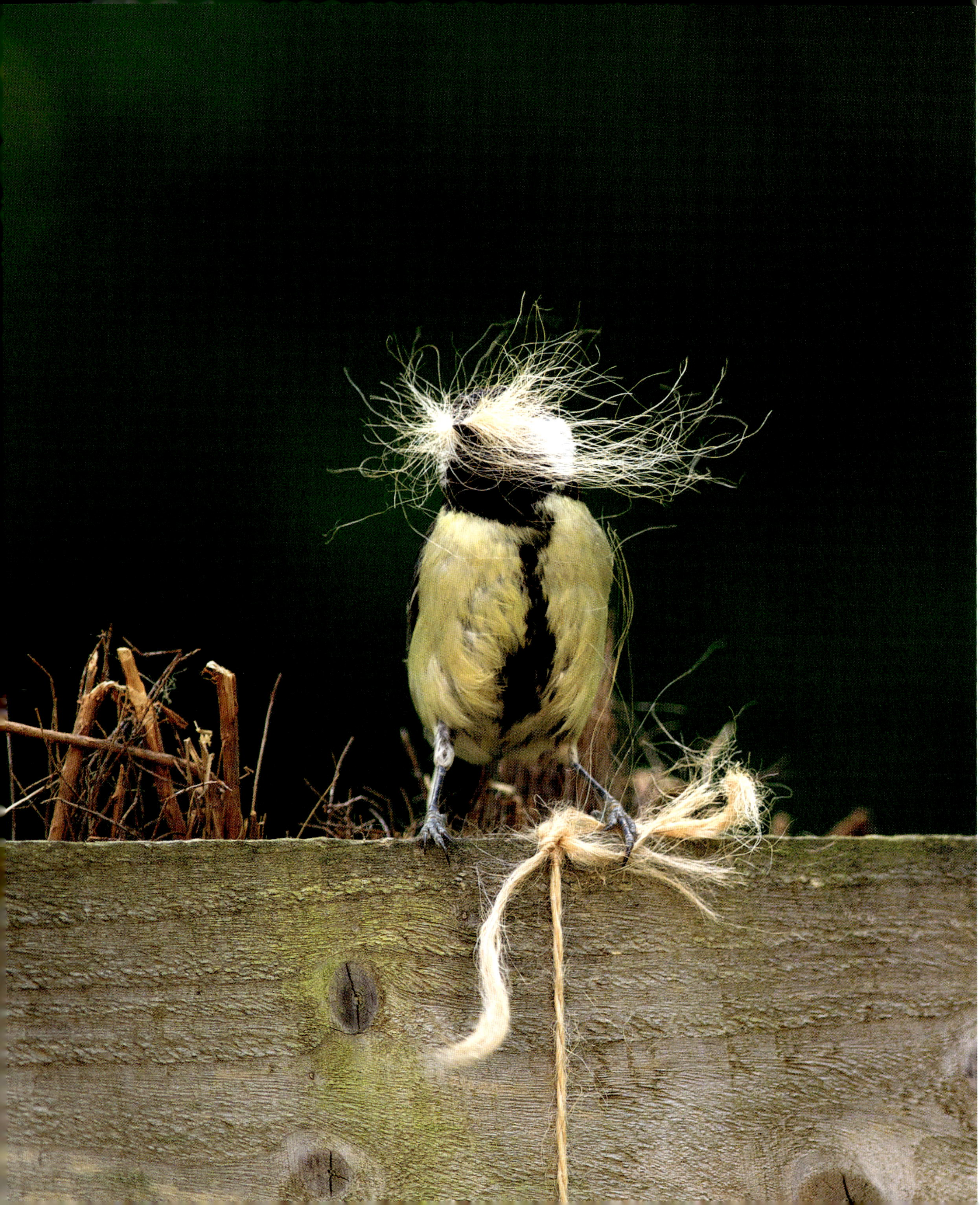

ANNE LATRÉMOLIÈRE <

Great Tit with a Beakful of Nesting Material

(Great Tit, *Parus major)*

Great Hill, near Tring, Hertfordshire, England

This photograph was taken from some distance in order not to disturb the tits taking string tied to our fence. The image shows the large number of strands that the birds would tease out while also taking care not to drop earlier strands. I wondered how well they could see while carrying their pickings.

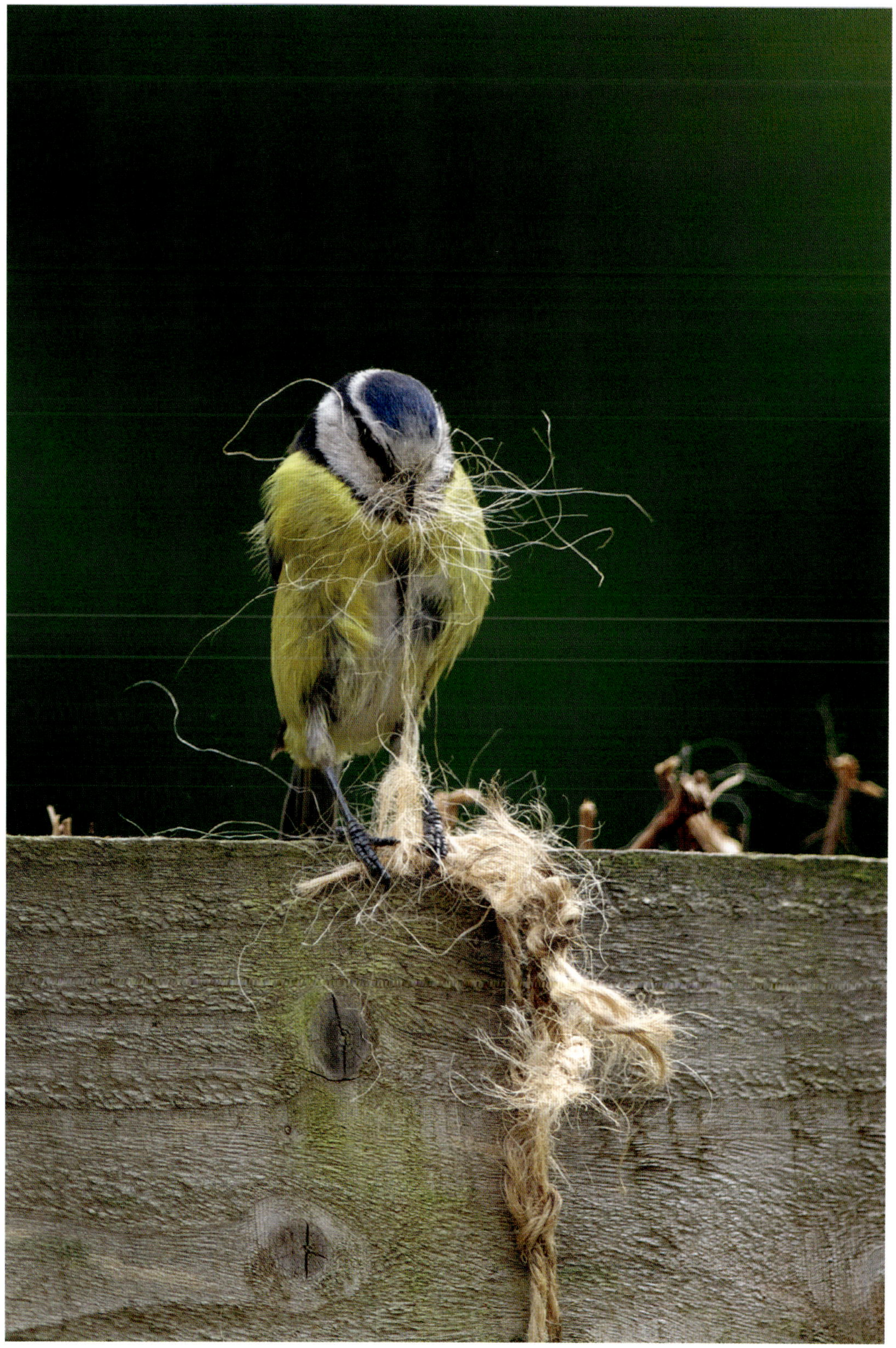

ANNE LATRÉMOLIÈRE >

Blue Tit Teasing out Nesting Material

(Blue Tit, *Cyanistes caeruleus)*

Great Hill, near Tring, Hertfordshire, England

I noticed a blue tit taking nesting material from the string tied to our fence. This image captures the effort and skill that it used to tease out the individual strands of the string.

WILD WOODS

SPONSORED BY
VISITWOODS

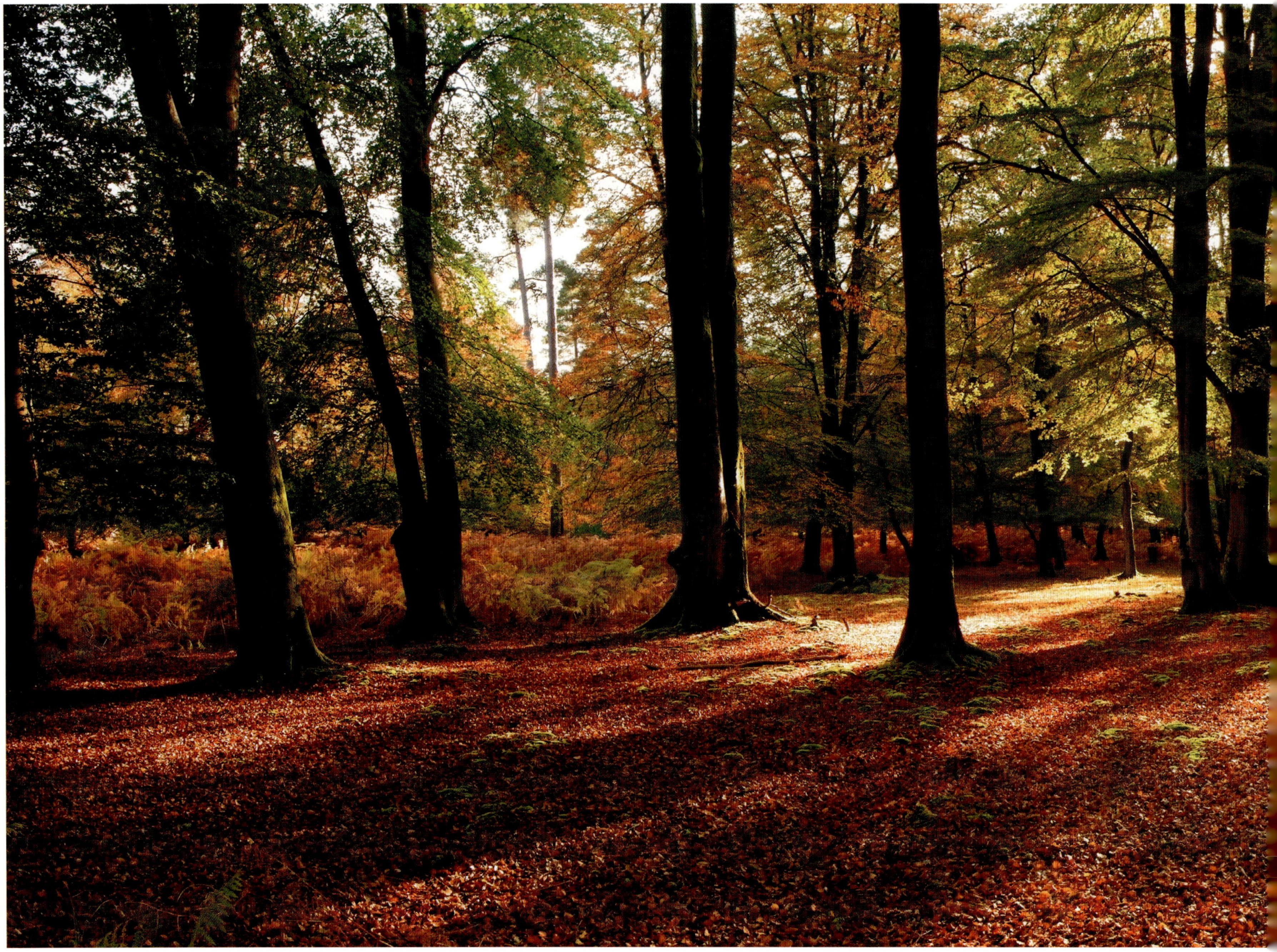

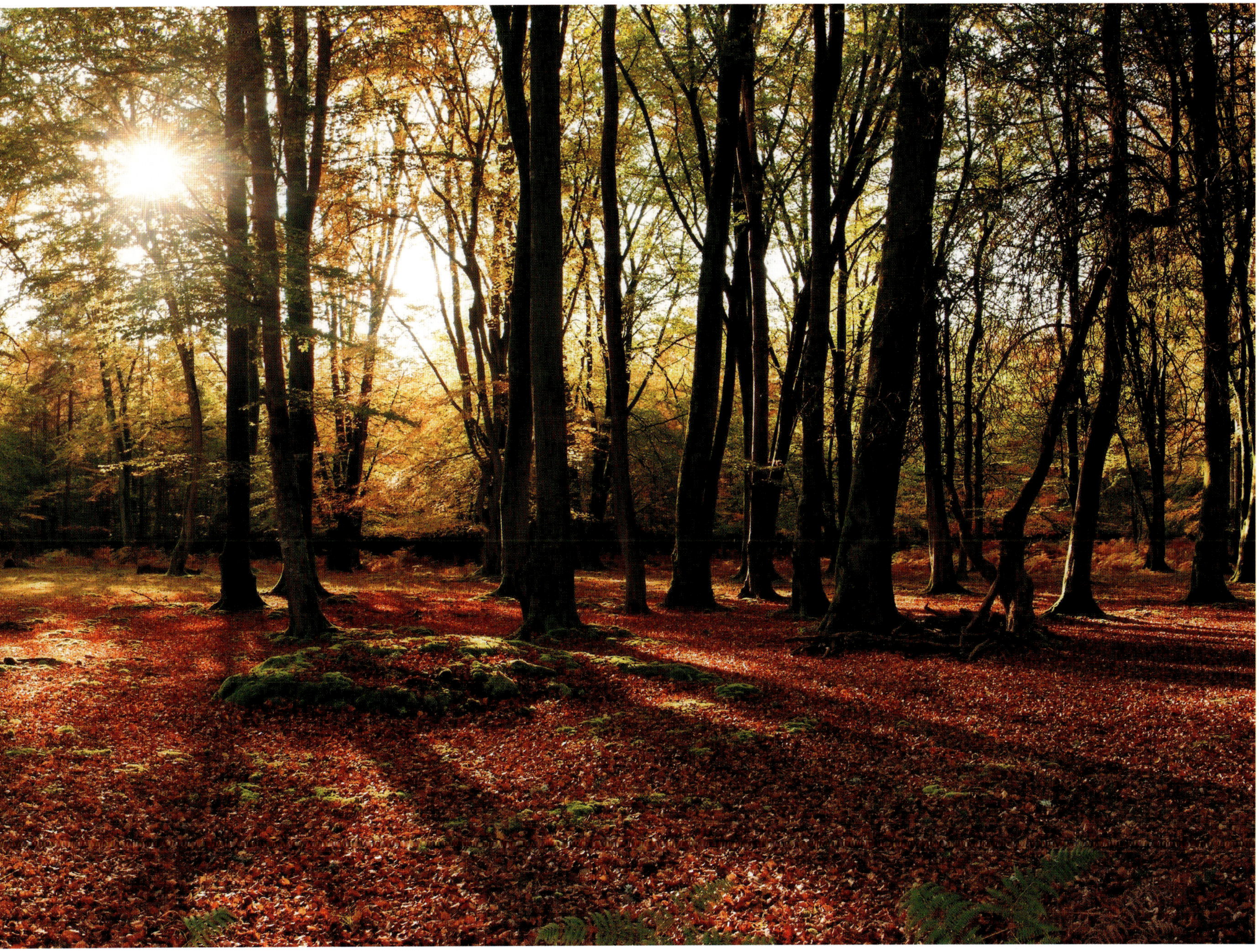

WILD WOODS WINNER

JEREMY WALKER

New Forest in Autumn
(Common Beech, *Fagus sylvatica)*
The New Forest, Hampshire, England

A classic autumnal afternoon, clear skies, a distinct nip in the air and dappled light breaking through the golden canopy to play on the forest floor. An absolute joy to photograph, but you have to work quickly as the patterns and dapples are constantly on the move.

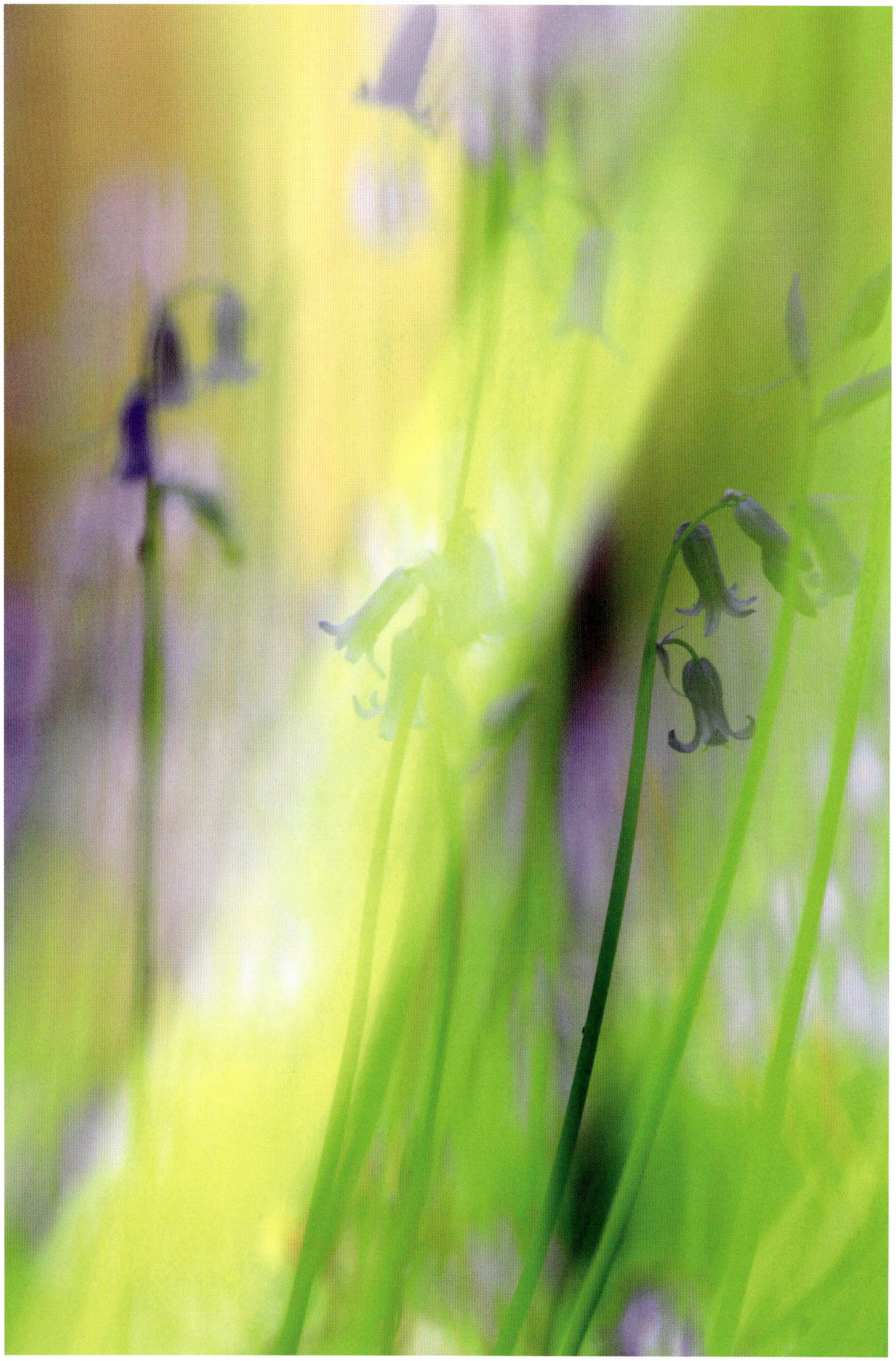

IAN WADE < HIGHLY COMMENDED

Beautiful Bluebells
(Bluebell, *Hyacinthoides non-scripta)*
Wrington, Somerset, England

I wanted to photograph bluebells in a different light; to show the bluebell in a creative painting style. Using my zoom lens I picked out one bluebell and blurred the rest into a dreamy, almost painted effect of colour and form...

ANTHONY MURDEN > HIGHLY COMMENDED

Autumn Trees, New Abbey Woods
New Abbey, Dumfries and Galloway, Scotland

On arriving in Dumfries and Galloway, one of the first locations that caught my eye in terms of the autumn colours was New Abbey. On the last day of my trip I got the break in the weather and the light that I had been waiting for and captured the season in all its glory.

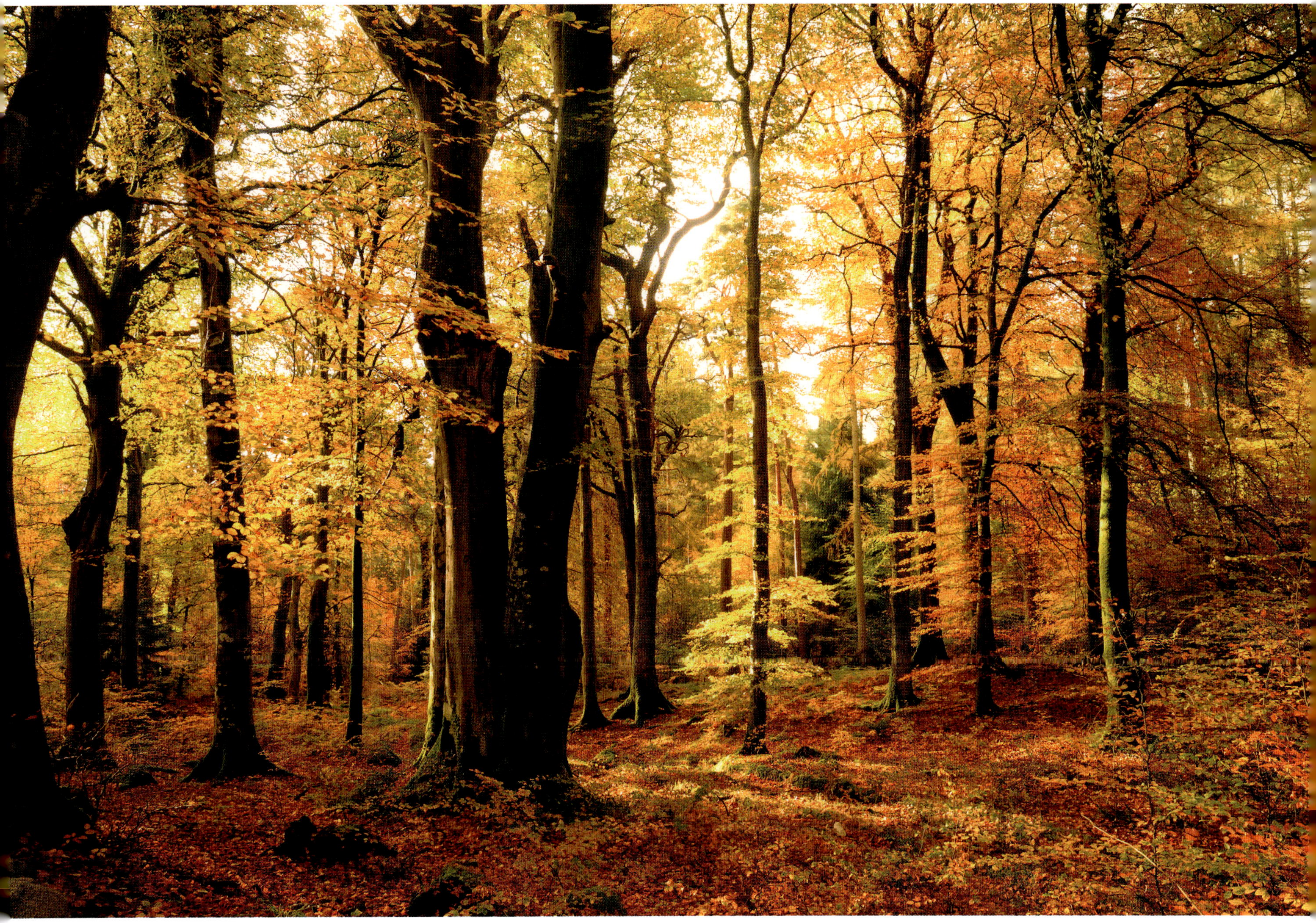

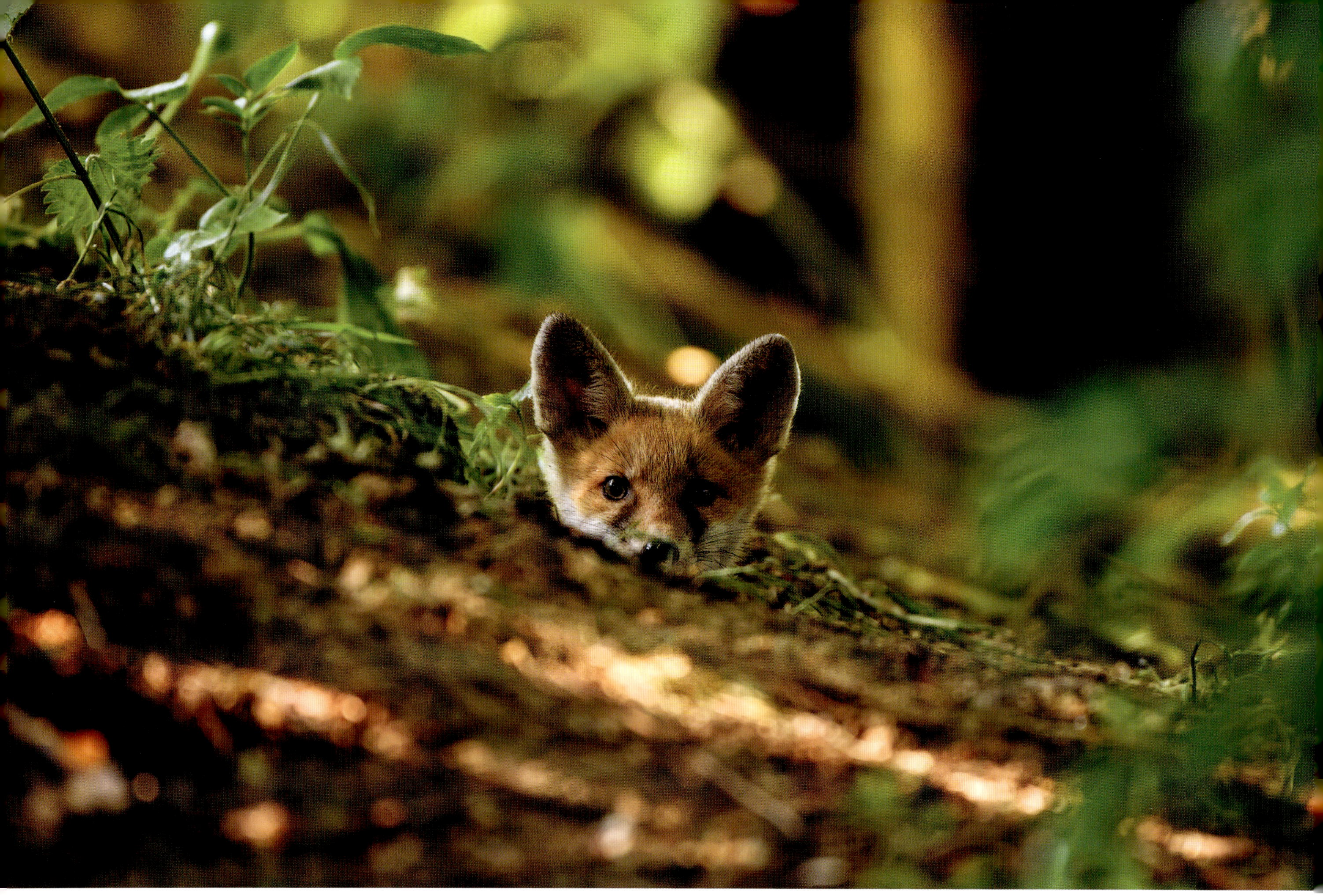

CALUM DICKSON HIGHLY COMMENDED

Fox Hole
(Red Fox, *Vulpes vulpes*)
Epping Forest, Essex, England

I had been watching the behaviour of this fox cub for days. Each time it emerged from the hole it momentarily checked that the coast was clear. There was never a warning when it was about to emerge, so it was a case of lying on the forest floor behind a tree with a finger on the shutter. Shooting in the forest meant little light, high ISO and a slow shutter speed: tricky conditions.

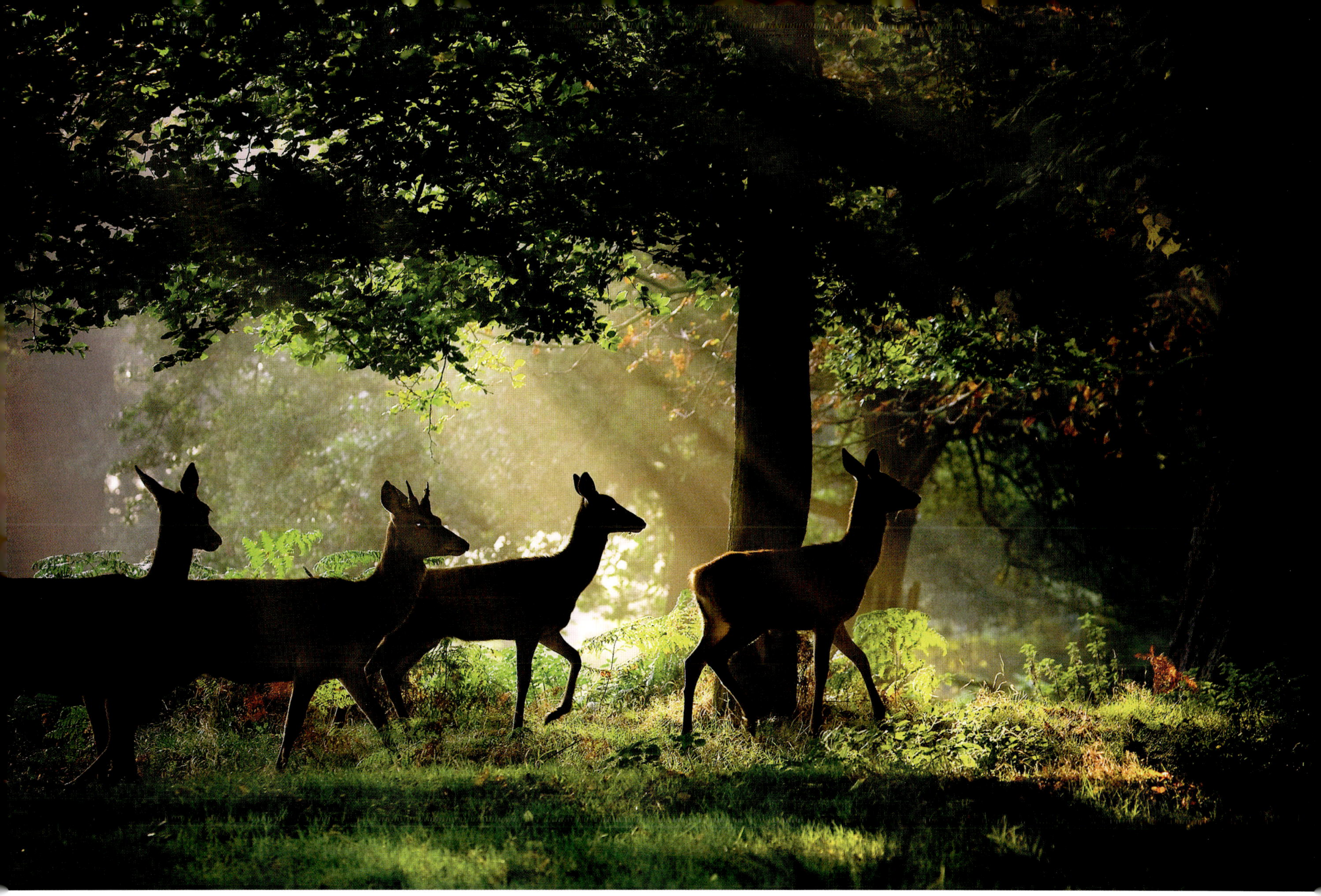

ALEX SABERI HIGHLY COMMENDED

Fairytale 1
(Red Deer, *Cervus elaphus)*
Richmond Park National Nature Reserve, Surrey, England

Almost right at the end of the morning's shoot I saw these stunning deer pass by. It was a magical moment because of the light, the mist and the colour. One of my all-time favourites I have taken in Richmond Park.

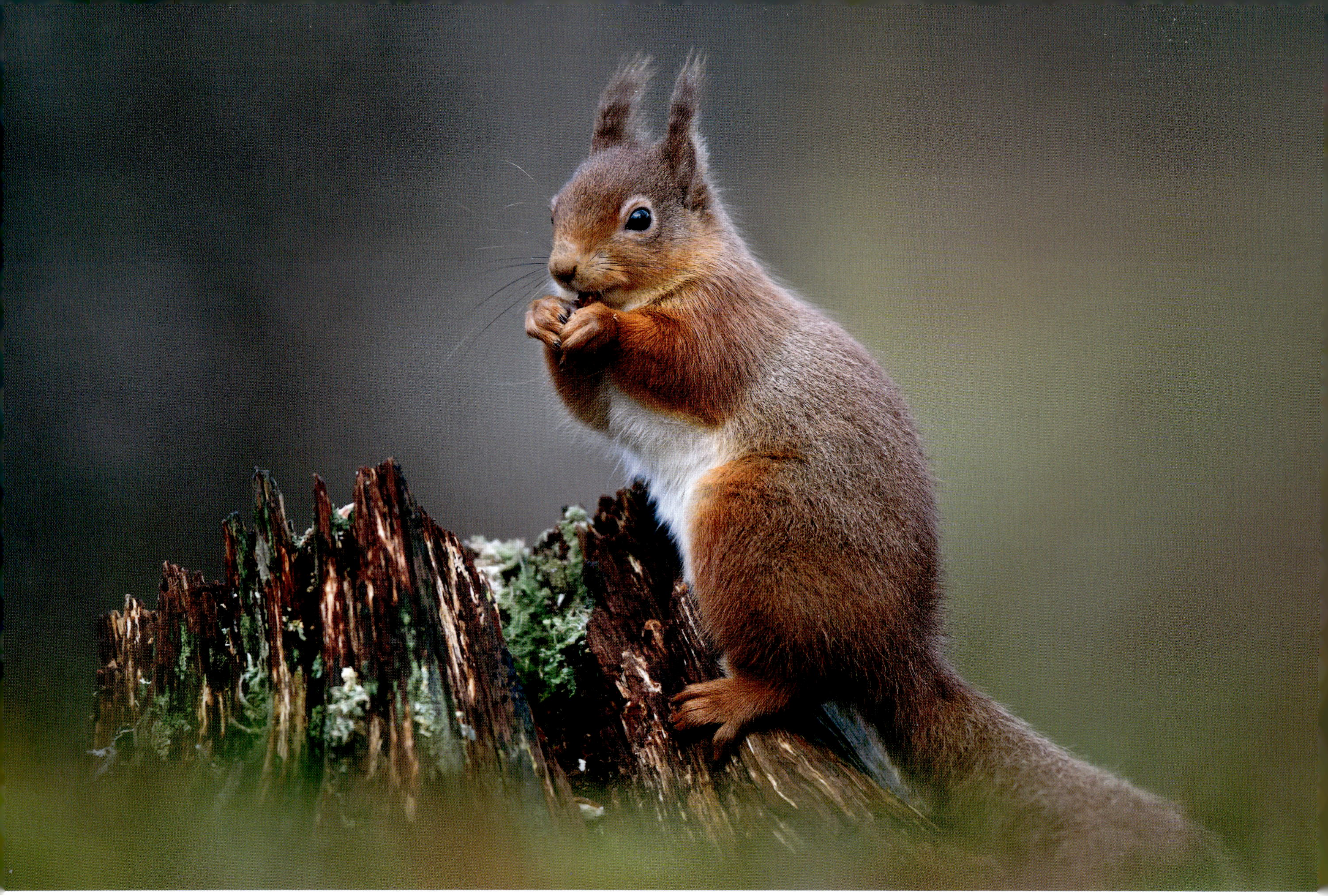

CALUM DICKSON ∧ HIGHLY COMMENDED

Squirrel Mush
(Red Squirrel, *Sciurus vulgaris*)
Aviemore, Highland, Scotland

The squirrel was feeding in the heather around the tree stump in very poor weather conditions. The lichen-covered tree stump caught my eye and I wanted to capture the squirrel against this with everything else out of focus. I set up low and waited for the squirrel to go on to the stump. Shooting wide open I knew the heather would turn to mush. Getting the image sharp was the challenge.

DREW BUCKLEY > HIGHLY COMMENDED

Water Under the Bridge
River Brathay near Clappersgate, Cumbria, England

Being out in autumn it's hard not to fall in love with the natural wonder that's going on all around you. Here, stunning amber foliage carpeted the riverbanks, encompassing the whole scene in an extremely vivid colour tapestry. Capturing the sheer beauty of this magical season in an almost timeless British image was my uppermost aim.

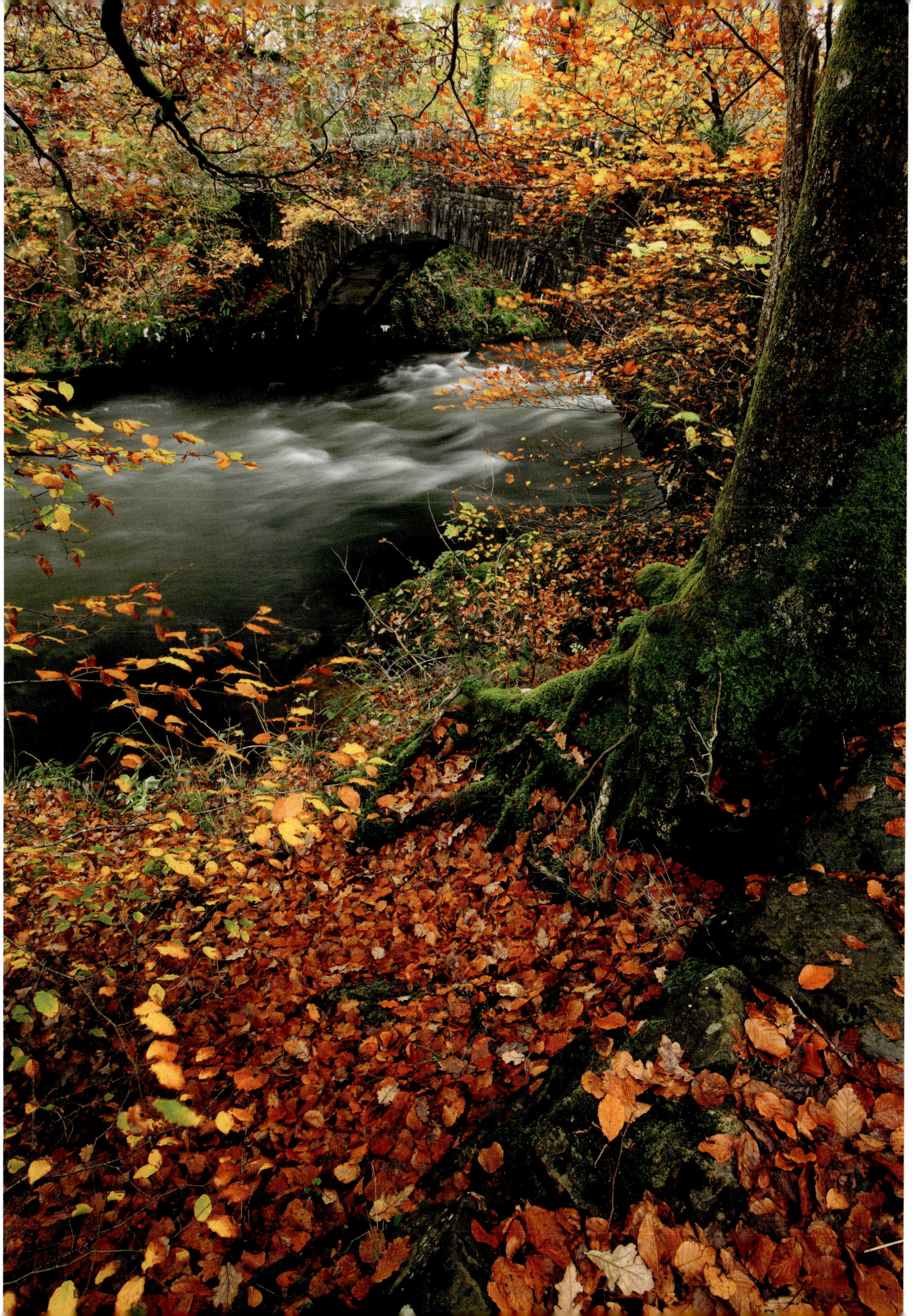

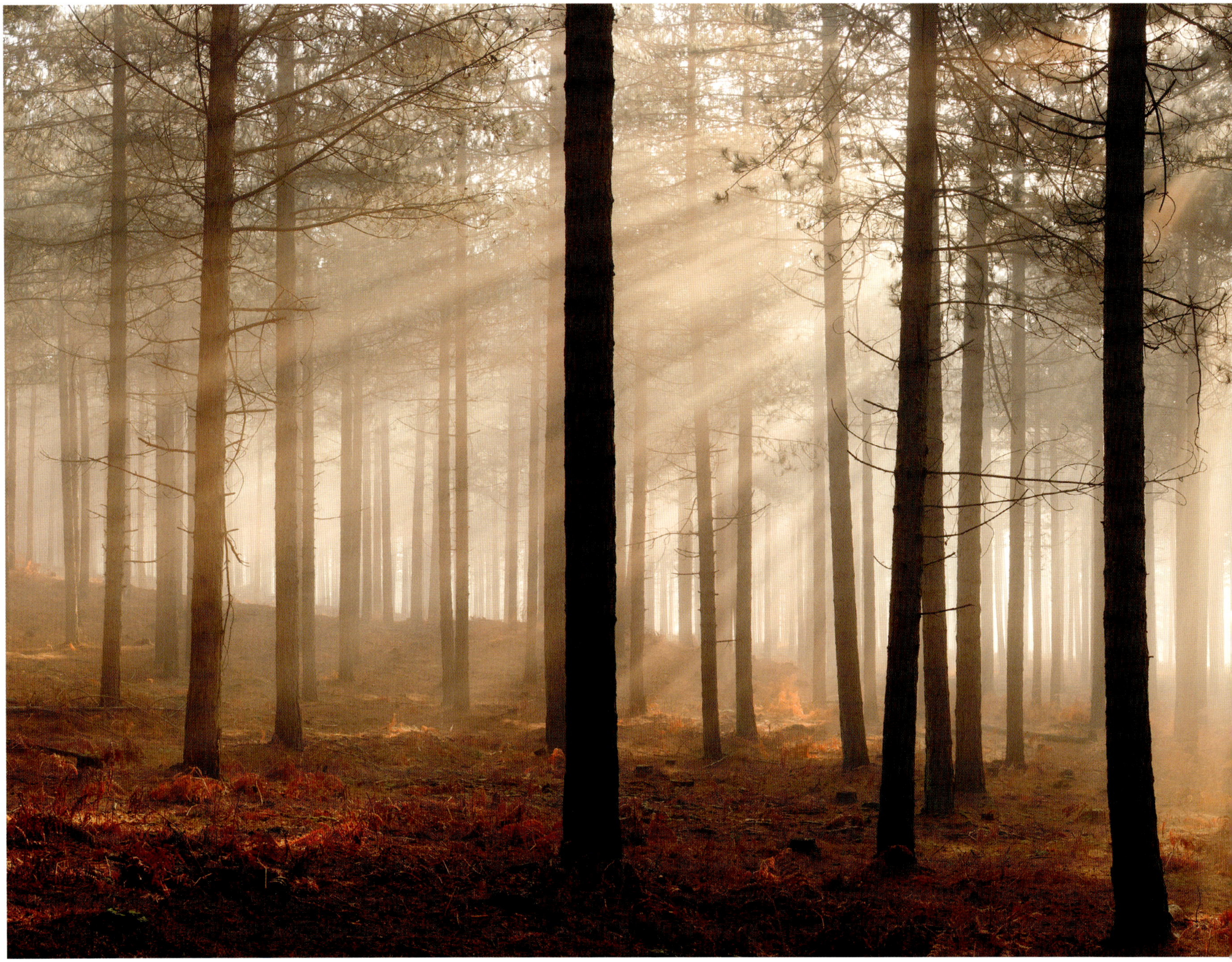

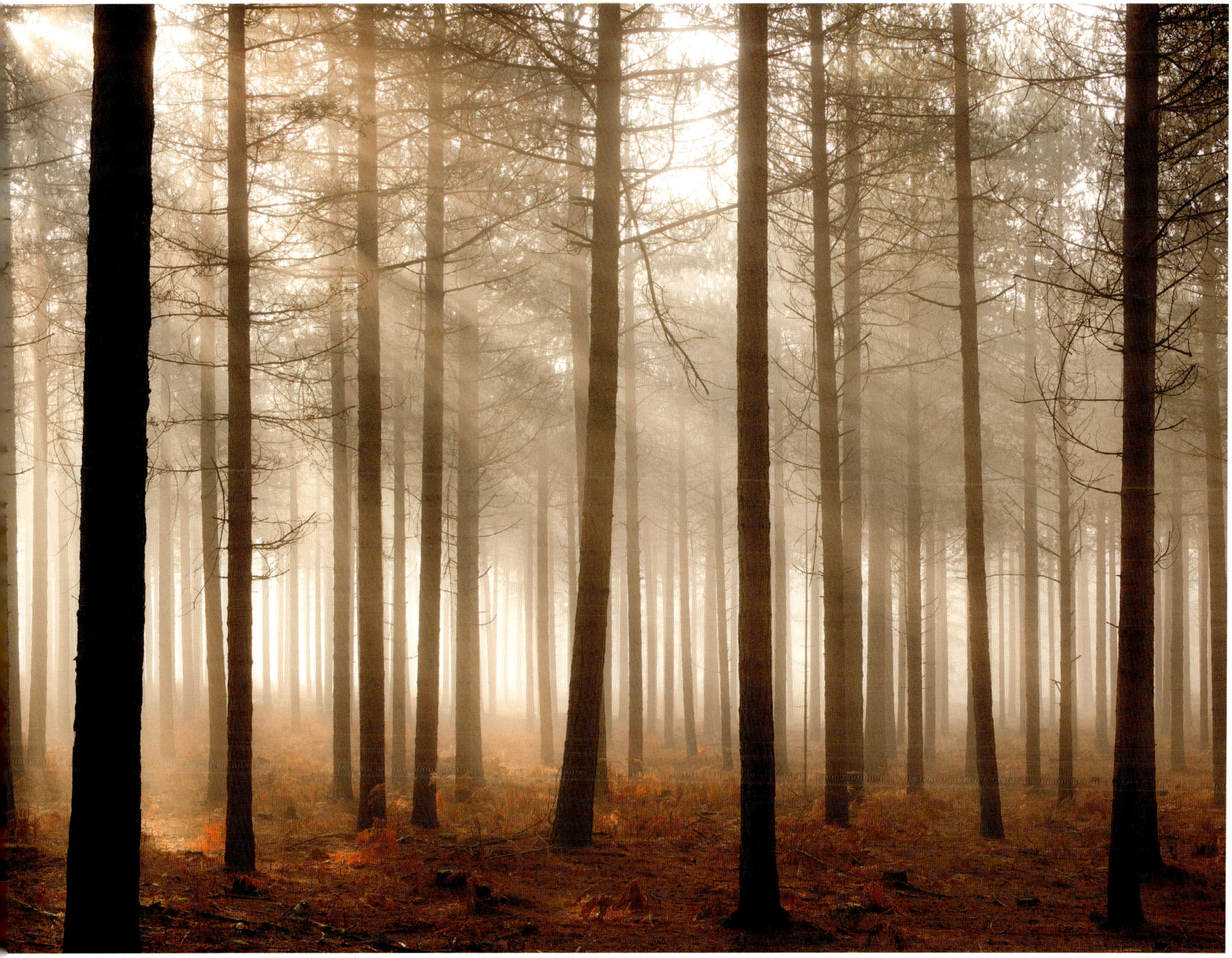

JEREMY WALKER HIGHLY COMMENDED

Morning Mist
(Pine, *Pinus)*
Wareham Forest, Dorset, England

A magical but rare morning when all the elements fall into place. Just the right amount of mist, the sun high and strong enough to penetrate the forest floor and a clean and simple composition to make the image work.

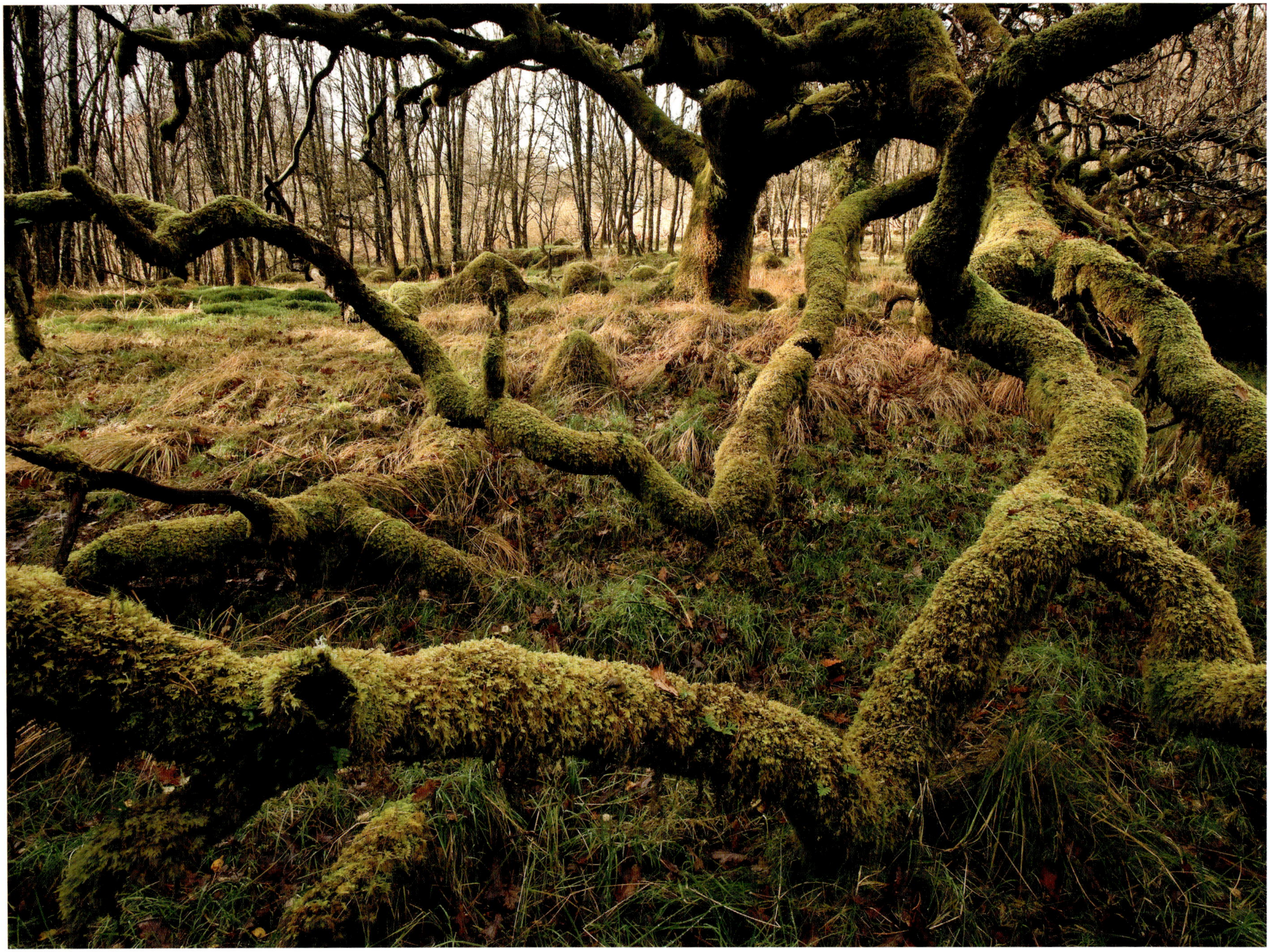

IAN PAUL HASKELL HIGHLY COMMENDED

Old Atlantic Oak
(Sessile Oak, *Quercus petraea*)
Acharacle, near Strontian, Argyll, Scotland

The old Atlantic oaklands at Ariundle have some of the best wild woodlands that I have experienced in Scotland. This particular tree had an old and mythical feel to it, representative of some of the best fantasy novels.

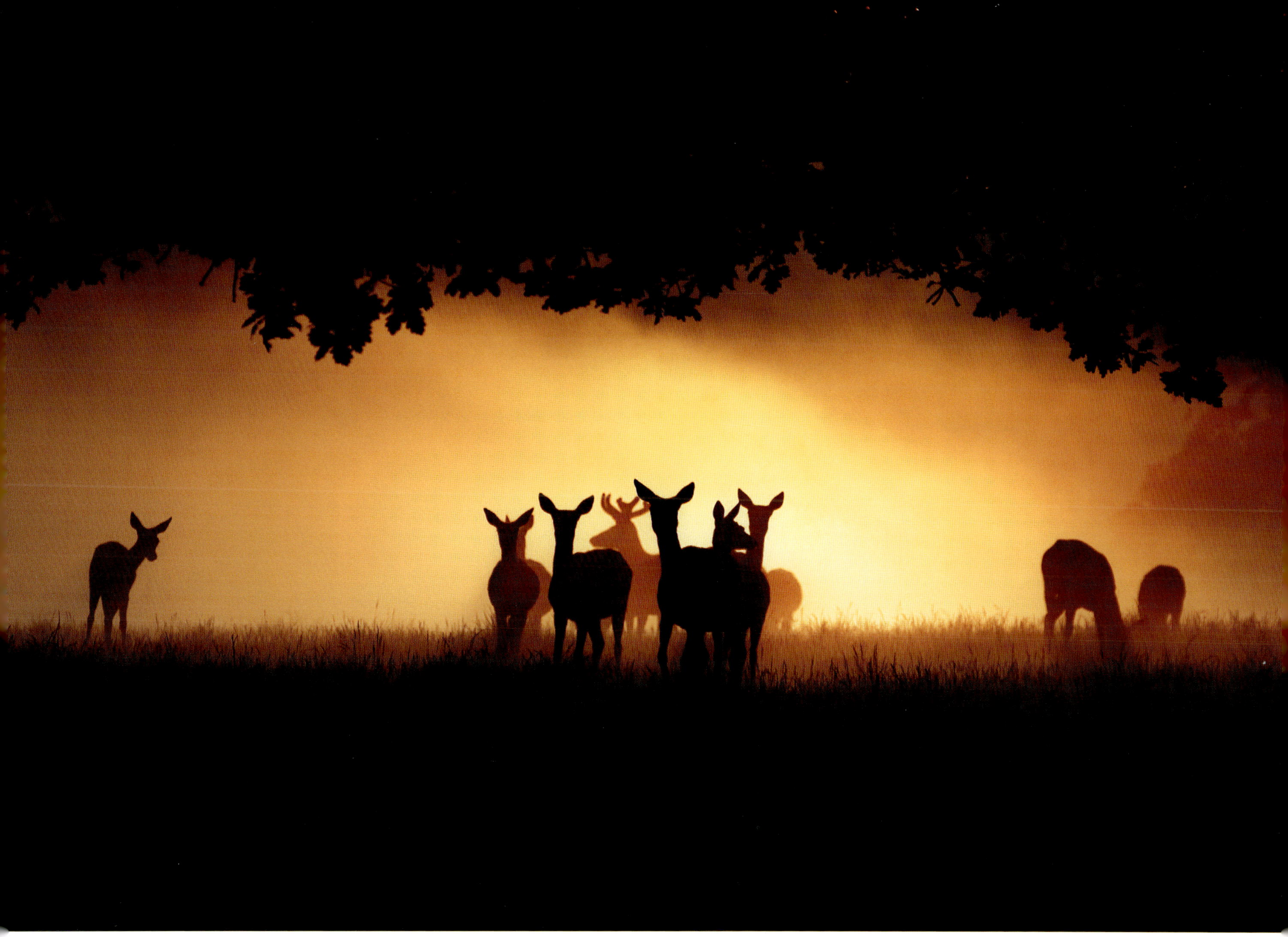

ALEX SABERI HIGHLY COMMENDED

Fairytale 2
(Red Deer, *Cervus elaphus)*
Richmond Park National Nature Reserve, Surrey, England

A rare spring/summer mist shot in Richmond Park. The conditions were perfect for me that morning in the park. The mist was just right and the formation of the young deer was great for this silhouette. One of my favourite shots.

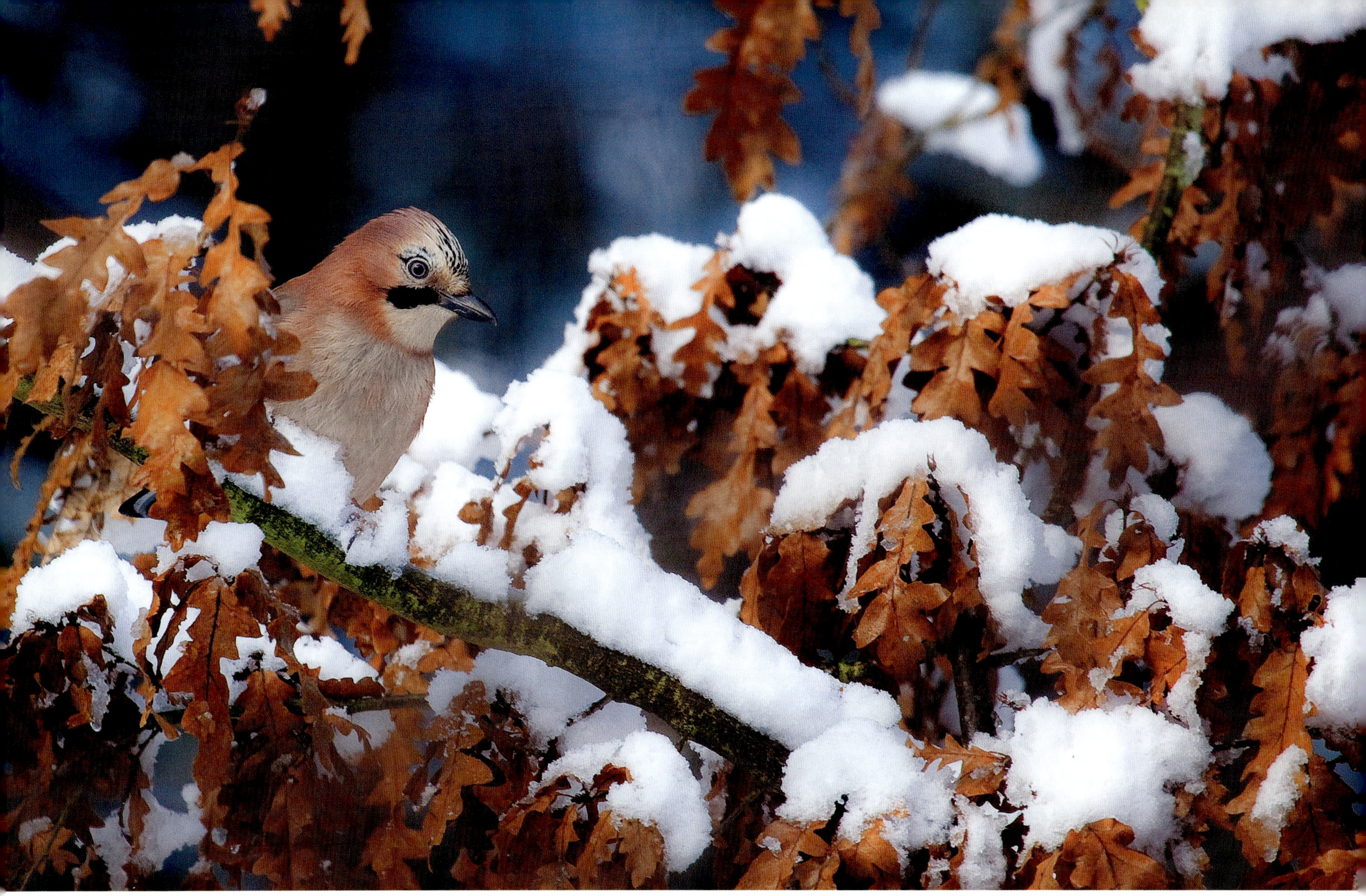

MARTIN WASE

Winter Roost
(Eurasian Jay, *Garrulus glandarius)*
Sandringham, Norfolk, England

The snow always gives any photographer the chance to capture nature in a different perspective. This jay was no exception when he came and sat in this snow-covered oak to see what I was up to. As usual he kept his distance and used his surroundings to his advantage, just peeping round the snow and leaves for a moment, long enough for me to snap him.

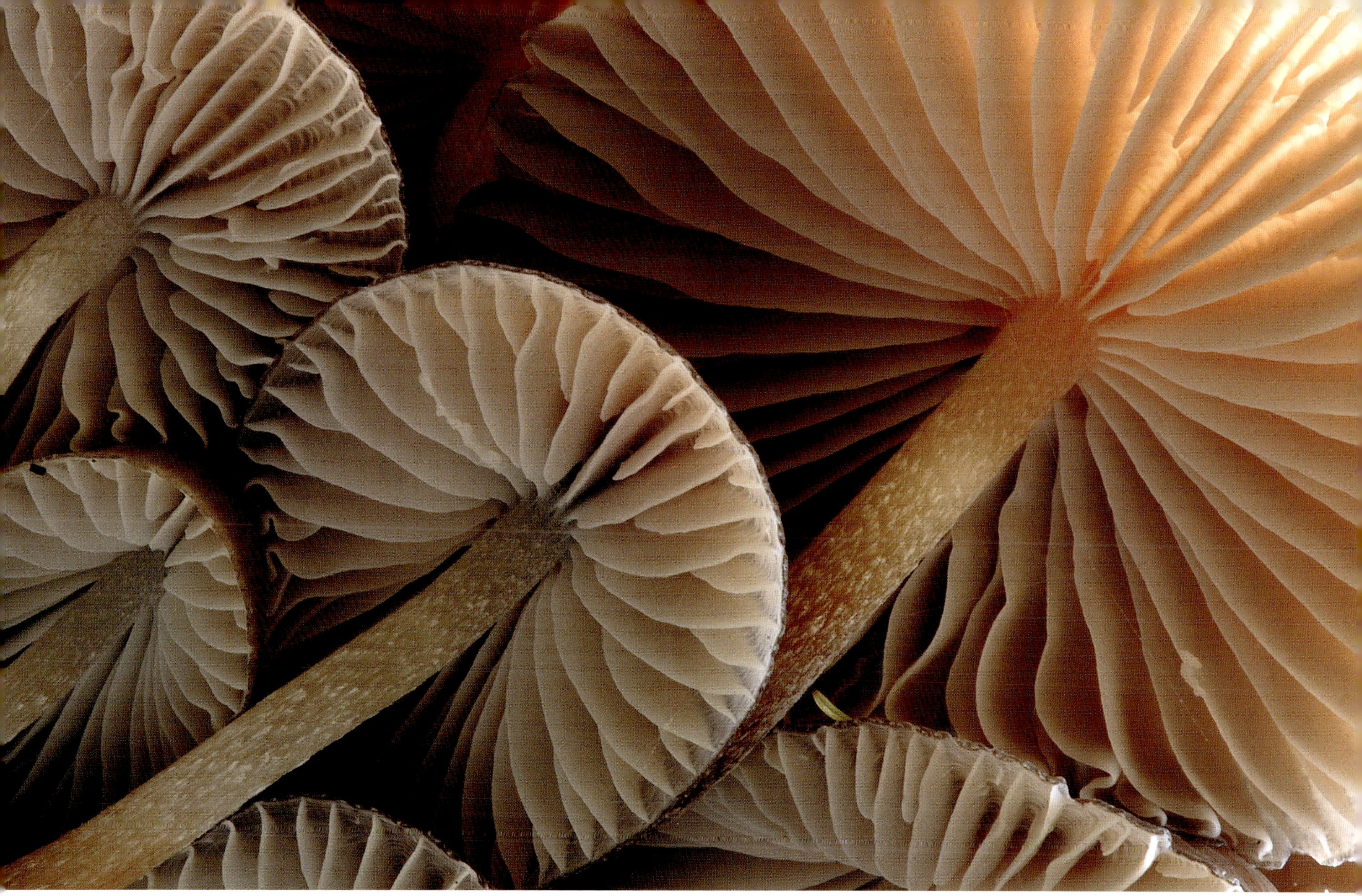

ROSS HODDINOTT

Gills
(Fungi, *Mycena)*
Shaugh Prior, Dartmoor, Devon, England

I discovered these tiny mushrooms sprouting from a tree stump while walking through the ancient woodland at Shaugh Prior, near Plymouth. I opted for a low shooting angle and filled the frame with the mushroom's gills for maximum impact. Due to where they were growing, it was tricky achieving a good angle. Hopefully the grubby knees proved worthwhile!

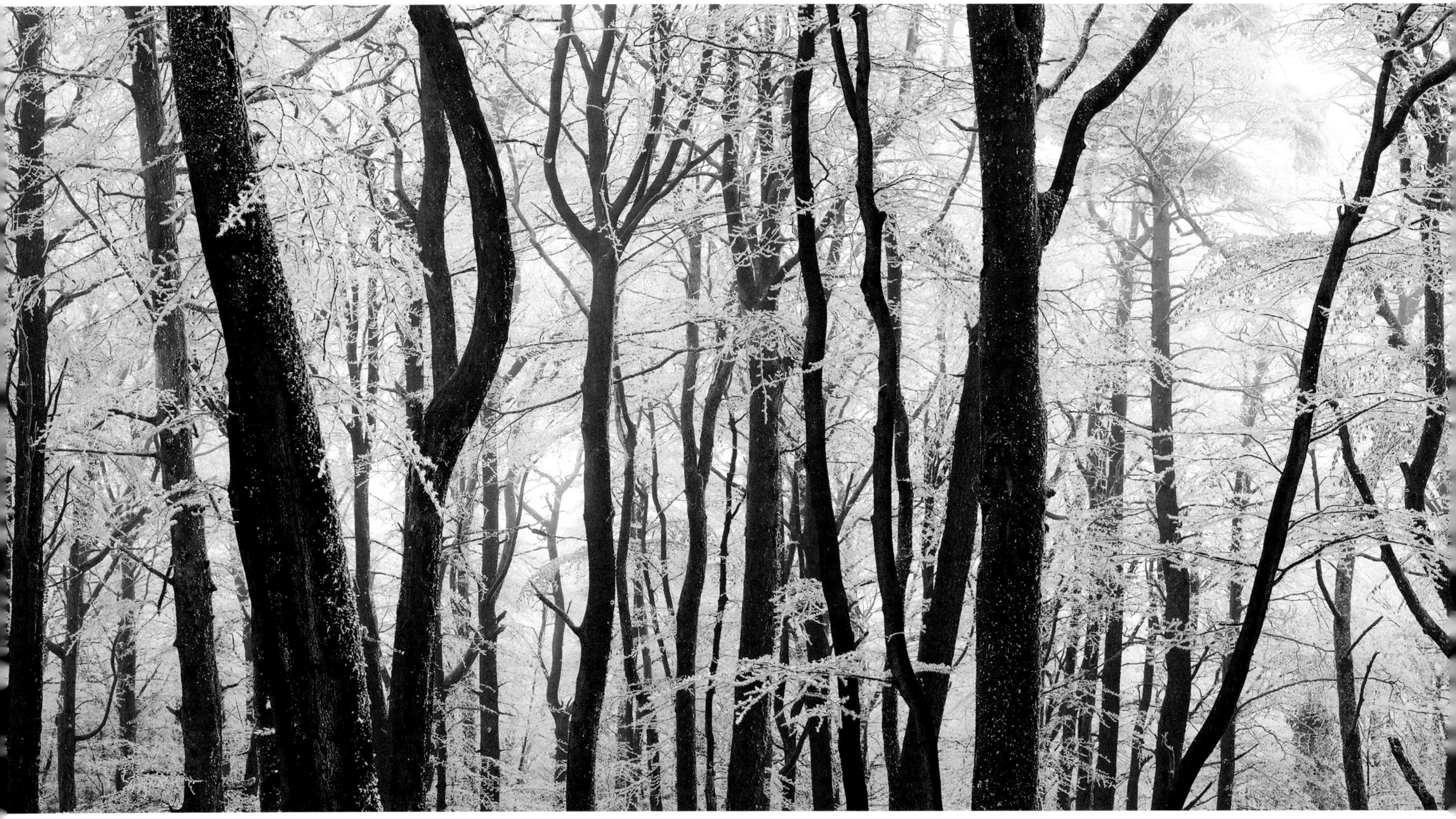

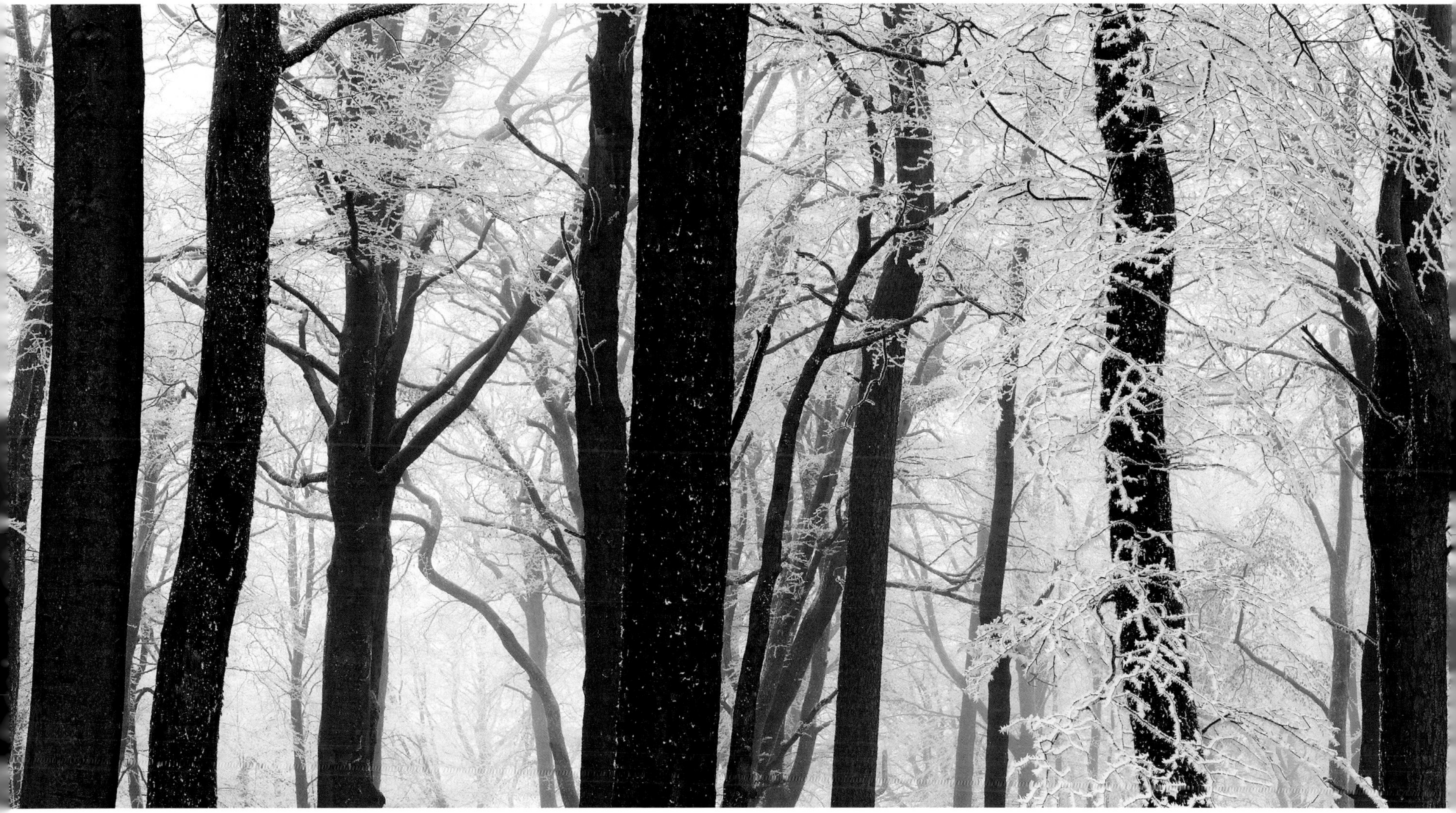

JEREMY WALKER

Fog Shrouded Woods
(Beech, *Fagus*; Pine, *Pinus* and other species)
Lambert's Castle, Dorset, England

The fog had effectively muffled the sound of everything. There was not a squeak or peep from any living creature, only the occasional rustle from the forest floor as ice fell from the trees. A very eerie working environment.

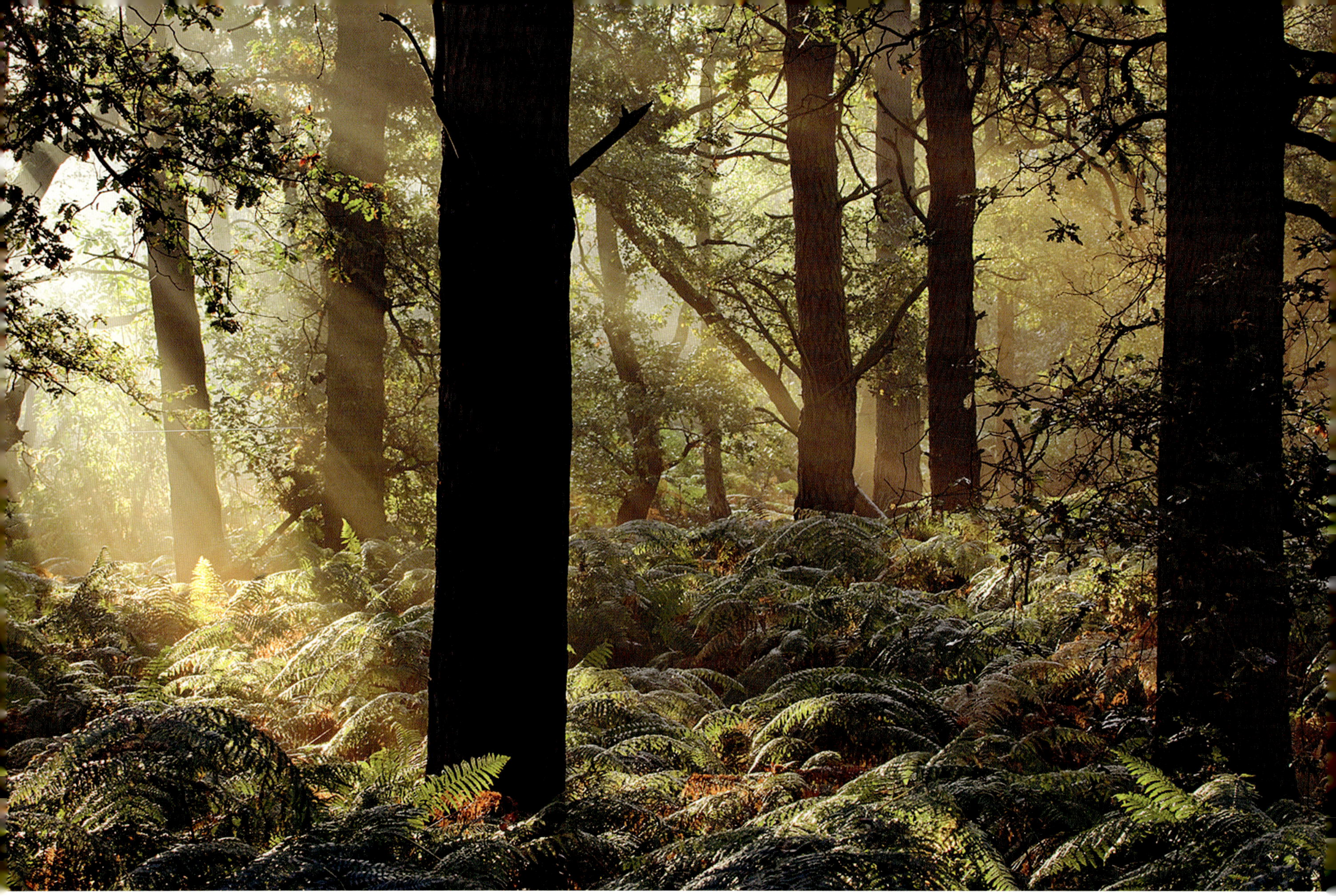

LEE BEEL ⋀

Sun Burns through Mist in Twigmoor Woods in Autumn
Twigmoor Woods, North Lincolnshire, England

The weather forecast warned of early-morning fog so I decided to visit Twigmoor Woods, which isn't far from my home. It was clear that the morning sun was rapidly burning through the fog when I arrived. I quickly found a spot to photograph and composed the shot to capture the sun's rays backlighting the trees and dispersing the fog. It was hard to concentrate on photography when faced with such a wonderful scene.

NEIL McINTYRE >

Crested Tit on Pine Tree
(Crested Tit, *Lophophans cristatus*)
Cairngorms National Park, Highland, Scotland

Crested tits are very addictive to photograph and I am lucky to have a good population on my doorstep in the Cairngorms area. For this shot I wanted the Cresty on top of a pine tree framed with out-of-focus branches from other pines. After a bit of a search I found this spot and with the lure of some food got the bird to land where I wanted.

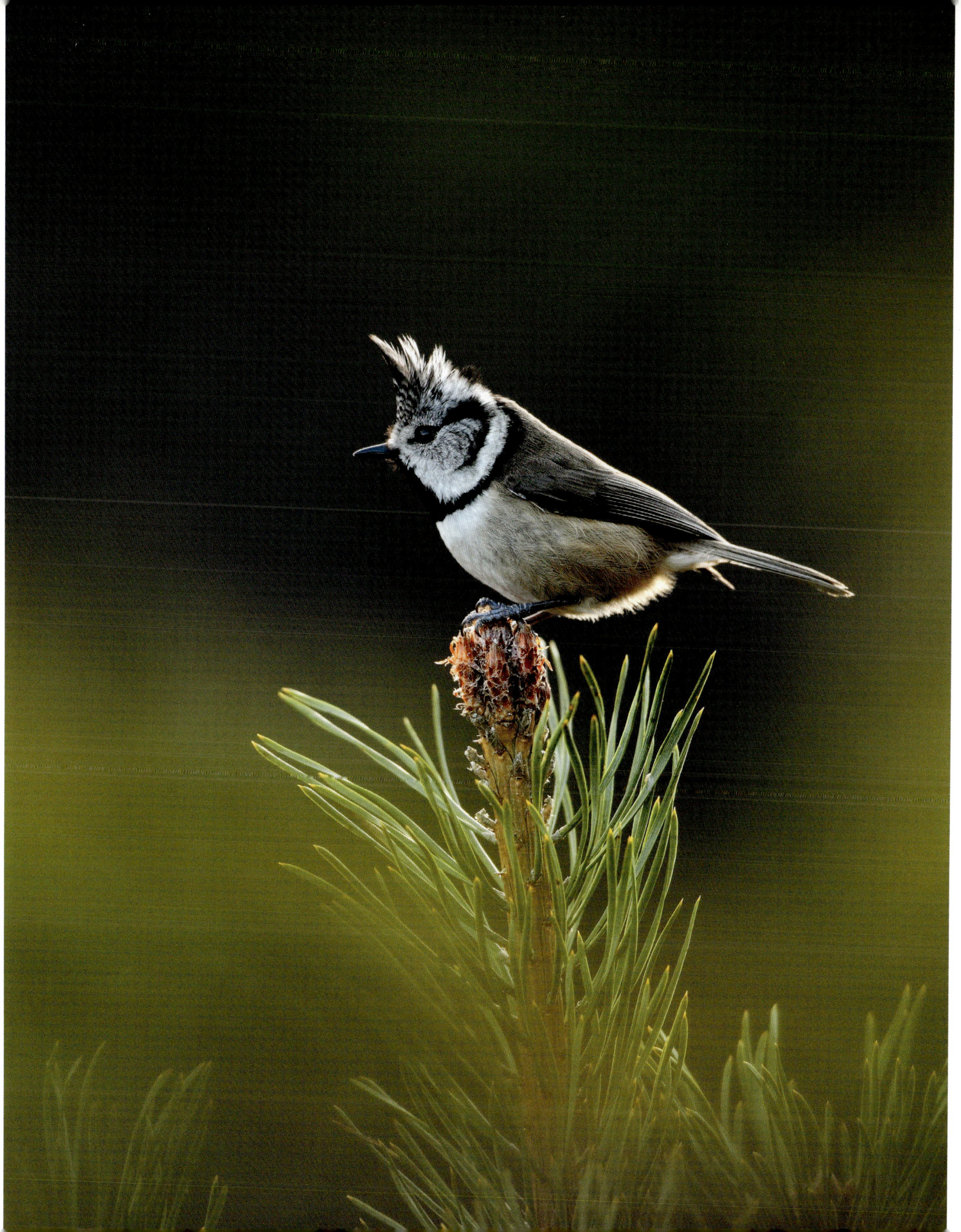

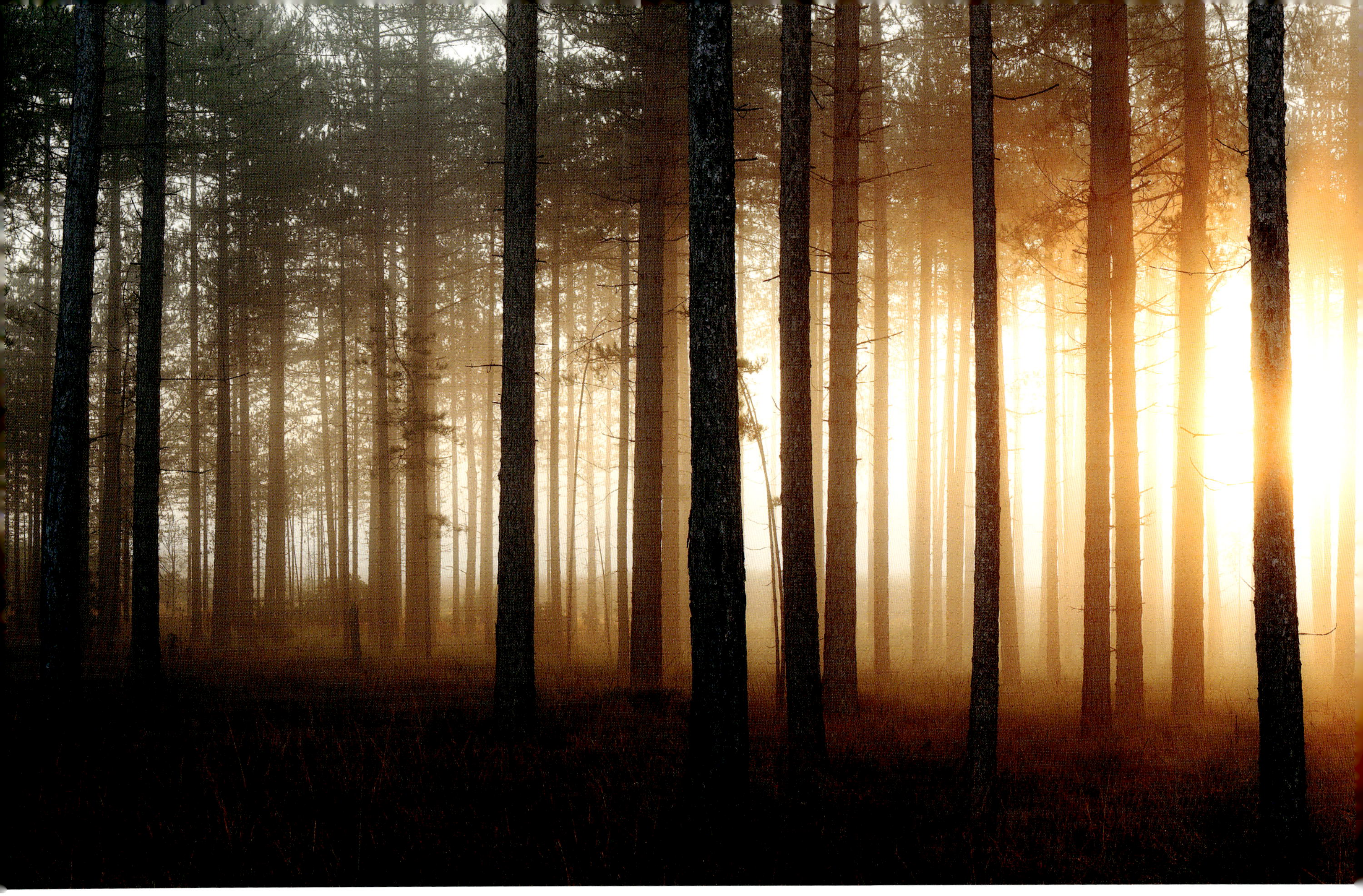

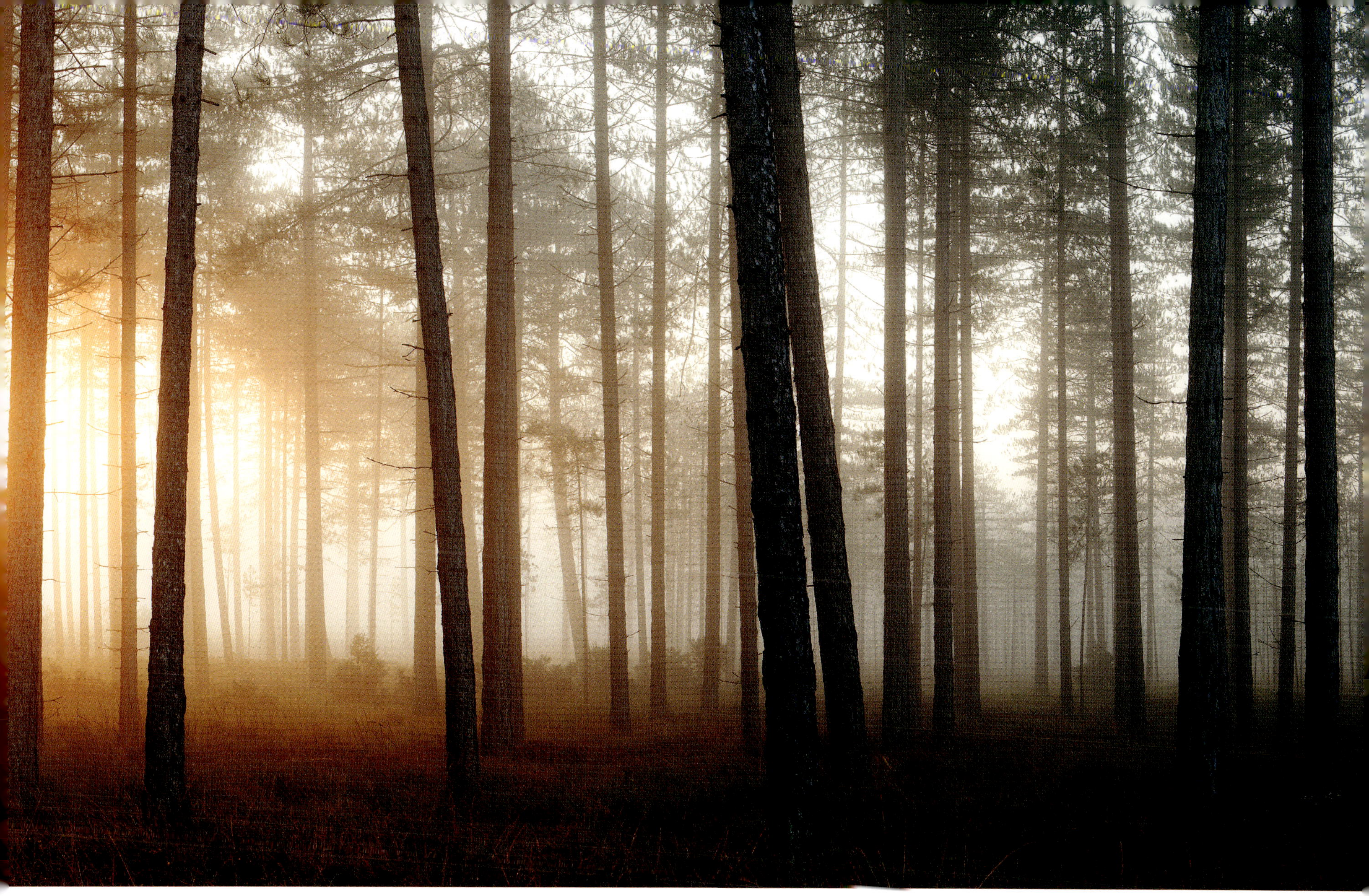

JEREMY WALKER

Misty Pine Forest at Dawn
(Pine, *Pinus)*
Wareham Forest, Dorset, England

Not the early morning image I was expecting to shoot. I was hoping for big heavenly beams of light but instead I had a soft ethereal glow slowly penetrating through the forest. I nearly dismissed the scene entirely but now I love the soft subtleness of the light and colours.

HABITAT

SPONSORED BY
COUNTRYSIDE JOBS SERVICE

HABITAT WINNER

ALEXANDER MUSTARD/2020VISION

Grey Seal in Underwater Cave
(Grey Seal, *Halichoerus grypus*)
Lundy Island, Devon, England

I saw a seal swim through this colourful cave and so I waited... and waited. About every 20 minutes a seal would come along and have a look, I suspect seeing if there were any fish sheltering here. This small cave was obviously an important hang-out for seals.

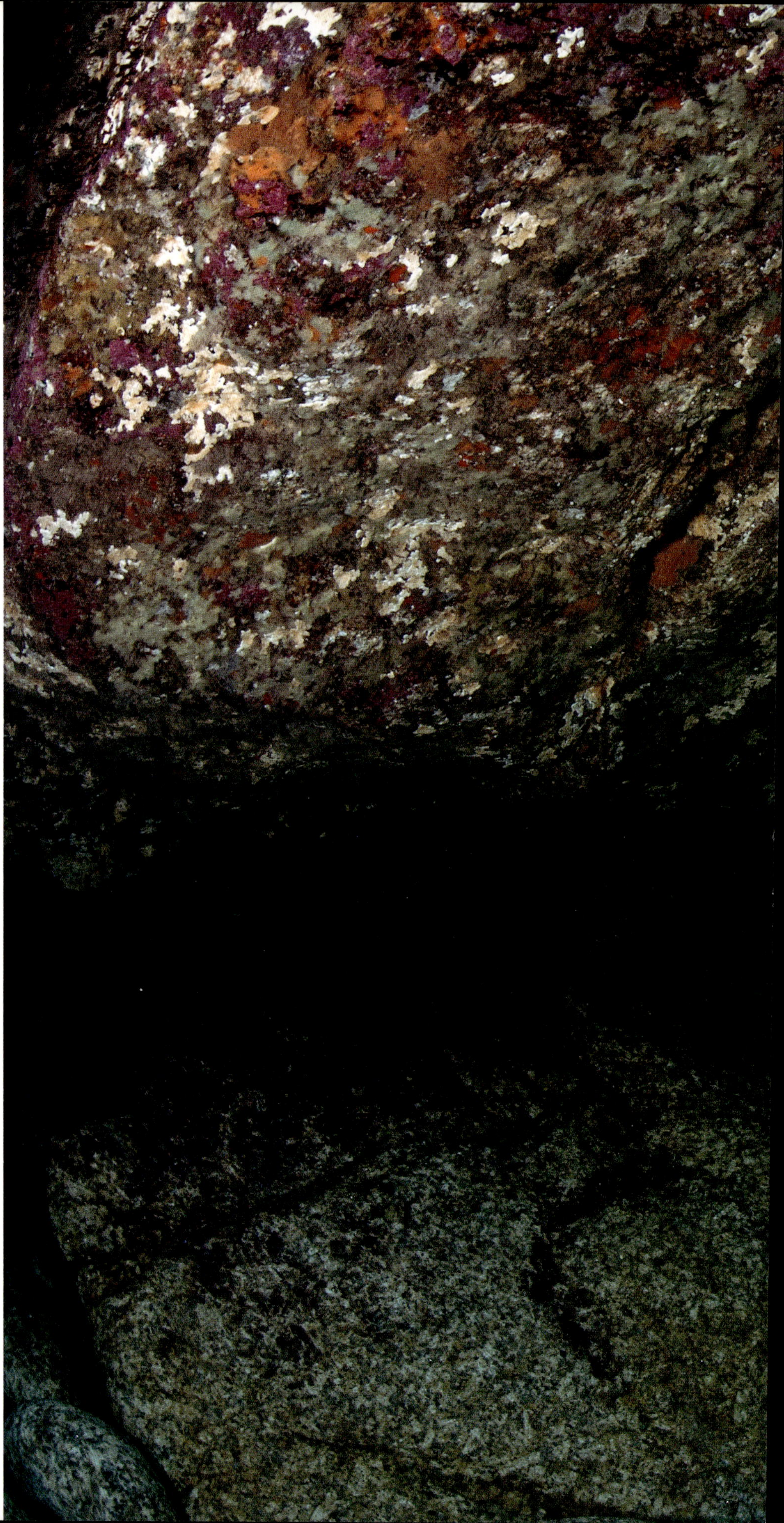

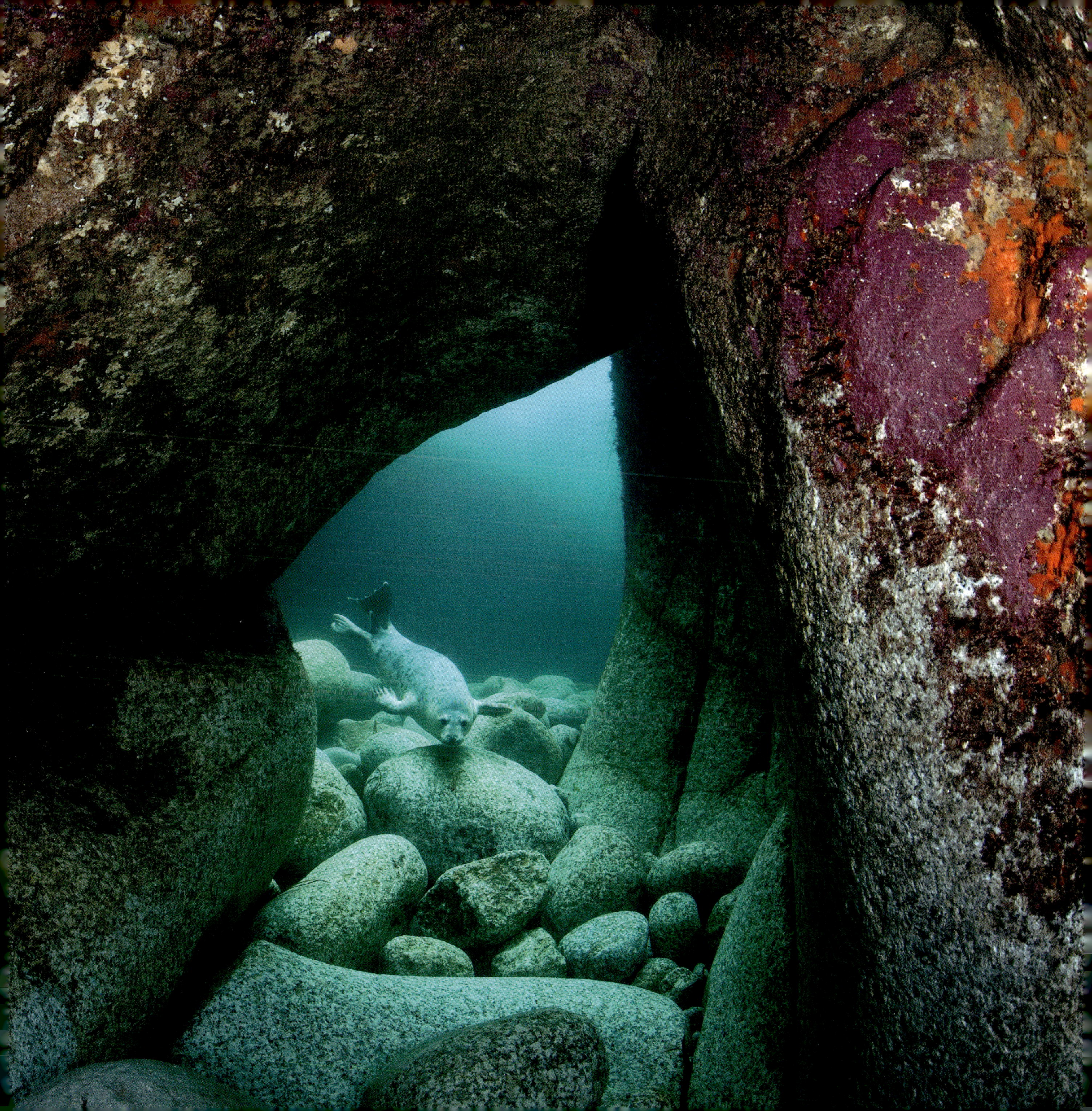

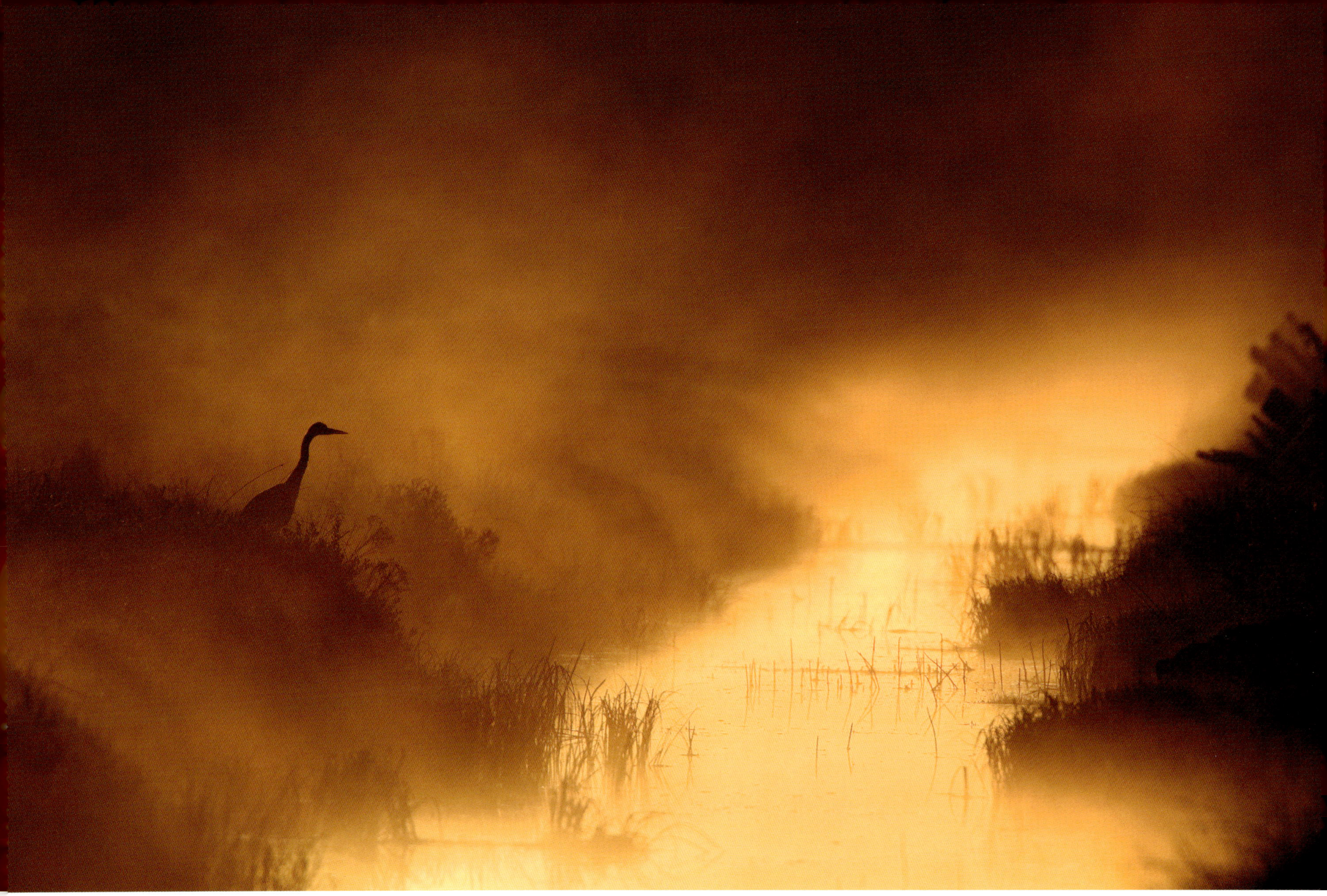

MARK HAMBLIN HIGHLY COMMENDED

Grey Heron Fishing Drainage Ditch at Dawn
(Grey Heron, *Ardea cinerea)*
Ballinlaggan Marsh, Highland, Scotland

Cold, clear nights with no wind will inevitably result in dawn mist over any body of water, providing my favourite conditions. Rarely are these images planned but there are several locations close to home that offer potential for such shots when all the elements come together. I have driven past this particular spot in these conditions on perhaps 50 occasions and just once there was a grey heron fishing in precisely the right spot as the sun rose.

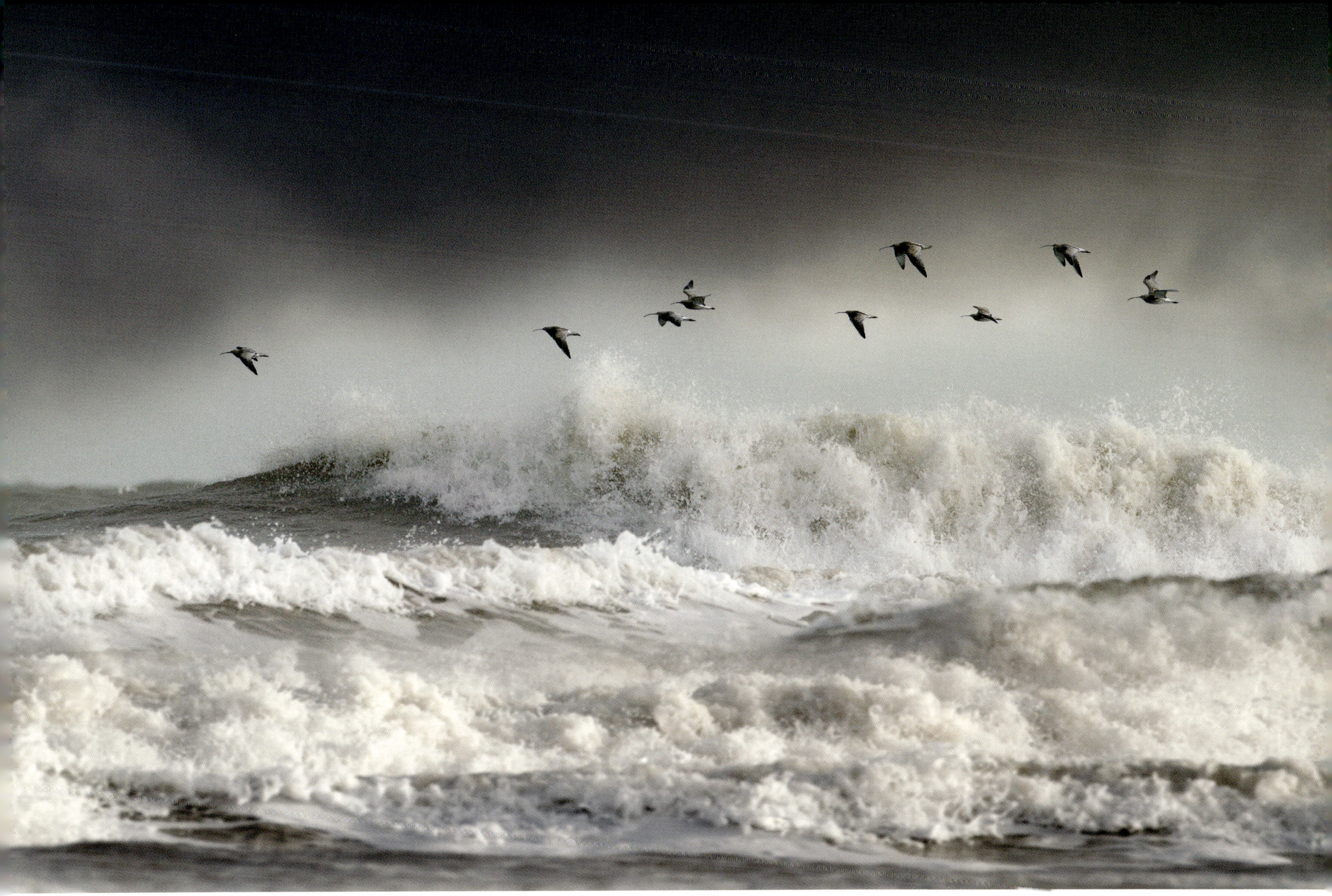

BEN HALL HIGHLY COMMENDED

Curlews in Flight over Stormy Sea
(Curlew, *Numenius arquata)*
Gronant, North Wales

I took this image whilst working on a commission to photograph coastal wildlife. I love to photograph in adverse weather conditions, and on the afternoon I took this a fierce storm was blowing in and churning the sea into a frenzy. I noticed a flock of curlew flying in from the right, so quickly panned with the birds and fired a sequence of images.

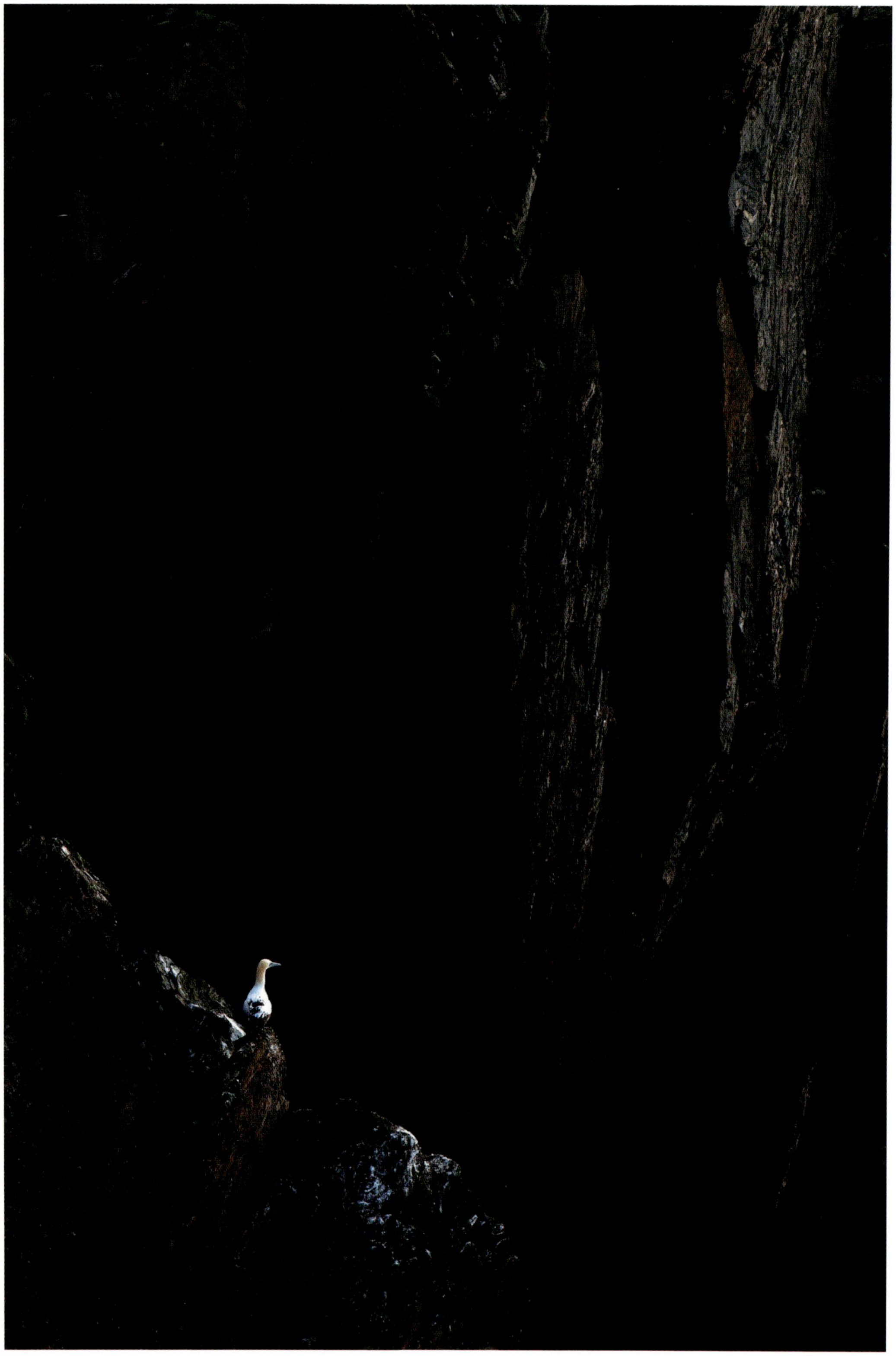

ANDREW PARKINSON < HIGHLY COMMENDED

Lone Gannet Among Precipitous Cliffs
(Northern Gannet, *Morus bassanus)*
Hermaness National Nature Reserve,
Shetland Isles, Scotland

Gannet colonies are so often overcrowded and busy that I wanted to make an image where a gannet was alone in its environment, like it is when it fishes in the open ocean. It took several weeks for the opportunity to reveal itself but I like the quiet, contemplative tone that such an image creates.

ANDREW PARKINSON > HIGHLY COMMENDED

A Gannet Lands on a Remote Scottish Gannetry
(Northern Gannet, *Morus bassanus)*
Hermaness National Nature Reserve,
Shetland Isles, Scotland

I always aspire to show birds and mammals within the context of the environment in which they live and few are as spectacular as Hermaness NNR on the northern tip of the Shetland Islands. I isolated this particularly interesting promontory and then waited in the cold until the right bird came in to land in the right place.

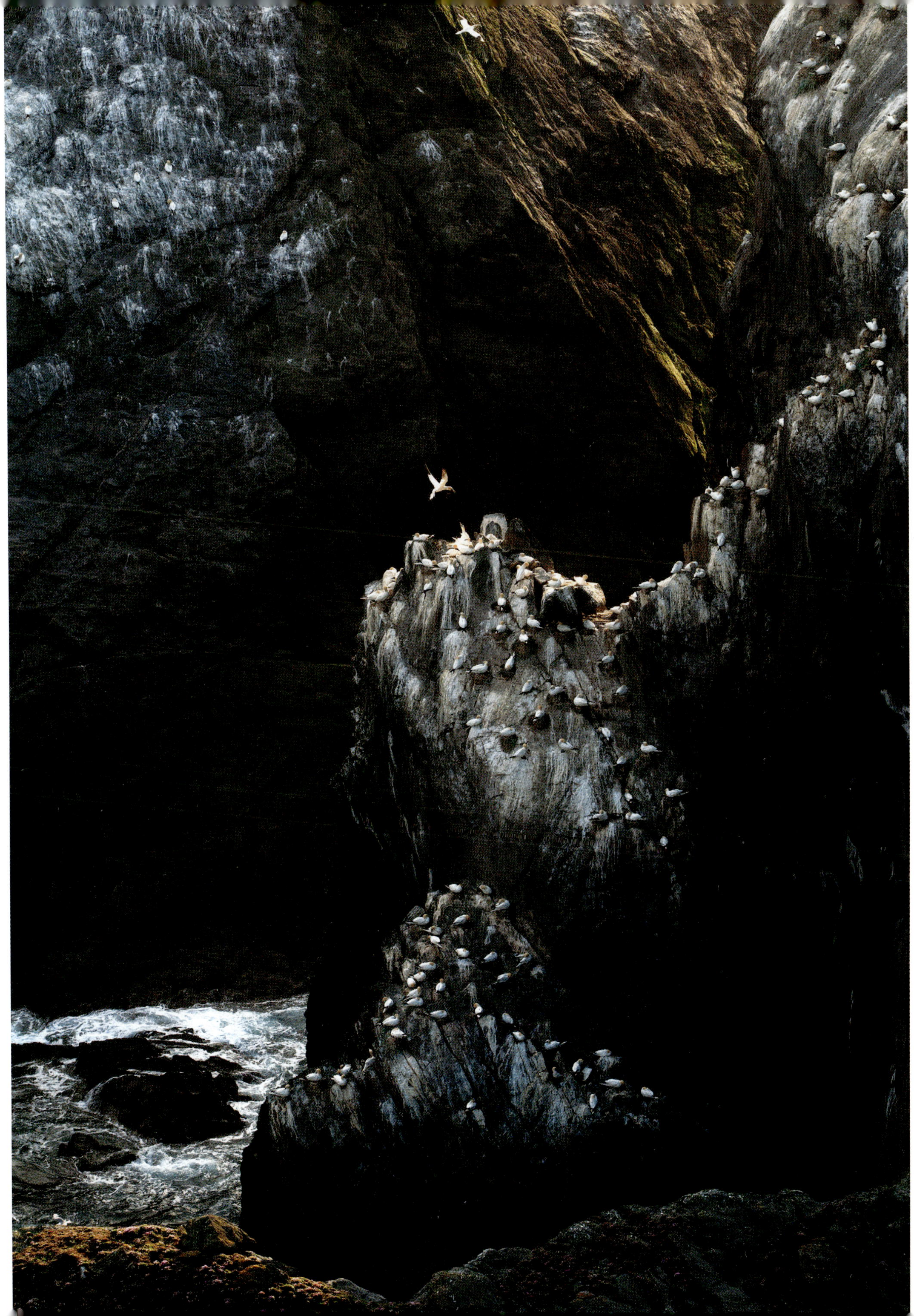

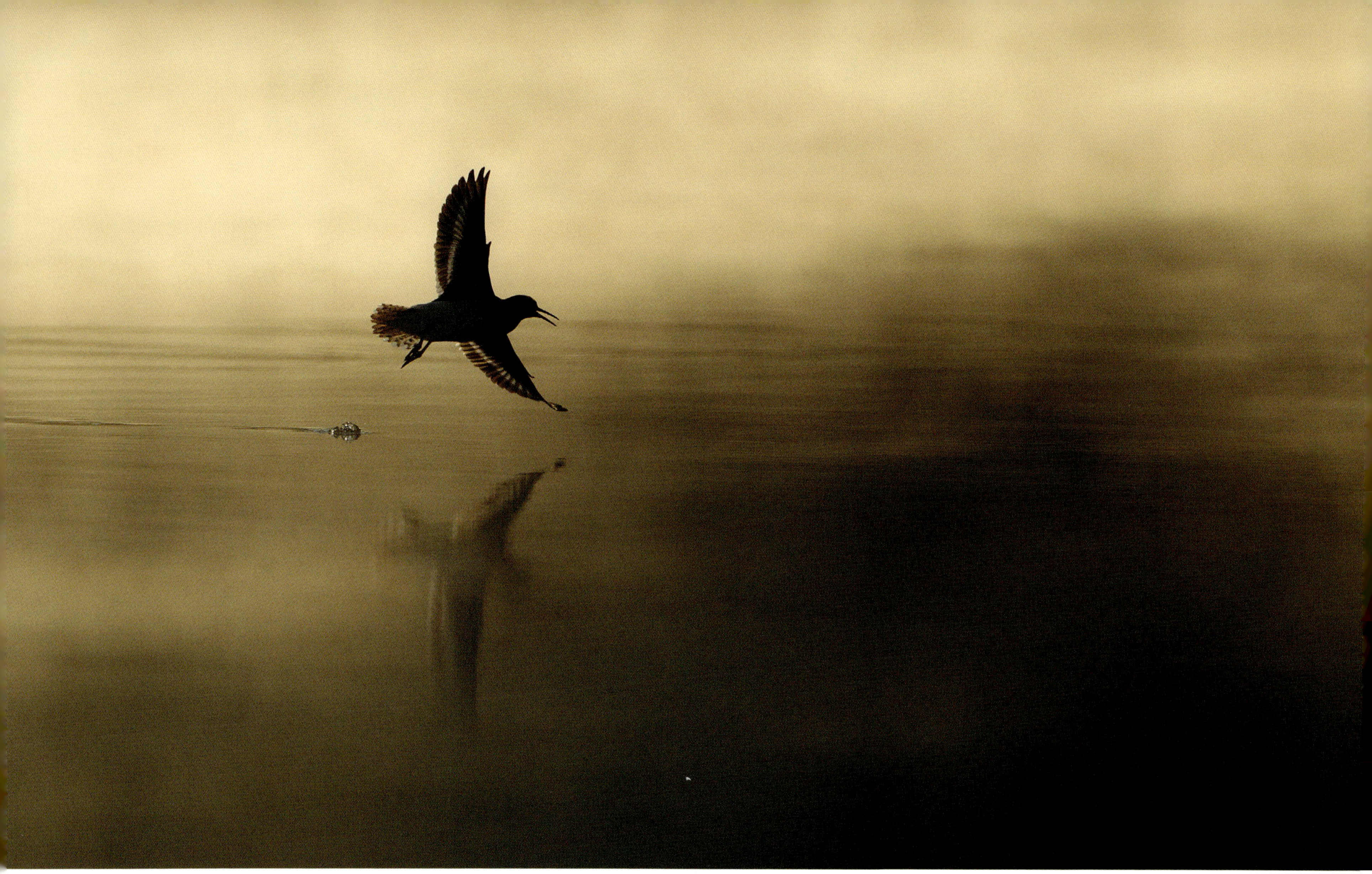

MARK HAMBLIN HIGHLY COMMENDED

Common Sandpiper in Flight over Misty Loch
(Common Sandpiper, *Actitis hypoleucos)*
Cairngorms National Park, Highland, Scotland

During late April and early May, common sandpipers are very active on their breeding grounds, calling and displaying to each other and defending their territories from rival birds. I had visited this small loch at dawn primarily for scenic images but was enthralled by the sandpipers, so instead turned my attention to trying to capture their display flight backlit against the dawn mist.

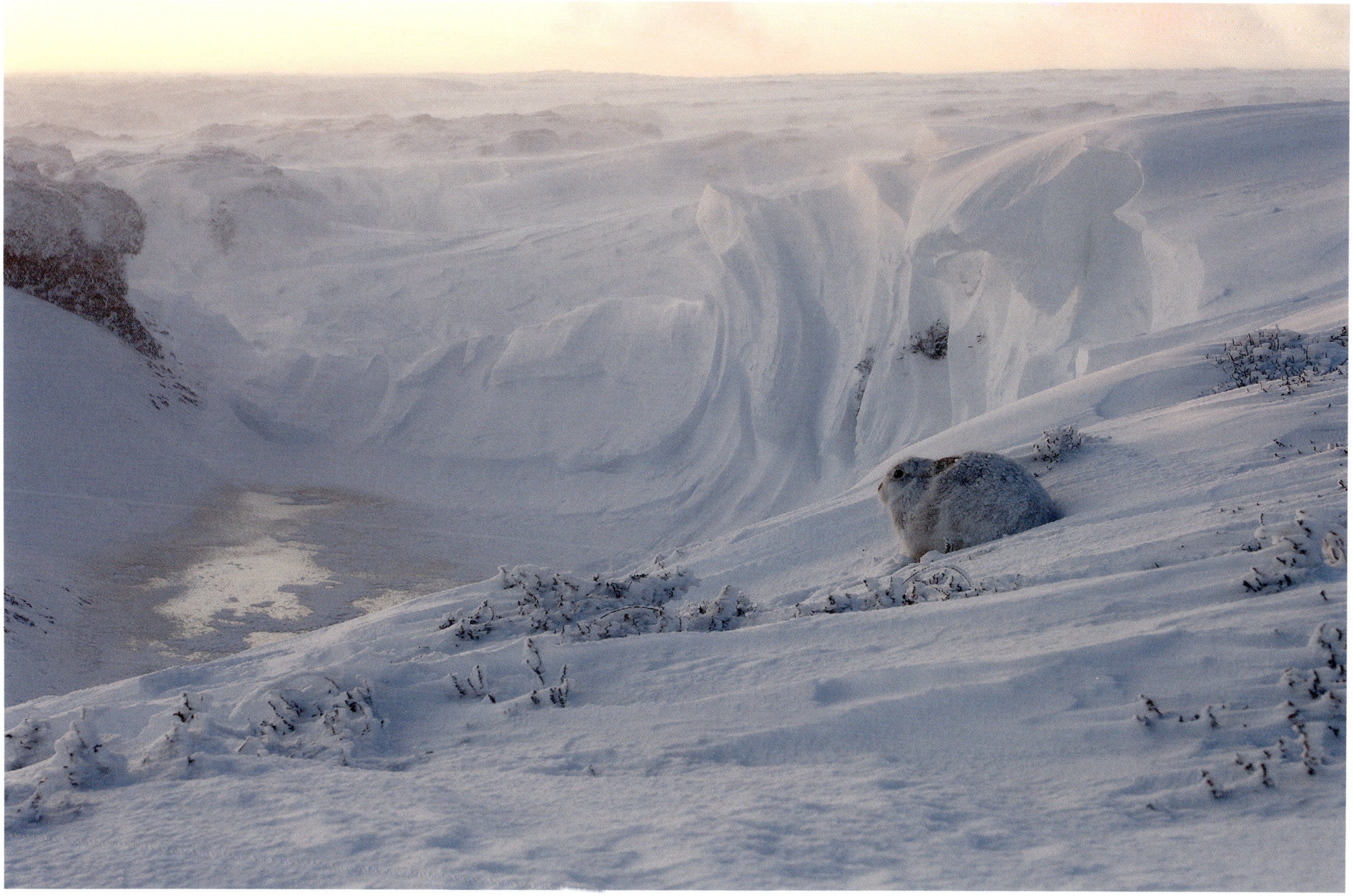

MARCIN ZAGORSKI HIGHLY COMMENDED

Mountain Hare
(Mountain Hare, *Lepus timidus)*
Kinder Scout, Peak District National Park, Derbyshire, England

I spent nearly five years photographing the mountain hare population around the Peak District area. My aim was to capture the spirit of this elusive animal together with the beauty of the harsh environment it inhabits. Hundreds of hours spent in the field eventually paid off when after a long-lasting blizzard a sudden change of weather unveiled this spectacular scenery. A split-second decision to change to a wide-angle lens was all that was needed to fulfil my ambition.

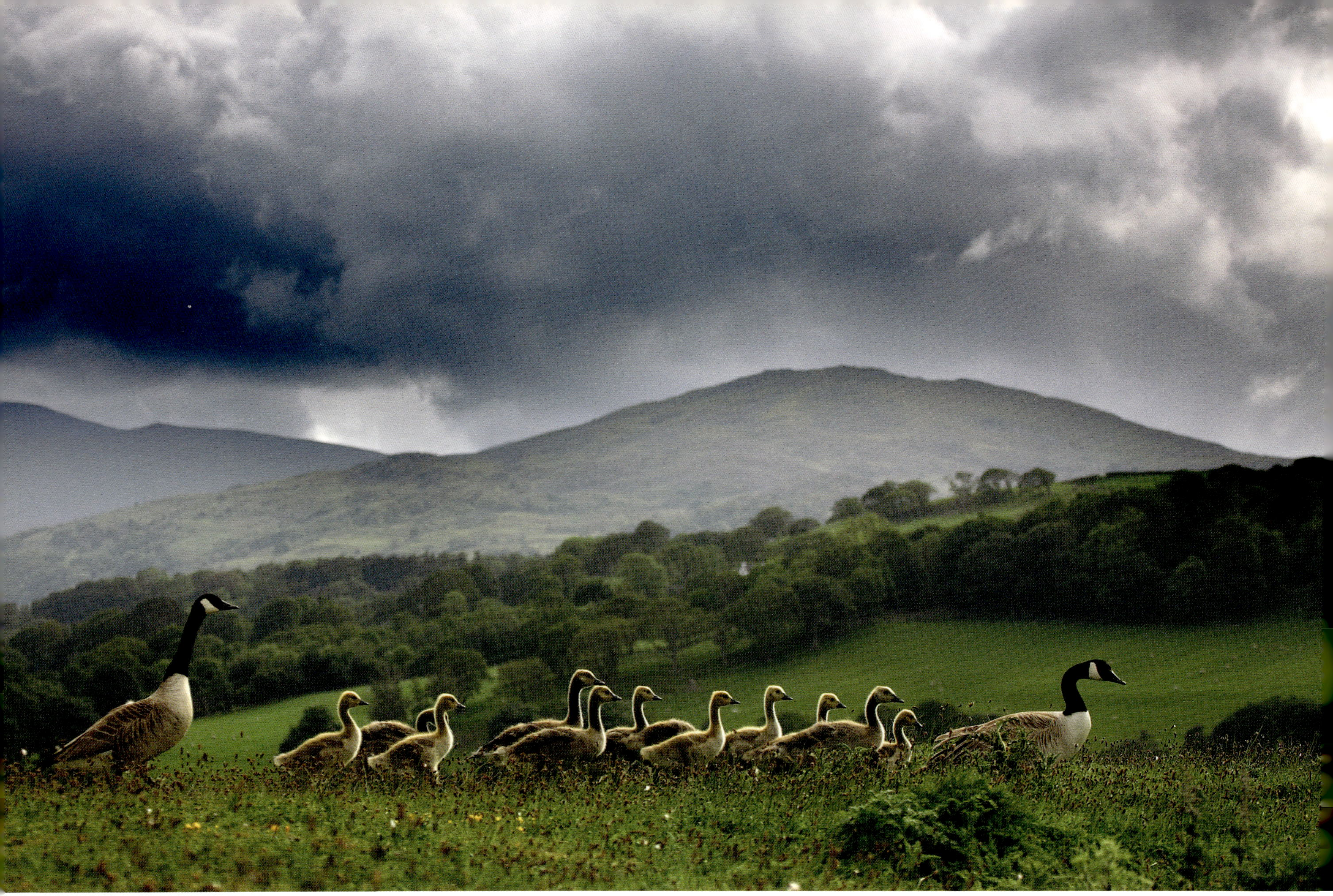

BEN HALL HIGHLY COMMENDED

Canada Geese family
(Canada Goose, *Branta canadensis*)
Conwy, North Wales

During a visit to Conwy RSPB reserve I noticed a family of Canada geese patrolling a grassy verge. My aim was to show the birds against the rolling hills of Snowdonia. I waited until they passed and used a zoom lens so as to include as much of the sky as possible. The subtle backlighting has created a rim of light around the young birds, and the dark sky has given just the right atmosphere.

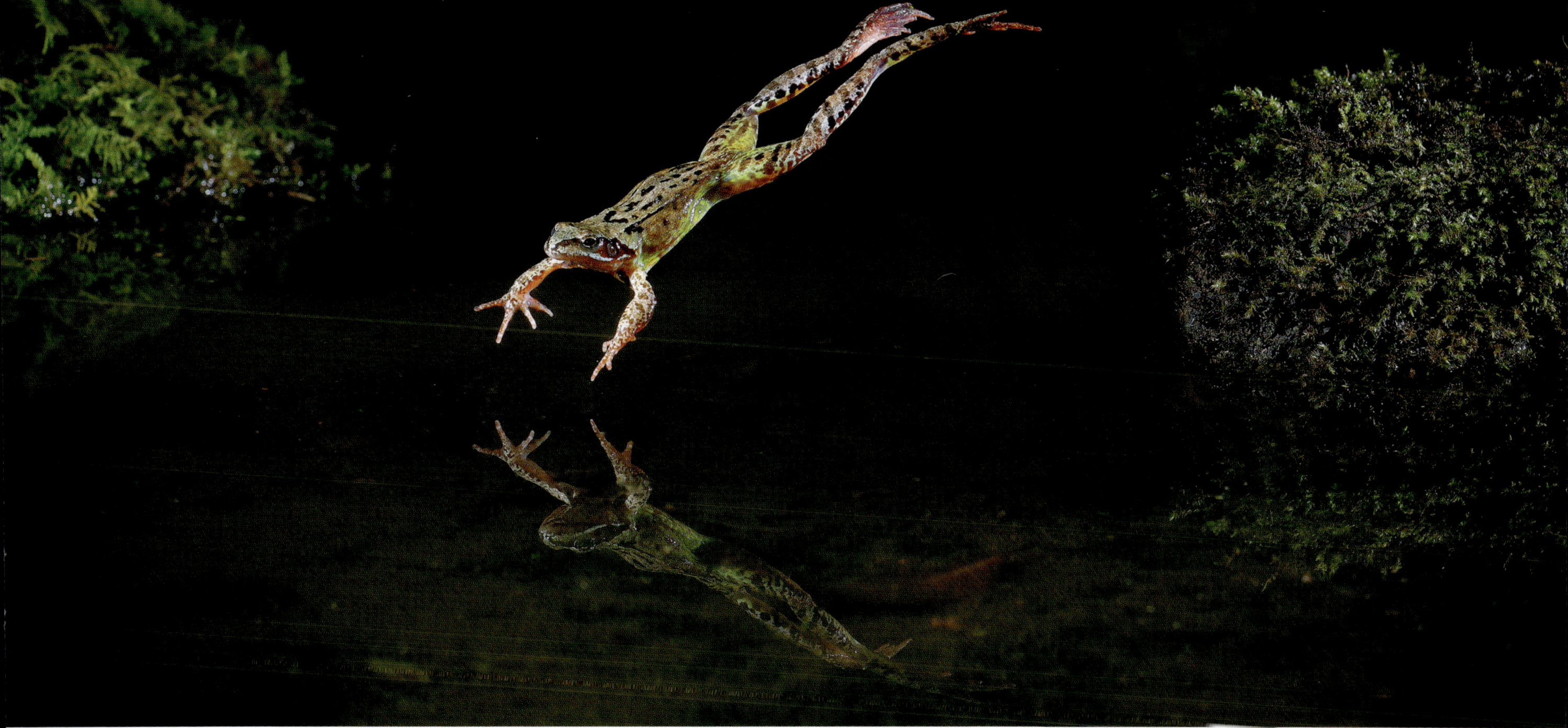

DALE SUTTON HIGHLY COMMENDED

Dive Dive Dive
(Common Frog, *Rana temporaria)*
Battle, East Sussex, England

Something I've always wanted to capture is a frog in mid dive, so I set the whole thing up in my garden pond. It was all there, all I had to do was record it, utilising a high-speed flash to freeze the frog before it left the stone it was sitting on. It was the positioning I had trouble with, getting it to look right in the image. The composition and lighting is the main aspect in all photography and it's got to look natural. I'm very pleased with the overall result and achieved what I set out to capture.

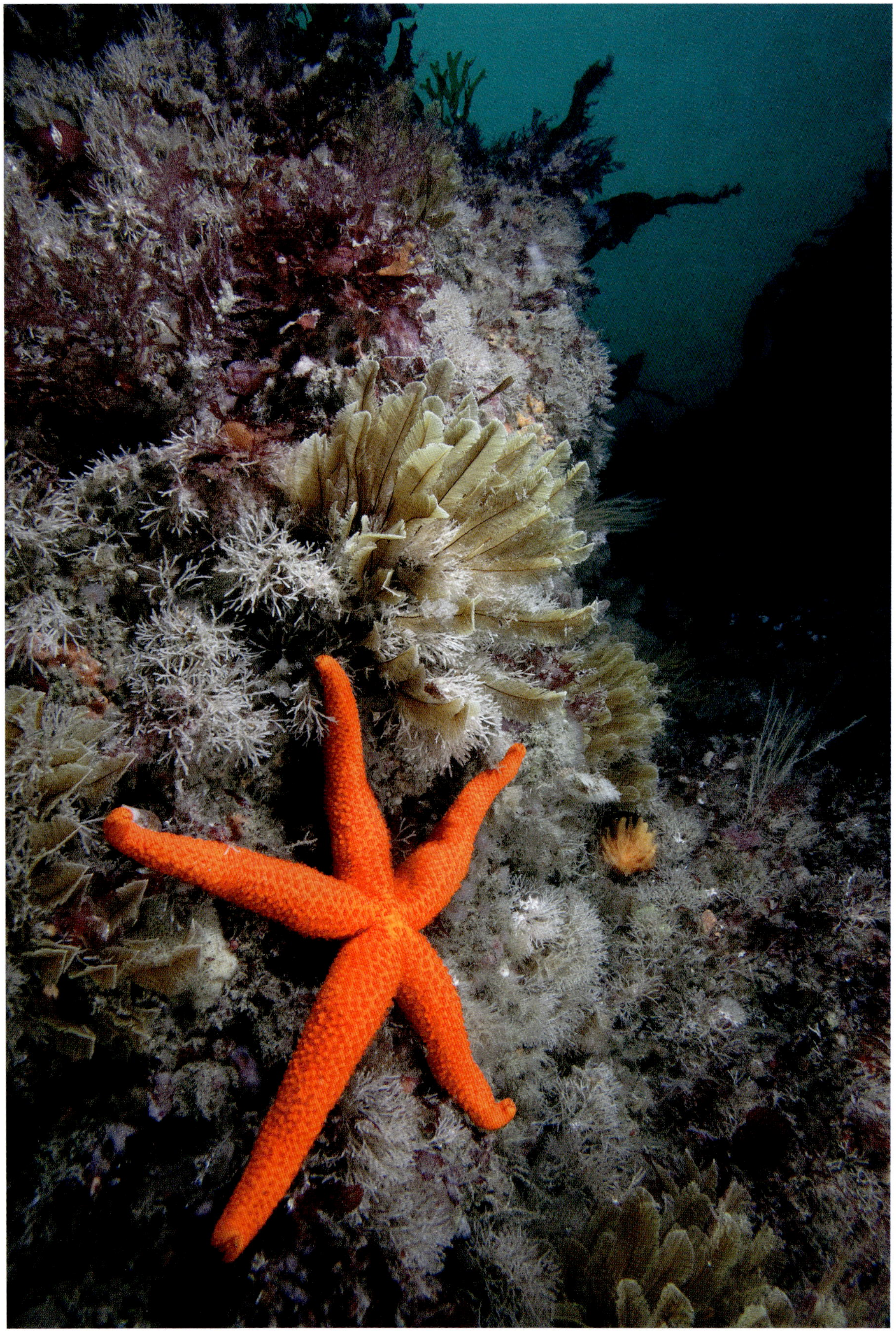

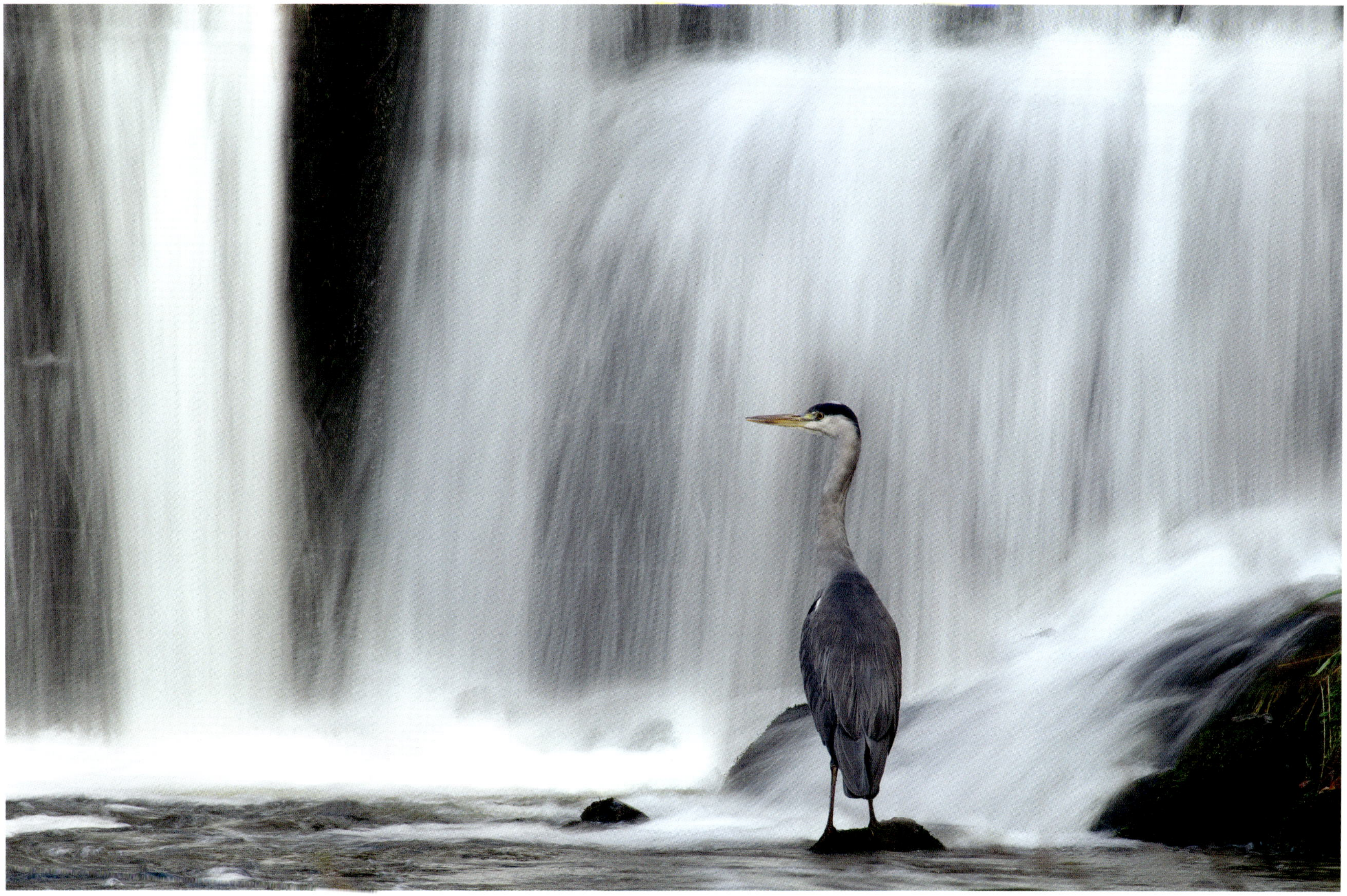

SUE DALY < HIGHLY COMMENDED

Soapy Starfish
(Soapy Starfish, *Echinaster sepositus)*
Pavlaison, Sark, British Channel Islands

Among the muted colours of the reef the bright orange of this starfish almost appeared to glow in the greeny-blue light 25m down. The attached life at this depth is almost entirely animal based; a rich carpet of bryozoans (the feather-like ones are obvious in the picture), hydroids and sponges cover the wall, giving way to kelp and other seaweeds as you look up towards the surface.

BEN HALL ∧ HIGHLY COMMENDED

Heron Perched Underneath Waterfall
(Grey Heron, *Ardea cinerea)*
Ambleside, Cumbria, England

I noticed this heron perched motionless on a mossy rock underneath a large waterfall. I immediately saw the potential for an environmental study and somehow managed to crawl down to the water's edge without being detected. I set the camera on a tripod and used a slow shutter speed to blur the movement of the water, capturing the bird in the context of its habitat.

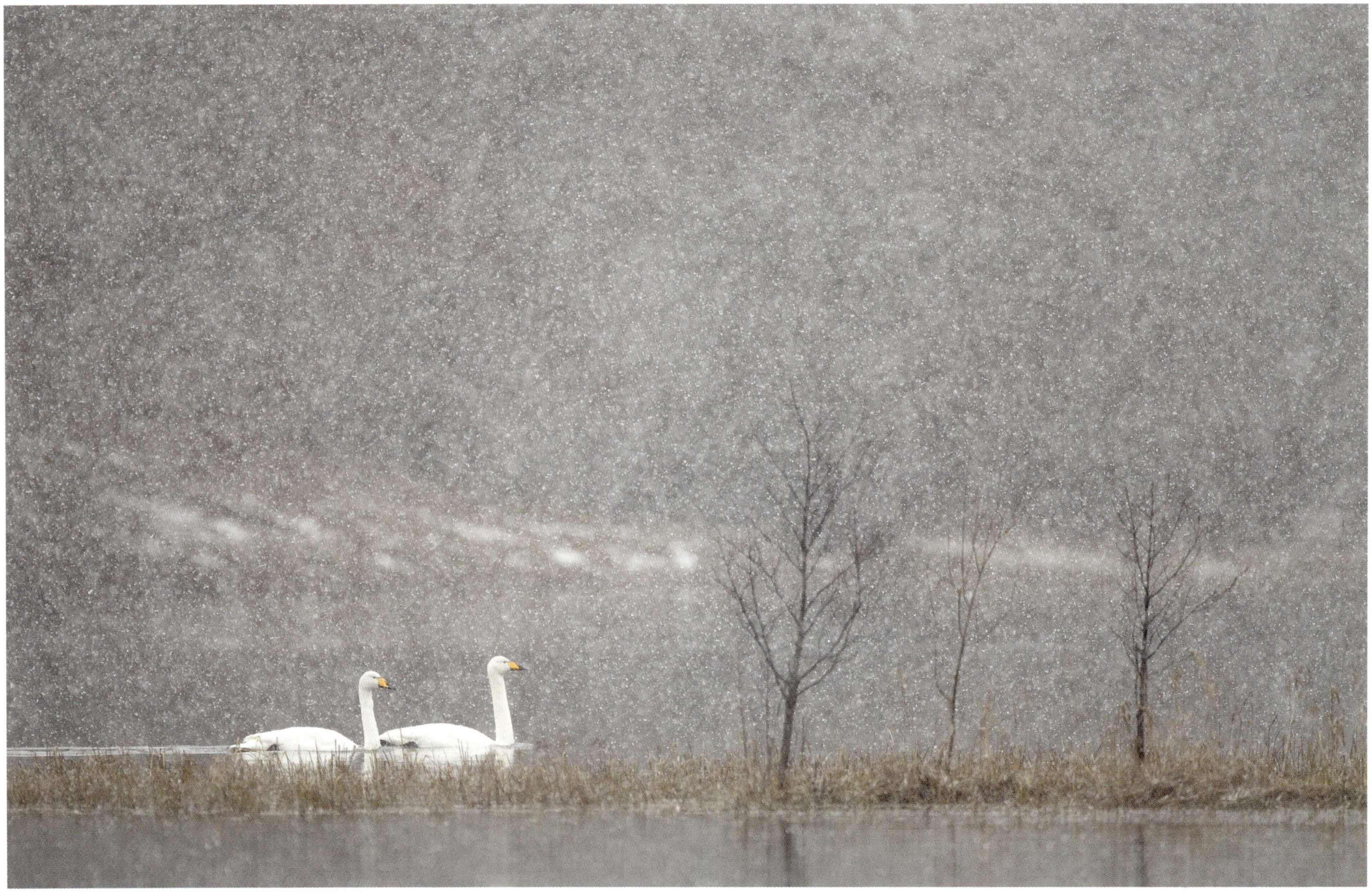

PETER CAIRNS/2020VISION HIGHLY COMMENDED

Whooper Swan Blizzard
(Whooper Swan, *Cygnus cygnus*)
Loch Insh, Cairngorms National Park, Highland, Scotland

As soon as the saucer-sized snowflakes started falling, I knew where I had to be. These whooper swans breed in Iceland but winter here and although not as approachable as whoopers elsewhere, they had gathered in a small area of the loch, giving me a fighting chance of a shot. I needed to line the swans up with the distant forest which provided a dark background for the falling snow. Waiting for the birds to come into position was agonising!

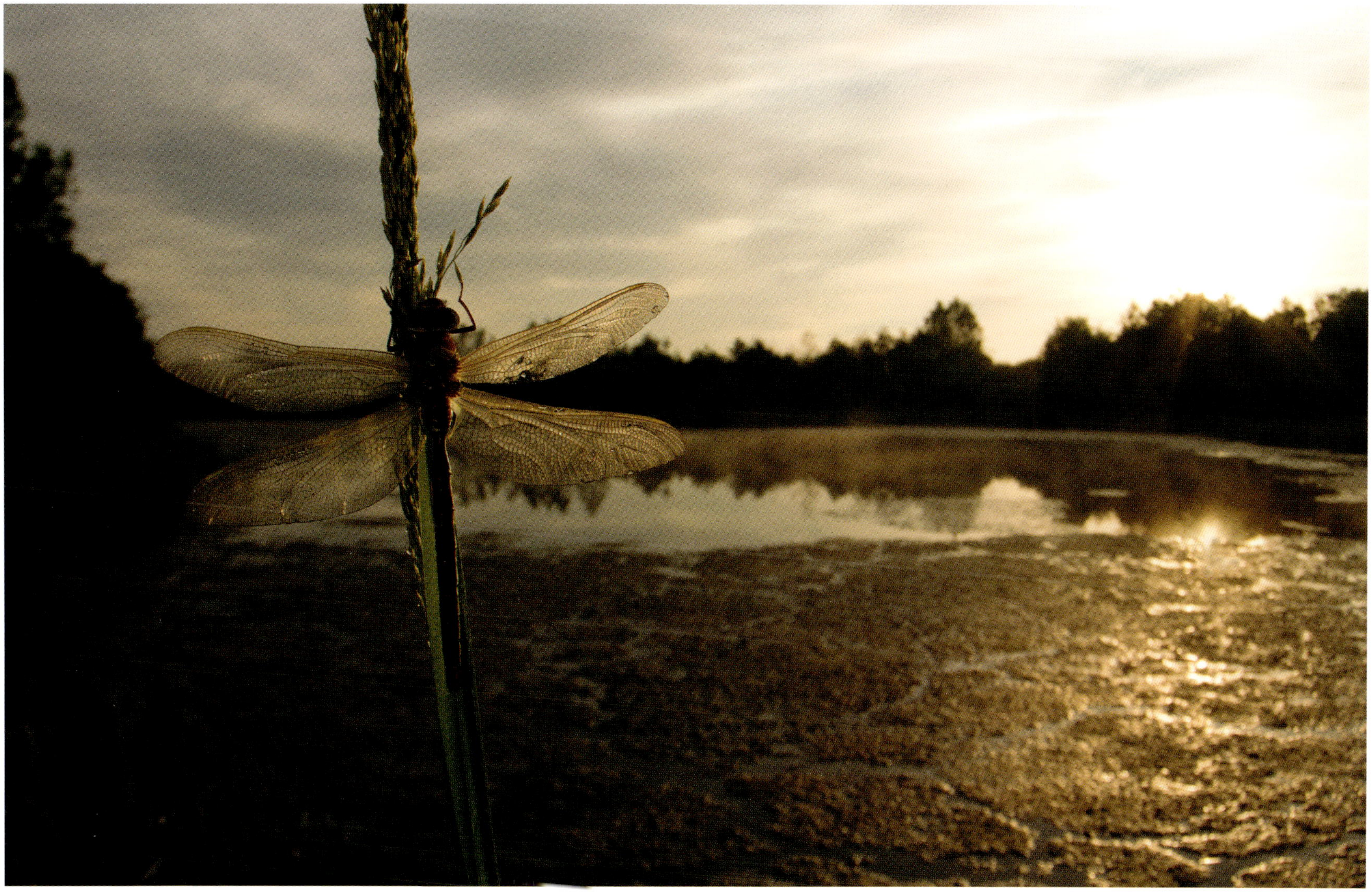

MATT COLE HIGHLY COMMENDED

An Early Morning Brown Hawker
(Brown Hawker Dragonfly, *Aeshna grandis)*
Oakthorpe, Leicestershire, England

An early start at a local nature reserve paid dividends when I found this brown hawker dragonfly. Being cool from the overnight drop in temperature it allowed me to approach, something this species does not generally do once it has warmed up. I was keen to show it in its environment and so I used a fisheye lens.

JAMIE HALL

Grass Snake Reflection
(Grass Snake, *Natrix natrix*)
Norfolk Broads, England

Living on the edge of the Norfolk Broads I spend much of my time exploring this habitat. One such morning I found this grass snake making its way along the edge of a small pond. Creeping up quietly and keeping low I managed to intercept it as it crossed over some pads to a warm bank on the edge of the pool.

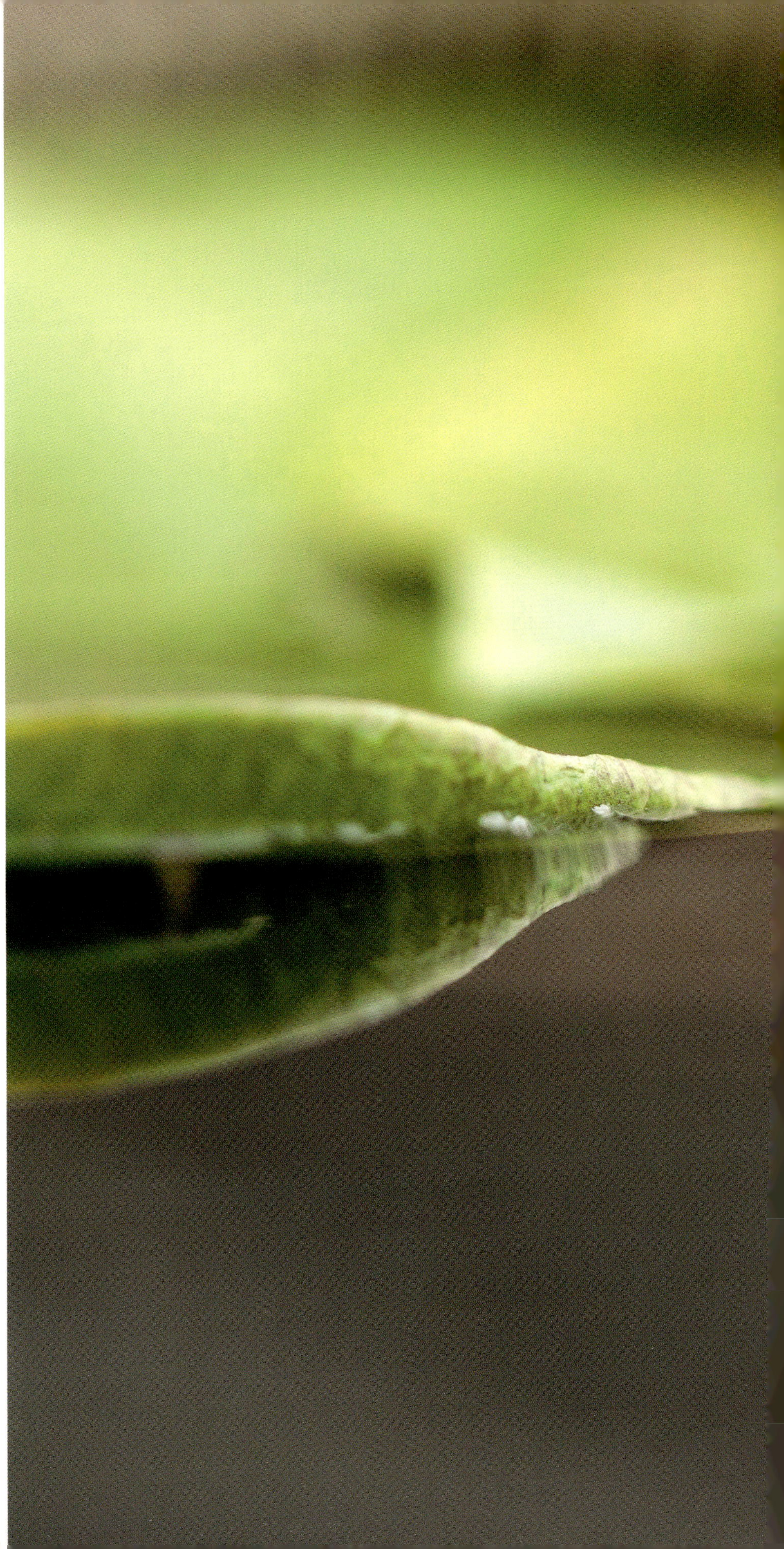

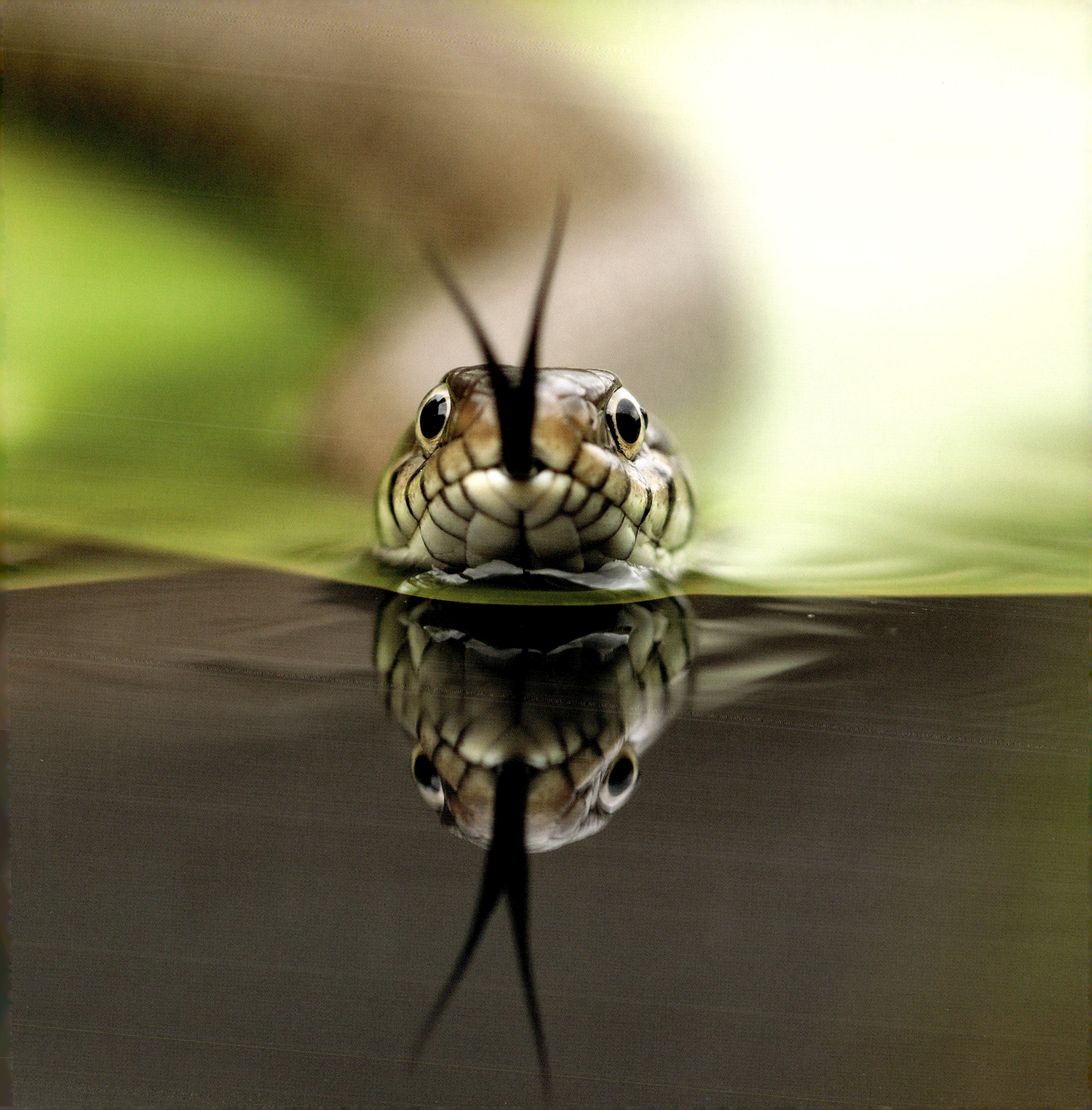

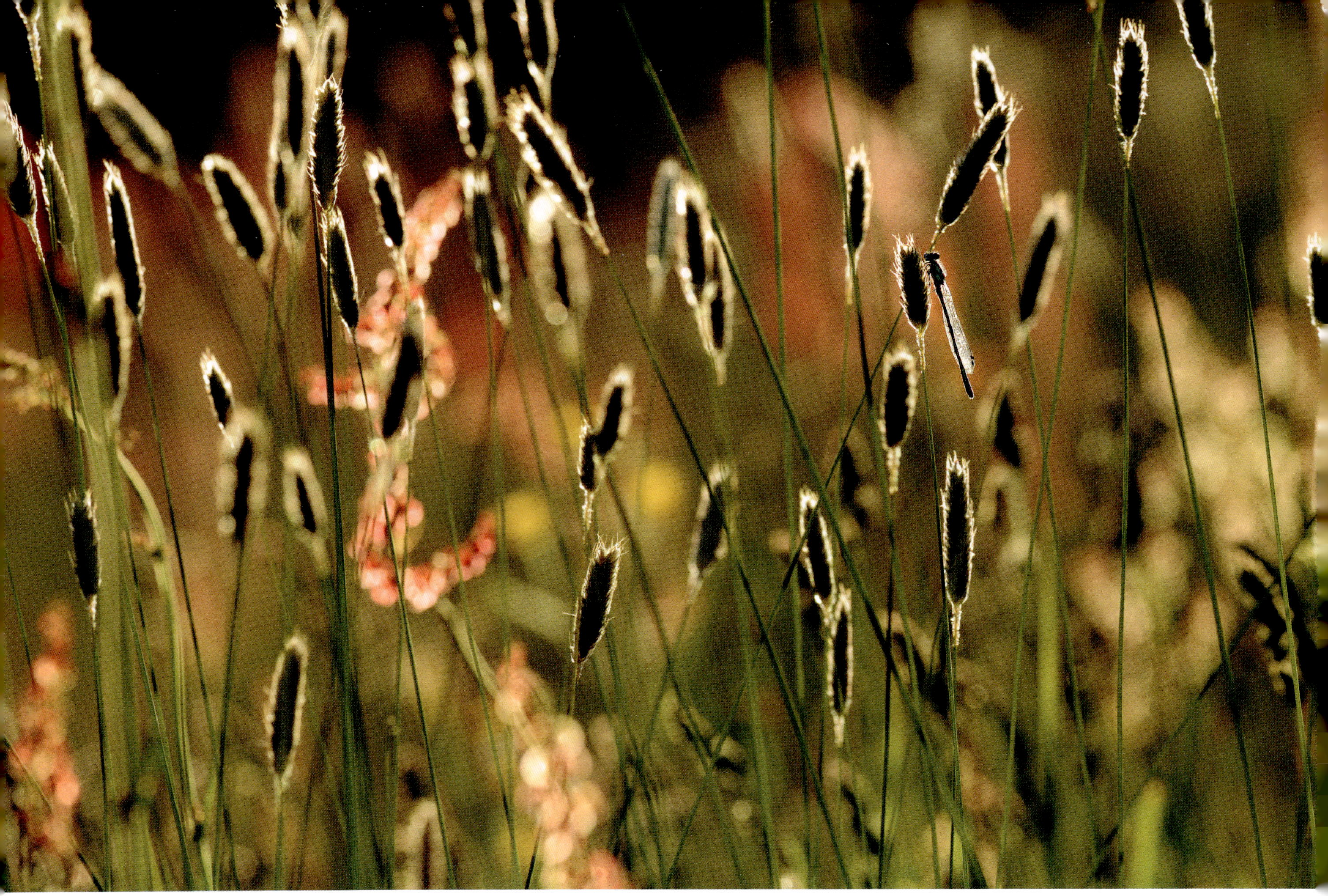

ALEX WINSER

Among the Grasses
(Common Blue Damselfly, *Enallagma cyathigerum)*
Chesworth Farm, Horsham, West Sussex, England

At roost among the long grasses at a local nature reserve this common blue damselfly is virtually invisible to most people. A slow approach was needed to ensure it wasn't disturbed. I decided to shoot with my long macro lens to compress the backlit scene and still show the species off in its habitat.

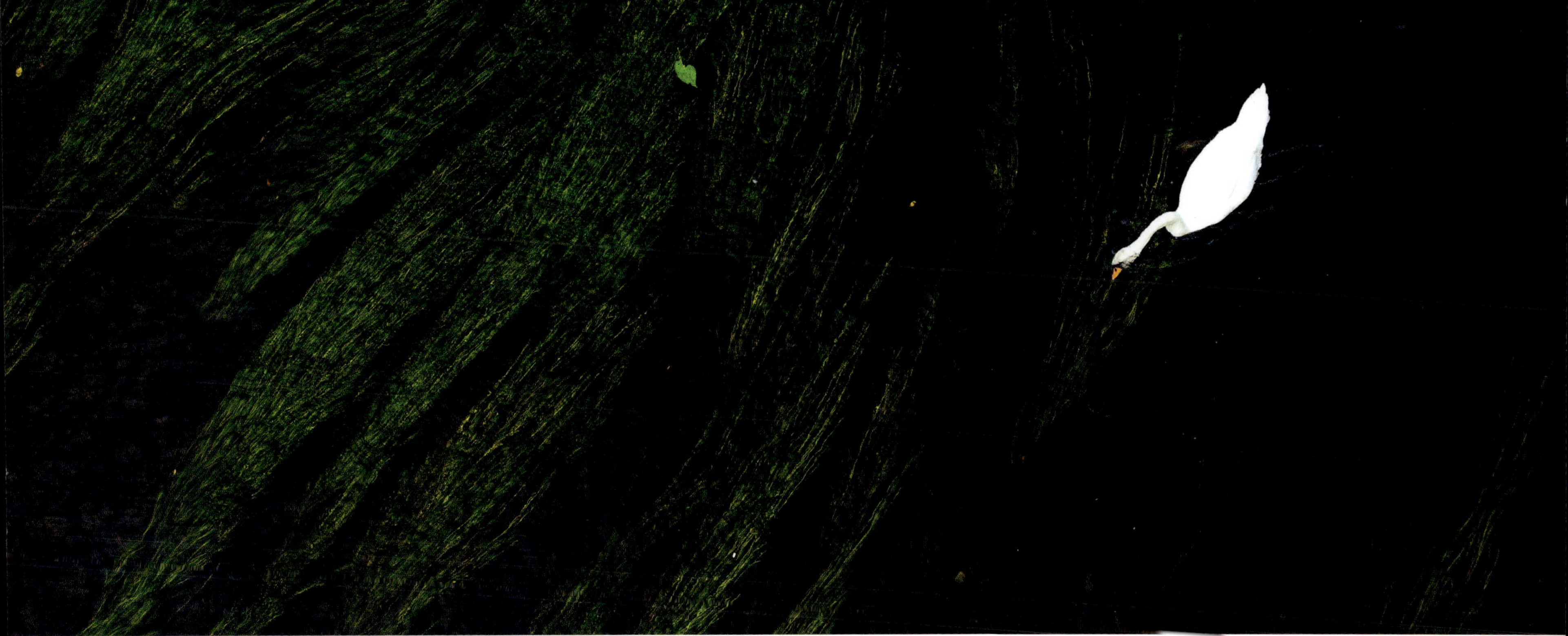

ANDREW FOX

Swan Feeding Amongst Reeds in the River Wye
(Mute Swan, *Cygnus olor*)
Hay-on-Wye, Herefordshire, England

The photograph was made looking down from a road bridge over the river.

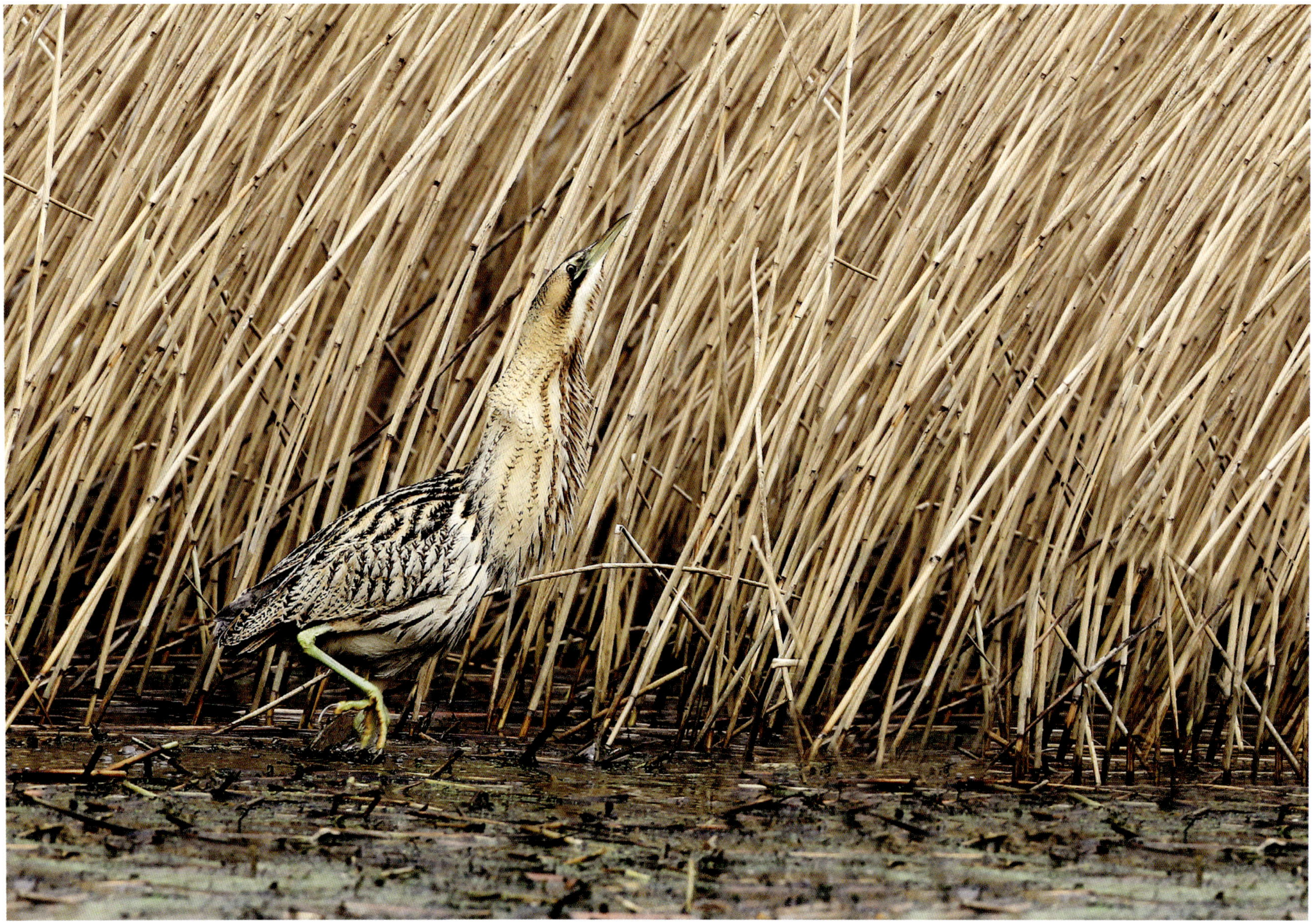

LESLIE HOLBURN

Bittern in the Reeds
(Bittern, *Botaurus stellaris*)
Northumberland, England

Now and again you hear whispers about a bittern showing and this was one of those occasions. Unfortunately I was away on holiday at the time so I had to wait until I got back before I could do anything about it. First time out and sure enough, there it was! Wow, what a privilege. To get photos was a massive bonus.

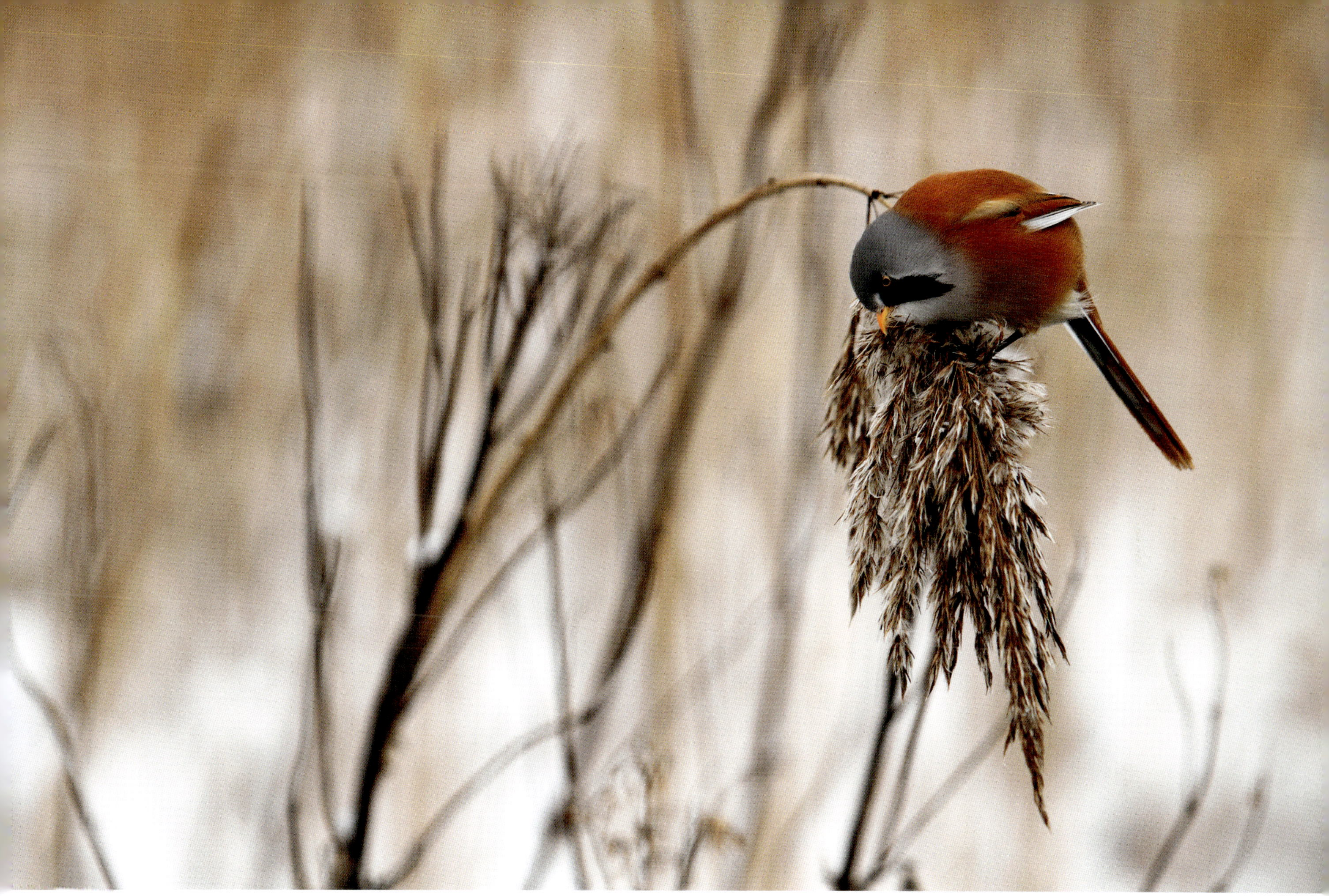

JAMES YAXLEY

Bearded Tit Feeding on Reed
(Bearded Tit, *Panurus biarmicus)*
RSPB Strumpshaw Fen, Norfolk, England

The reed bed is not looking at its finest but it is providing vital food for this beautiful little reedling. Conditions were very cloudy with limited light. I took the image using F5.6, in aperture priority mode, at 1/400sec. The bird was close enough to enable me to zoom out to 260mm to reveal an impressionistic interpretation of its wintry habitat of reed and snow.

BRITISH SEASONS

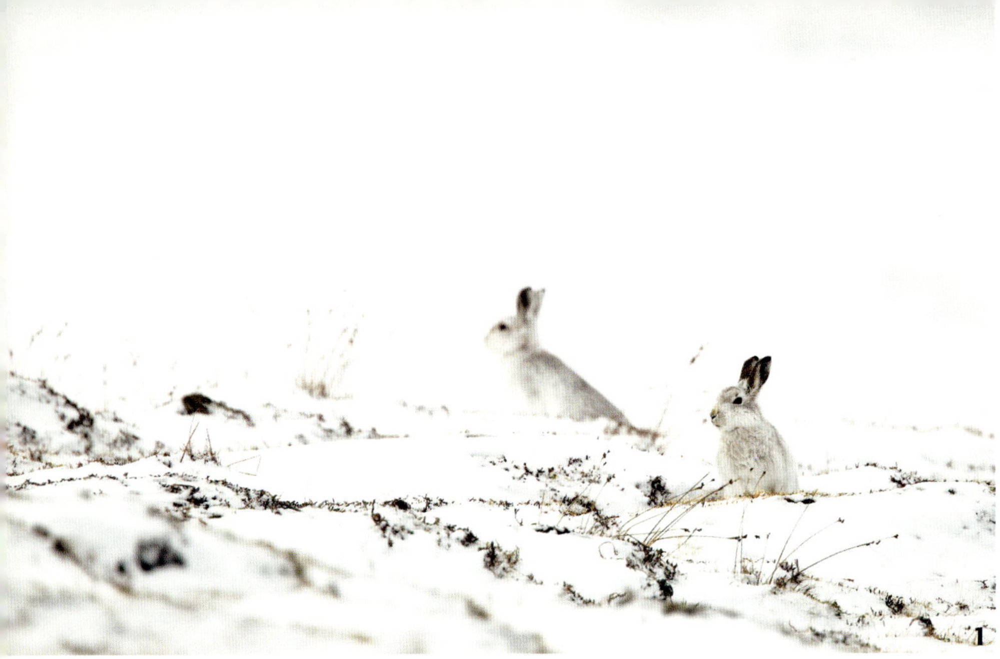

1

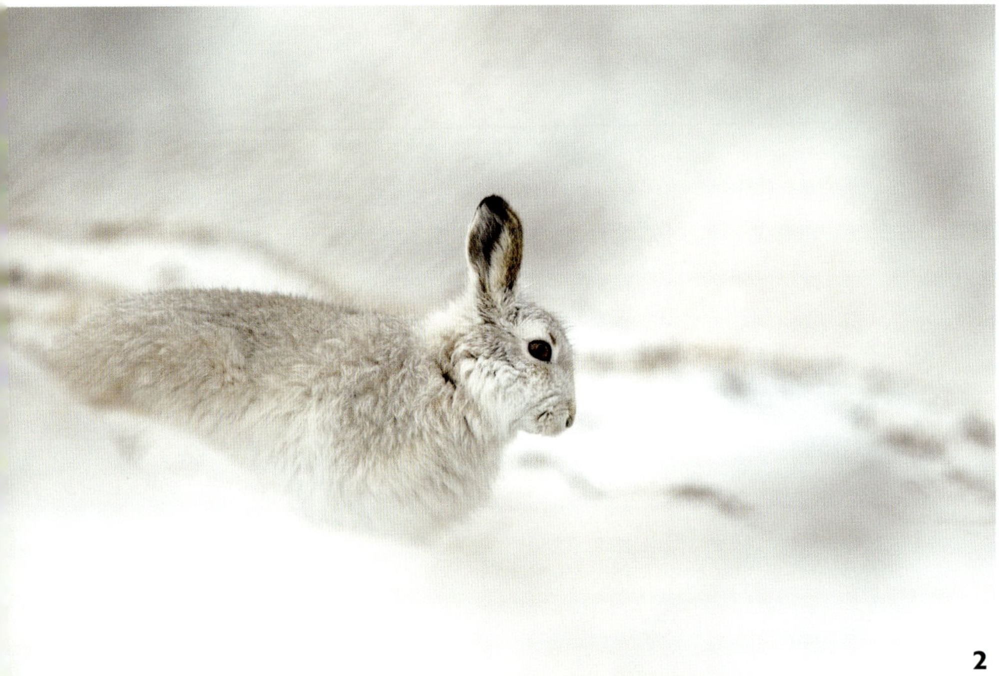

2

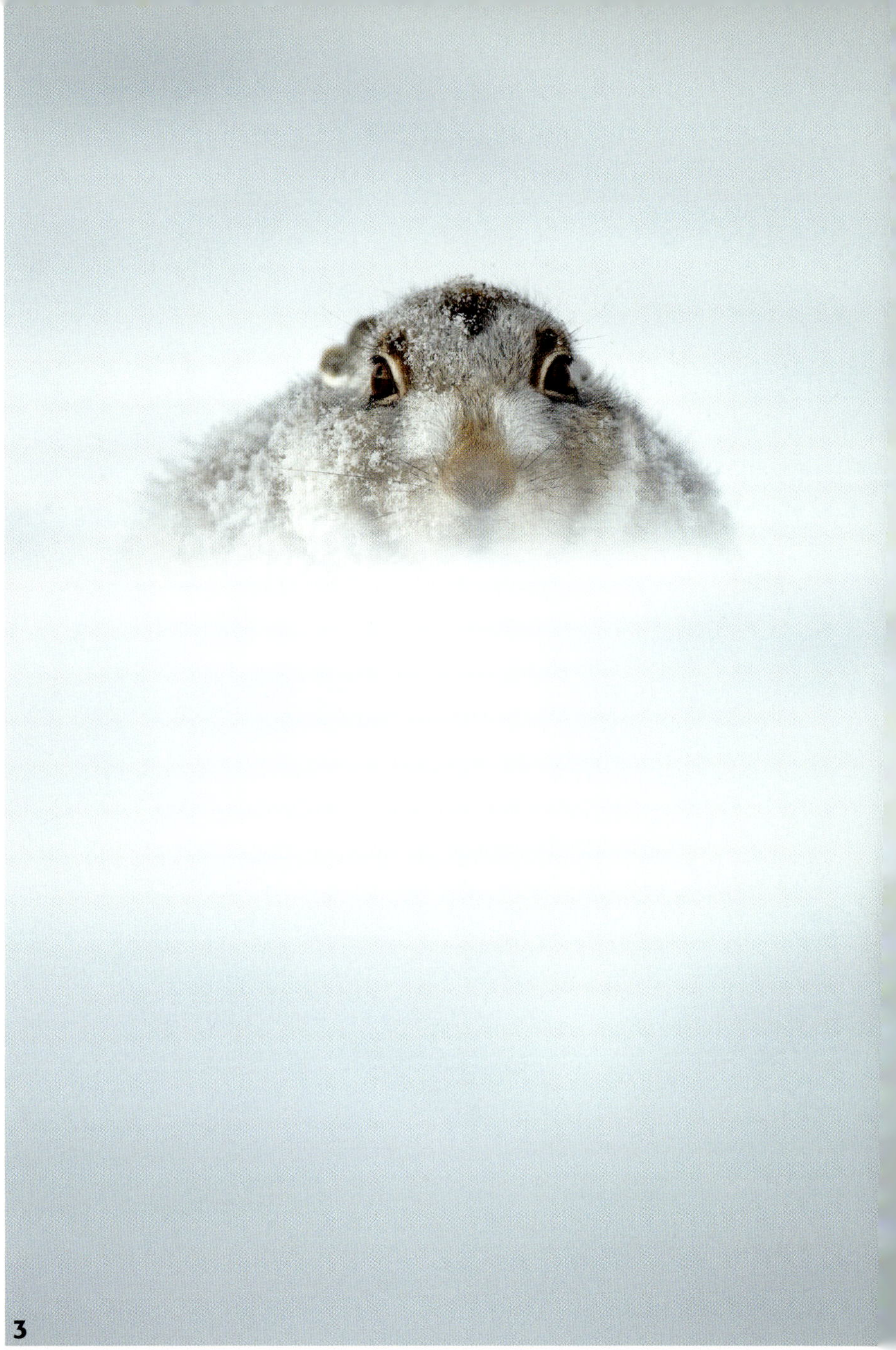

3

BRITISH SEASONS WINNER

JULES COX

WINTER HARE
(Mountain Hare, *Lepus timidus)*
Cairngorms National Park, Highland, Scotland

Last winter I worked on a project in the Scottish Highlands photographing the mountain hares that are native to the region.

1. Up on the Hill
During periods of snow cover, mountain hares gather in groups on leeward hill slopes to feed or shelter where shallow snow permits scraping to reveal underlying heather. I deliberately shot these hares small in the frame to give a sense of scale to the upland environment which they inhabit.

2. The Snow Hare
A mountain hare at rest, perfectly camouflaged in its pure white winter pelage. Hares are active at night, preferring to rest during the day. A careful approach can sometimes result in a privileged encounter such as this.

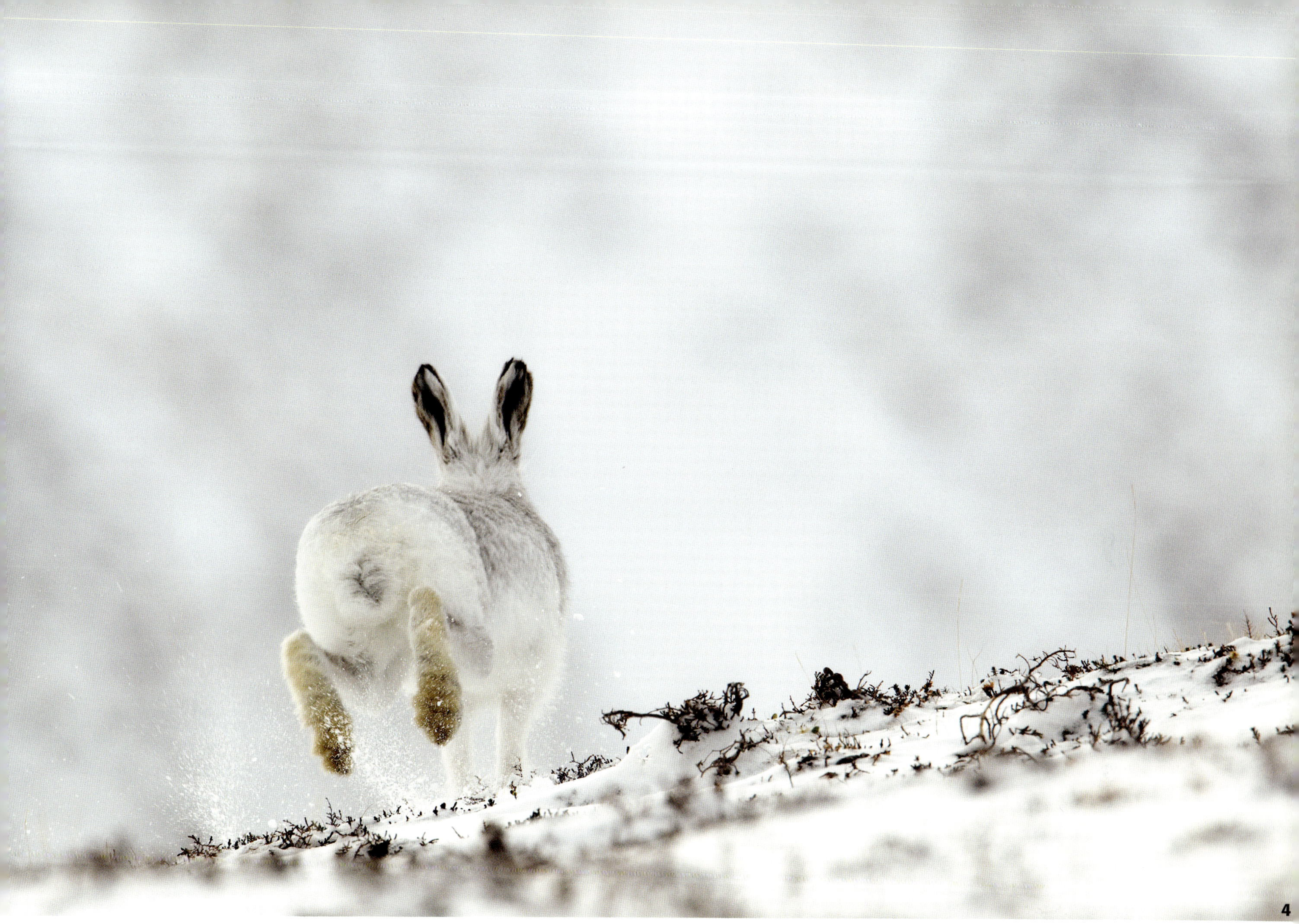

4

3. Snowed Under
A mountain hare shelters in a snow blizzard, perfectly camouflaged in her pure white winter coat. I found this old female hare high on a Scottish hillside not far from the path, sitting out the adverse weather conditions. She was perfectly relaxed, allowing me to photograph her even as the snow threatened to bury her completely. As I made my way back down the hill out of the blizzard I looked back and could just about see her still sitting there through the snowfall. The encounter left me marvelling at what hardy mammals mountain hares are.

4. Hop It
For anyone who has ever tried to get close to a mountain hare this will be a familiar sight. Shot by some grouse moor managers because they sometimes host ticks which in turn carry viruses to red grouse, hares are naturally wary of man. Perfectly adapted for the dangers presented by their natural predators, their powerful hind legs allow them to sprint away at phenomenal speed if being pursued. As a result of shooting and habitat loss, mountain hares are now a much rarer sight on some Scottish moors where once they were abundant.

LIVING LANDSCAPE: CONNECTIVITY

SPONSORED BY

THE WILDLIFE TRUSTS

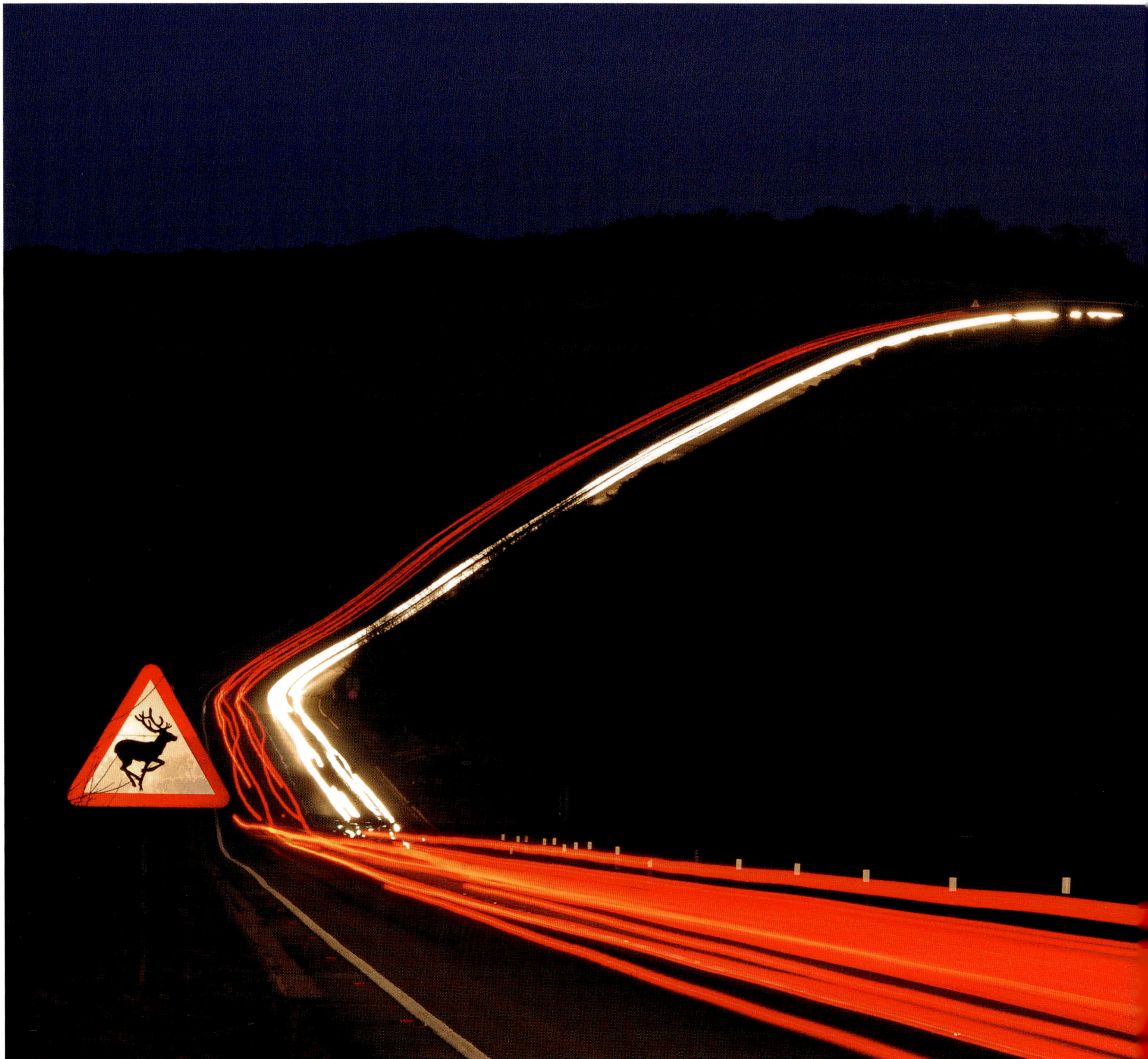

LIVING LANDSCAPE: CONNECTIVITY WINNER

STEPHEN WILLIAM POWLES

Culm Divided

Knowstone Moor, Devon, England

This is part of one of the largest expanses of Culm grassland in Britain. Culm grassland, found in North Devon and East Cornwall, is a unique and internationally threatened habitat. Sadly, the North Devon link road was put through the middle of this ecosystem. Taken at night, the image shows even more dramatically how the road has cut through this habitat. The deer sign further emphasises the conflict.

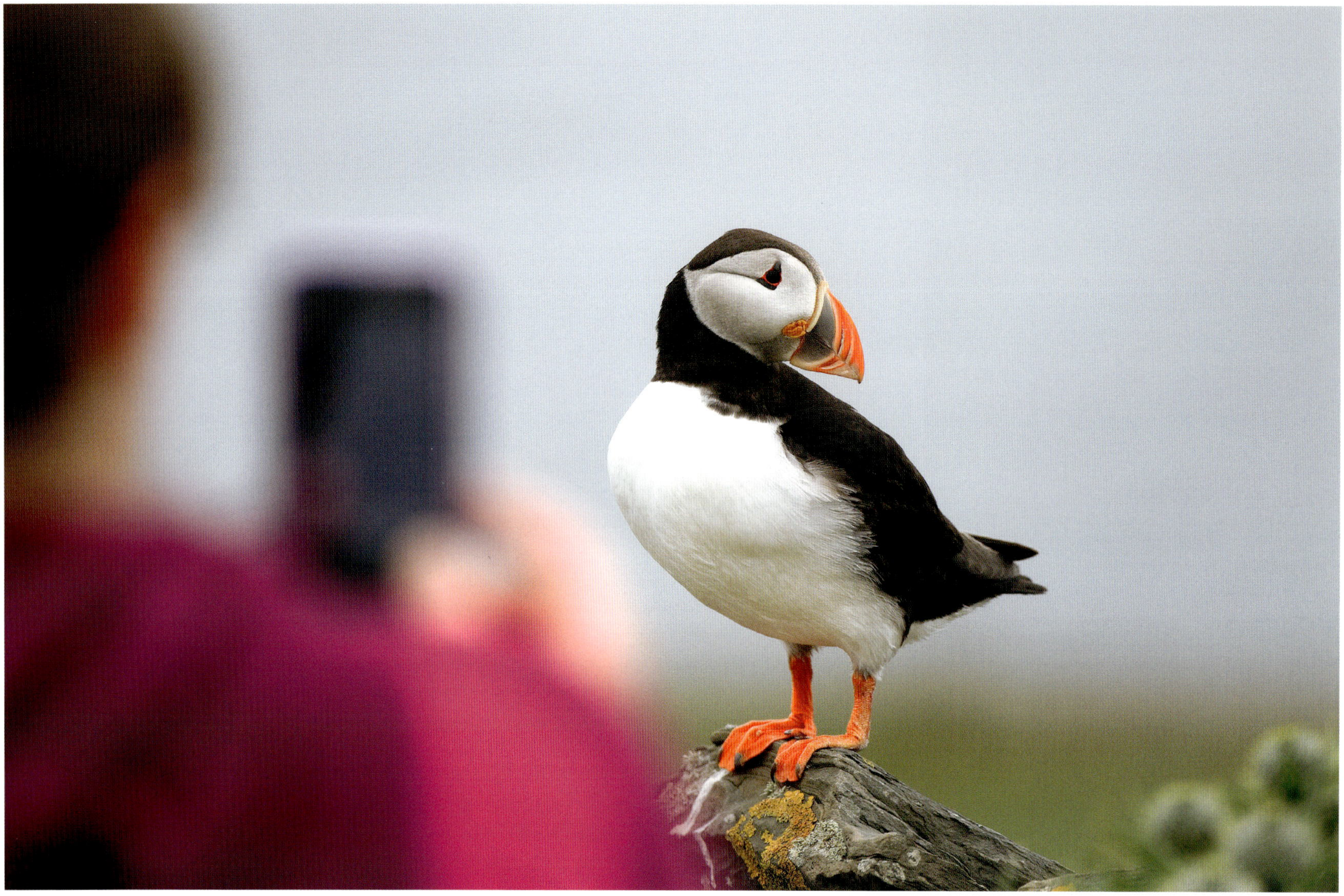

PETER CAIRNS/2020VISION HIGHLY COMMENDED

iPuffin
(Atlantic Puffin, *Fratercula arctica)*
Shetland Isles, Scotland

I'm always interested in the interaction between wildlife and people. When this girl produced her mobile phone in order to photograph an accommodating puffin, I wanted to try and portray not only the two species connecting but also the increasing role of technology in our relationship and understanding of nature.

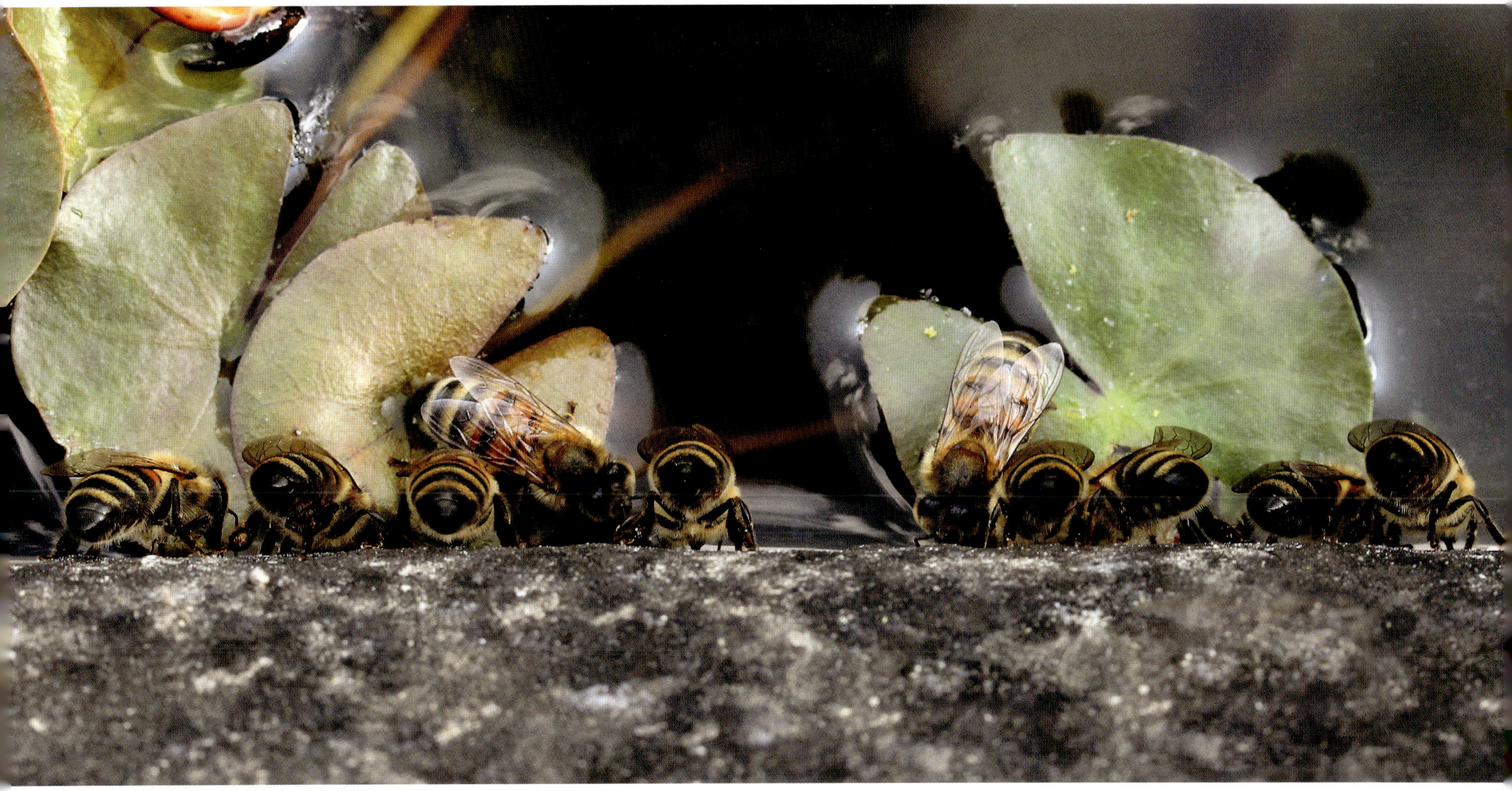

RACHEL PIPER HIGHLY COMMENDED

Bees Drinking from a Pond
(Honey Bee, *Apis mellifera*)
Ashridge House, Berkhamsted, Hertfordshire, England

I have always loved honey bees. This group were drinking water from a deep stone pond at my place of work, so my reason for leaning precariously over the edge was not obvious to those walking past! Ponds help sustain so much life, but in this case it was helping creatures that live on dry land, with lily leaves providing a ledge. Another good reason for building a pond.

DOCUMENTARY SERIES

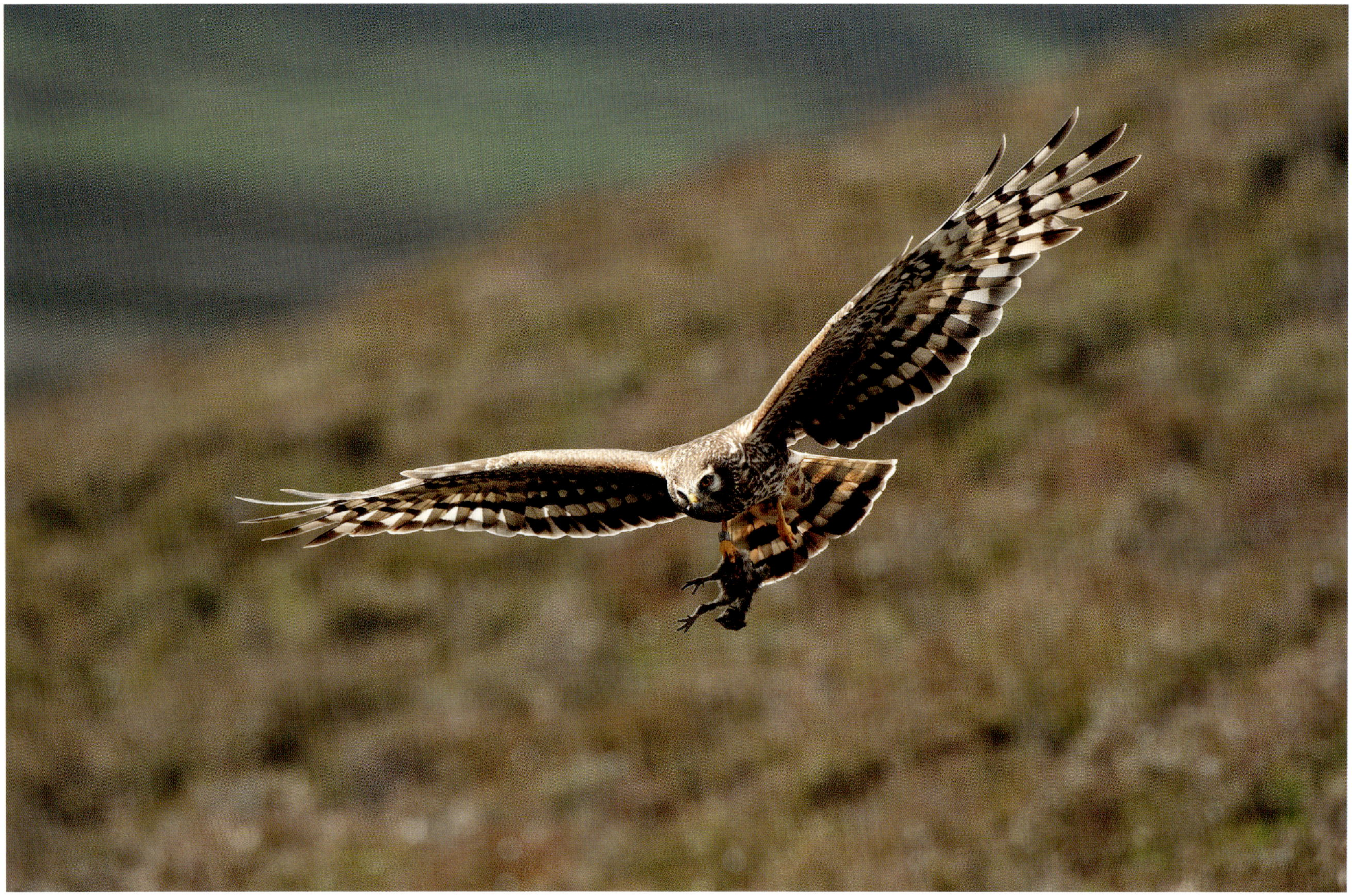

DOCUMENTARY SERIES WINNER

MARK HAMBLIN

A Fresh Approach to a Rural Tradition
(Hen Harrier, *Circus cyaneus*; Red Grouse, *Lagopus lagopus scoticus*)
Glen Tanar Estate, Aberdeenshire, Scotland

On many of Britain's upland estates there remains a precarious relationship between sport shooting and predators. Right at the forefront of this are two species – red grouse and hen harrier. The former is a highly prized game bird much coverted by shooting interests, the other a formidable predator of young grouse. This is an uneasy relationship that has been difficult to reconcile. One strategy put forward is to provide nesting harriers with an alternative food supply to lessen their impact on grouse stocks. My aim was to document this fresh approach to grouse management on Glen Tanar Estate in the Scottish Highlands.

A female hen harrier returns to its nest with a young grouse chick. Harriers and other moorland raptors can heavily reduce grouse stocks during the breeding season, which inevitably brings them into direct conflict with game shooting interests.

Red grouse is an important game bird whose numbers are maximised on sporting estates through careful management to provide high yields for paying clients during the shooting season, which begins on 12 August.

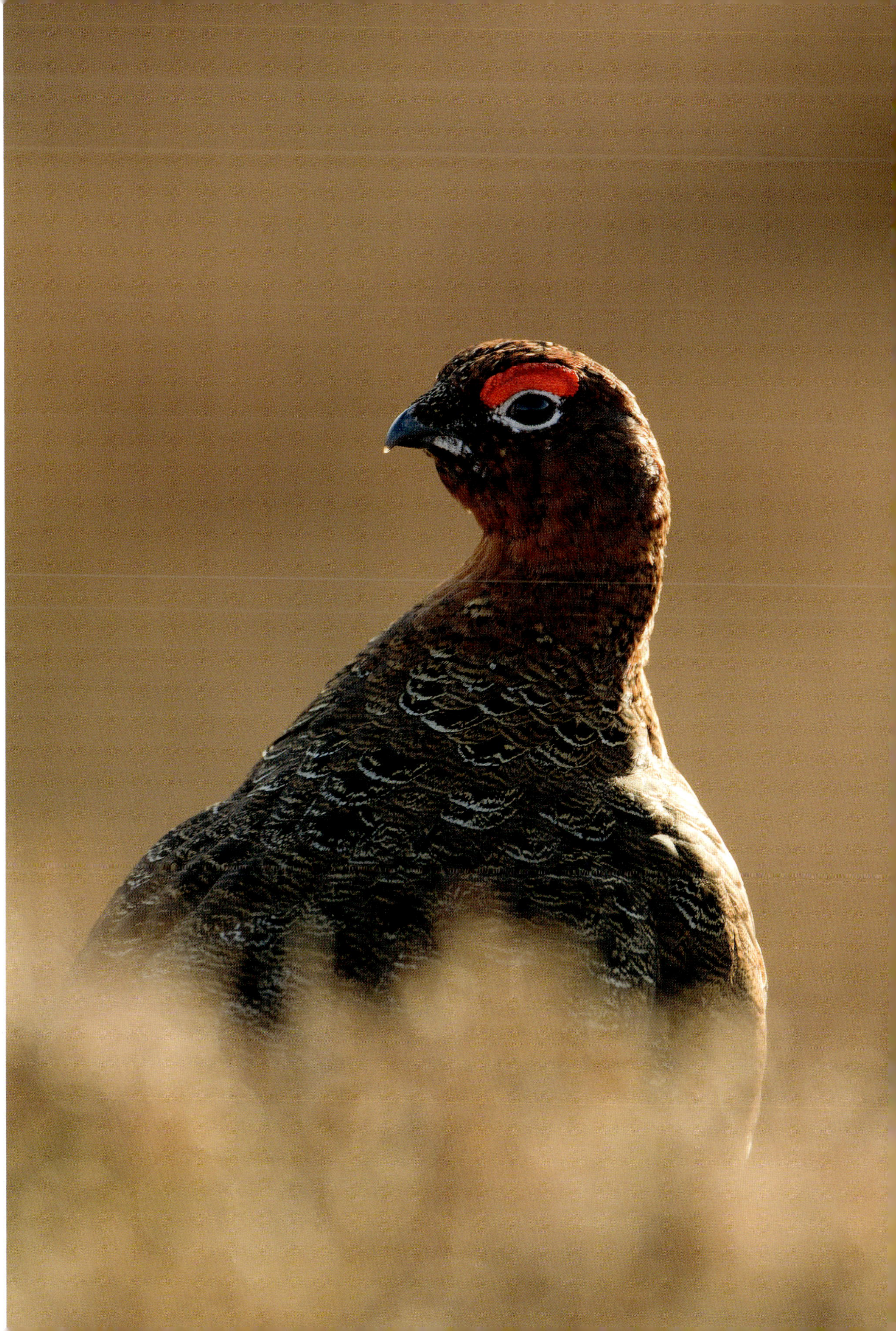

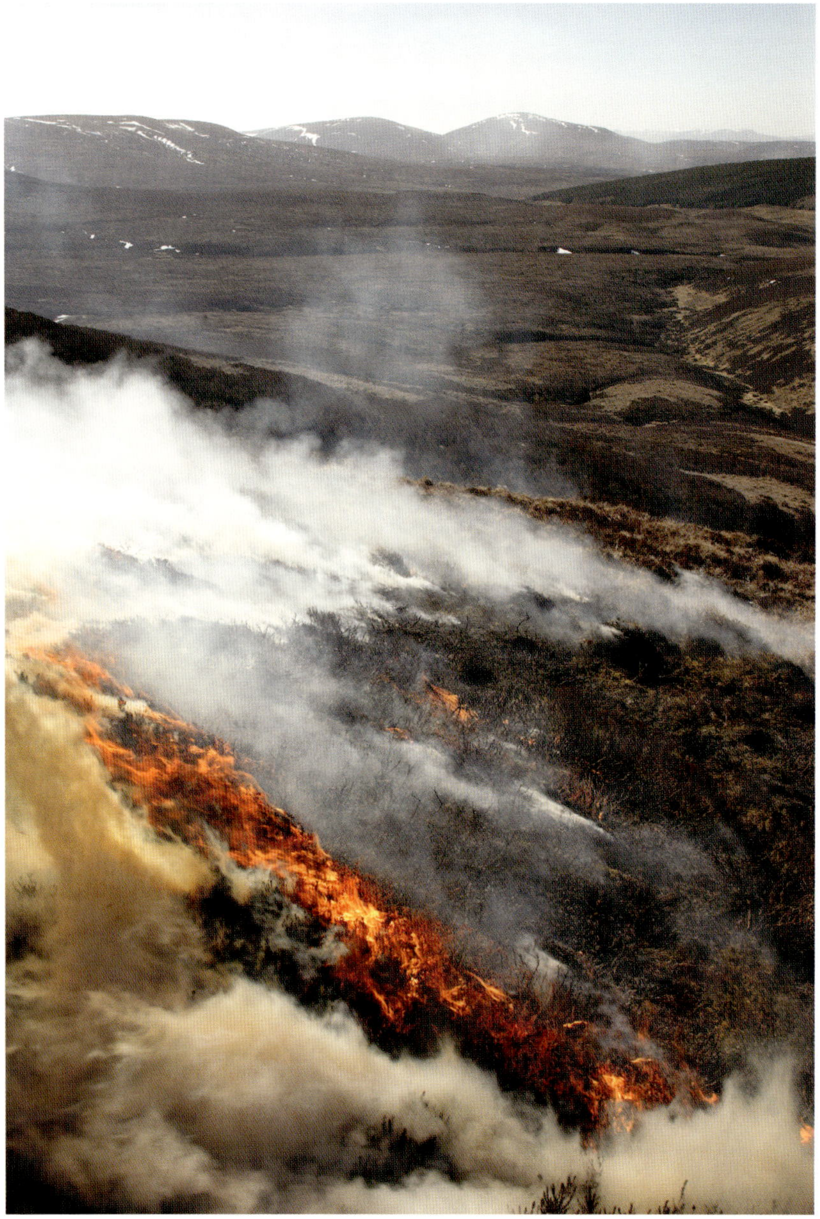

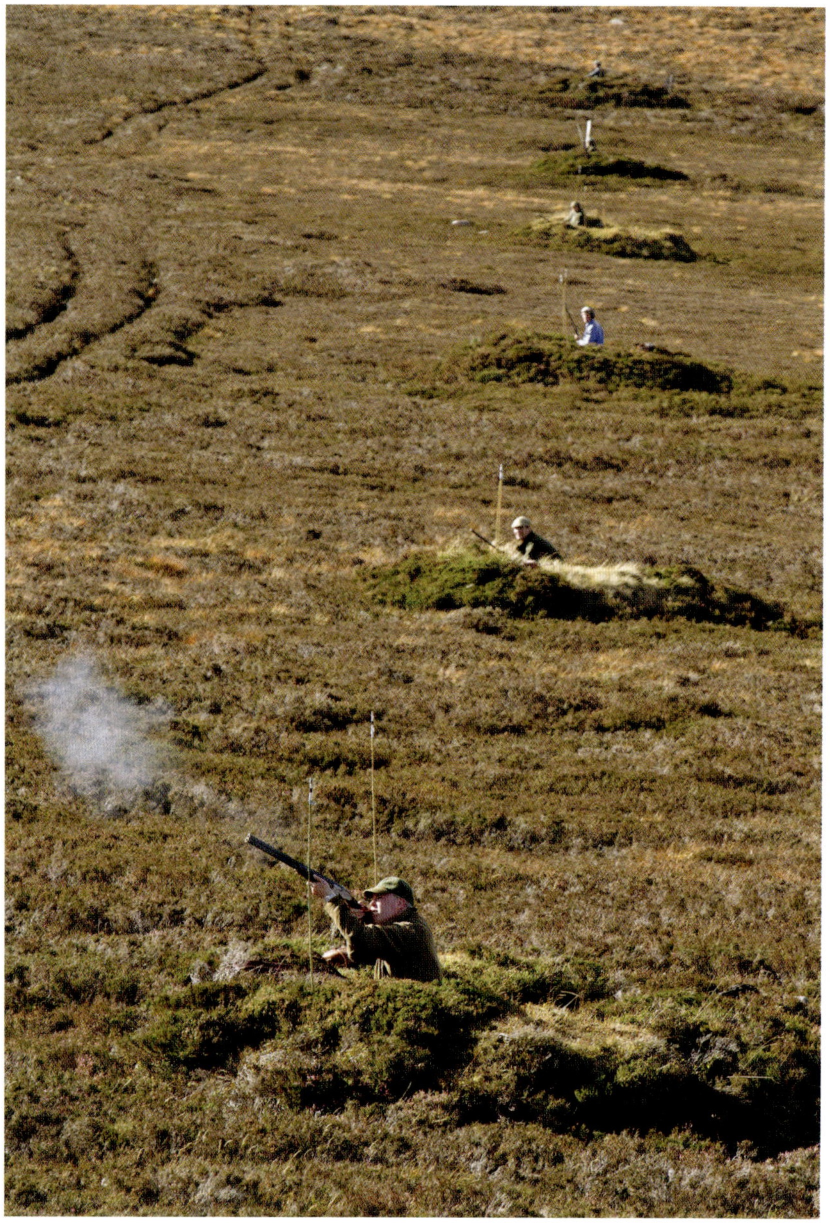

Muirburn (heather burning) on an upland sporting estate to burn off old growth and promote fresh heather shoots on which red grouse feed. Burning takes place on a 12-year cycle to generate a mosaic of heather patches of different ages as part of red grouse management.

Clients pay handsomely to shoot red grouse but if there are low numbers of game birds then there are less grouse shooting days, which, it is argued, has significant implications for rural economies and local employment.

An alternative approach to reducing the impact of grouse predation by harriers is to provide them with another food source. On Glen Tanar Estate, during the 2011 season, the single pair of breeding harriers were provided with chunks of venison that the female took to feed her chicks, which resulted in a reduction in the number of grouse taken.

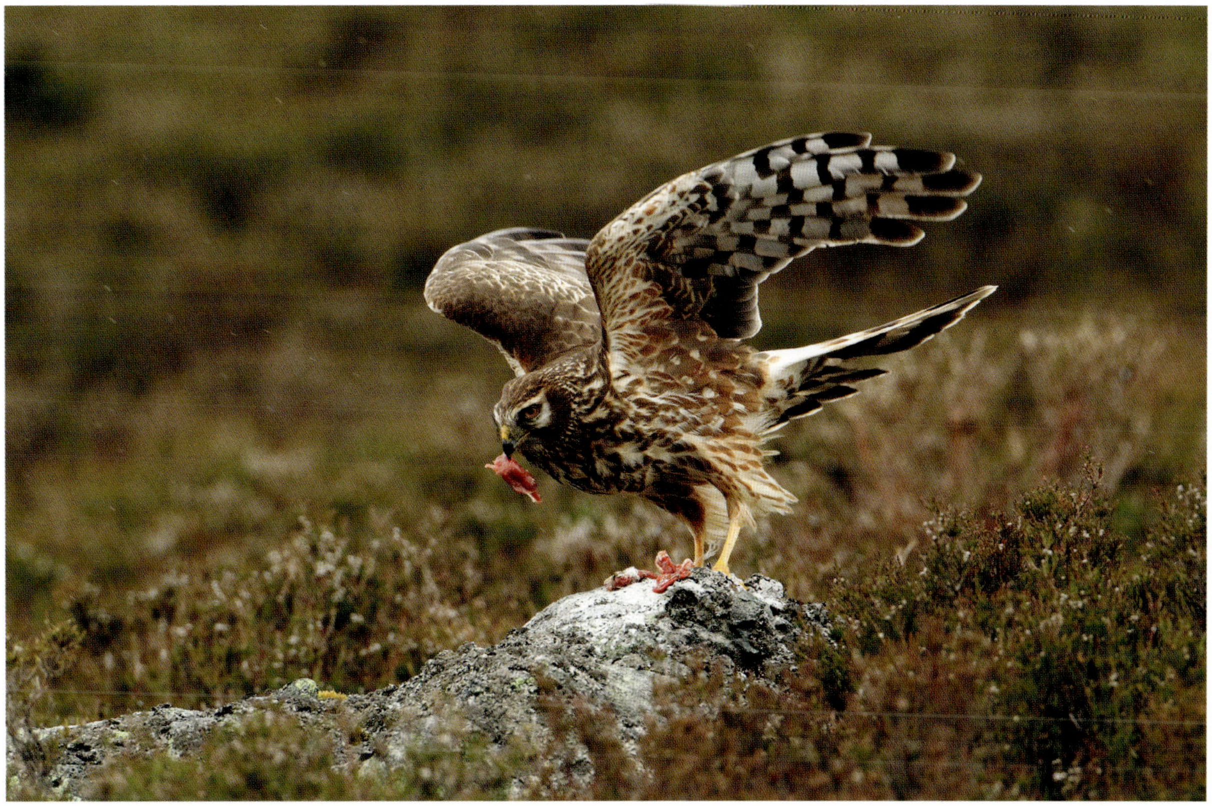

In 2011, Glen Tanar Estate recorded their highest bag (yield) of red grouse for many years, which also coincided with the successful fledging of four hen harrier chicks, one of which was fitted with a radio transmitter to track its movements.

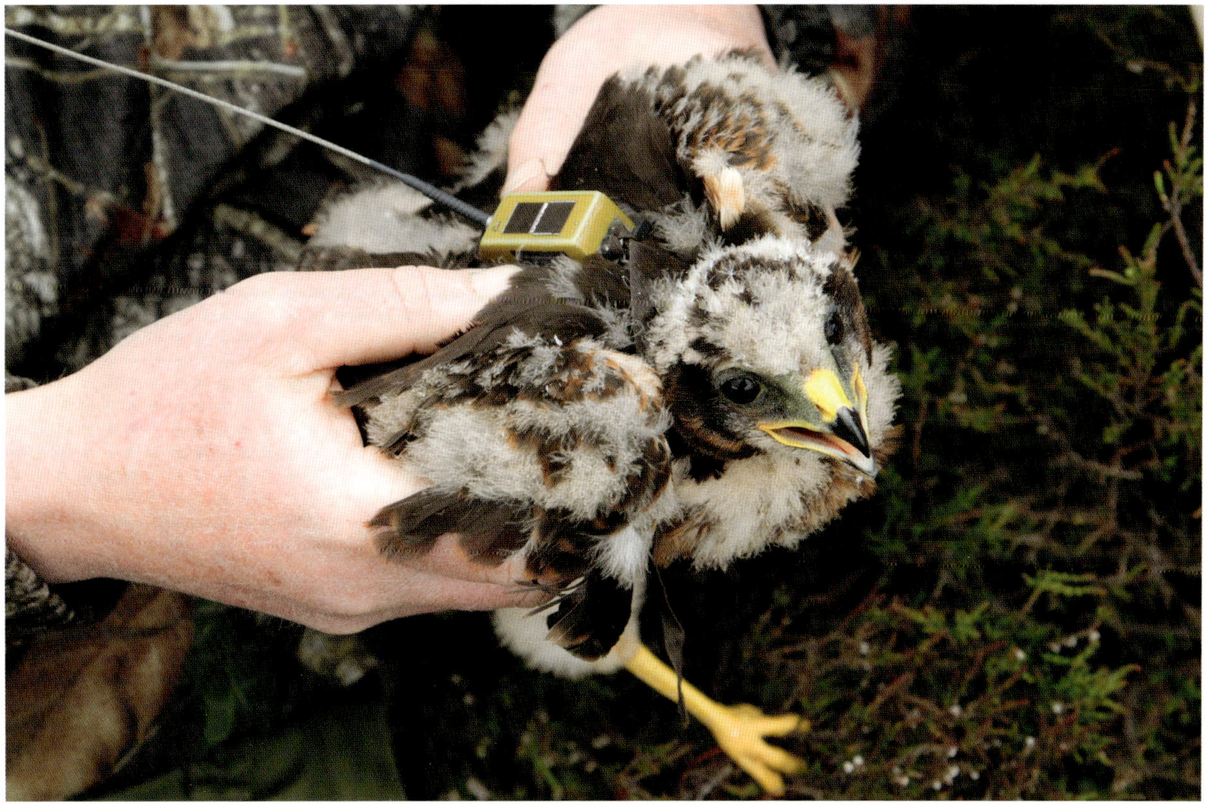

The Vanishing Shore

(Salmon, *Salmo salar*)

St Cyrus, Angus, Scotland

A massive decline in wild Scottish salmon numbers has led to the near demise of the wild salmon netting industry. I wanted to photograph the last of the salmon netters and the few returning salmon they caught. The project started in 2007. This year the Scottish government awarded massive grants to some salmon netsmen to expand their operations, which caused outrage among those concerned with salmon conservation. They believe it will further hasten the decline of the salmon. In the image below, Stuart Ogston holds up a small trout that was the only catch taken from the nets. Wild salmon numbers are in precipitous decline.

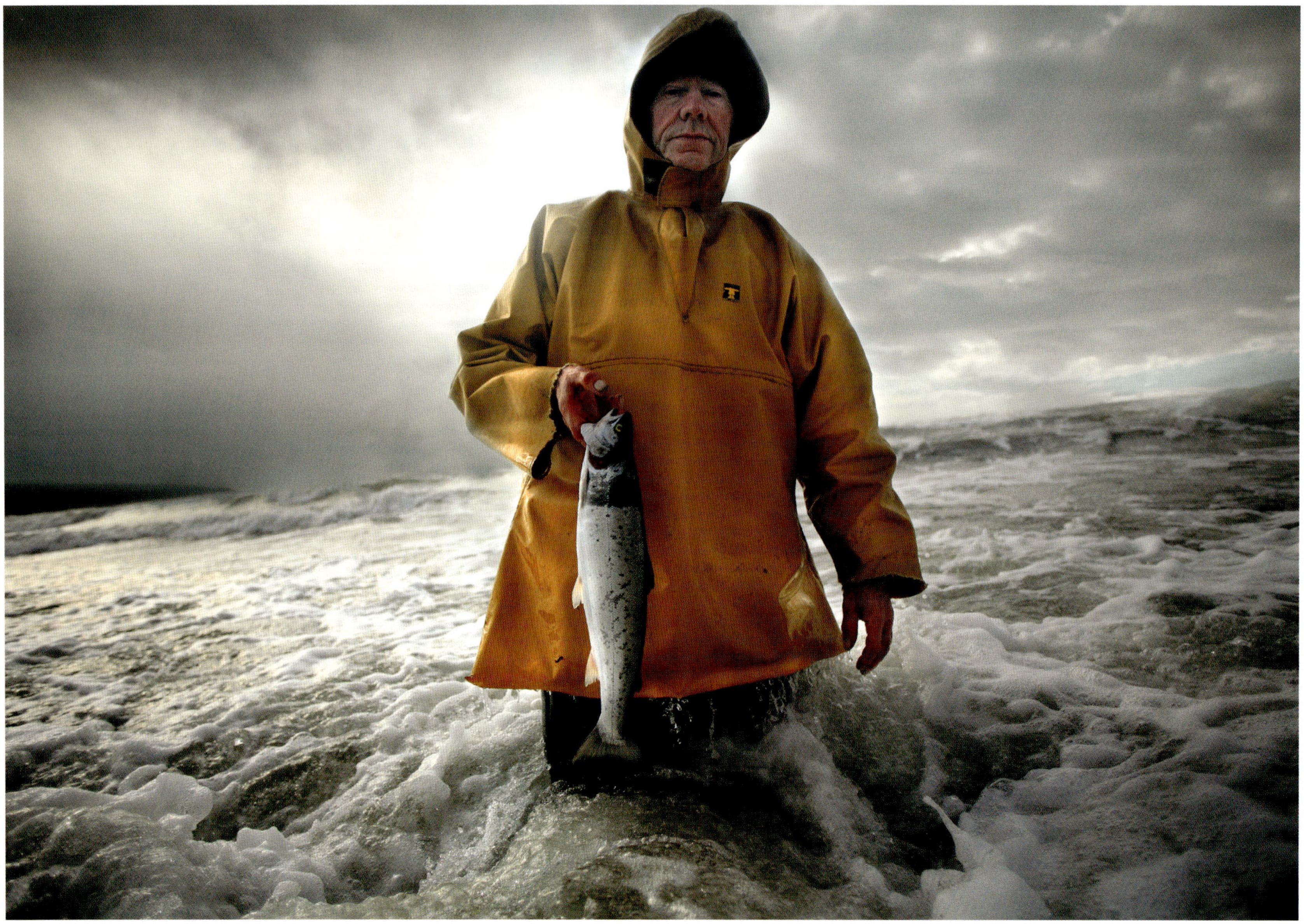

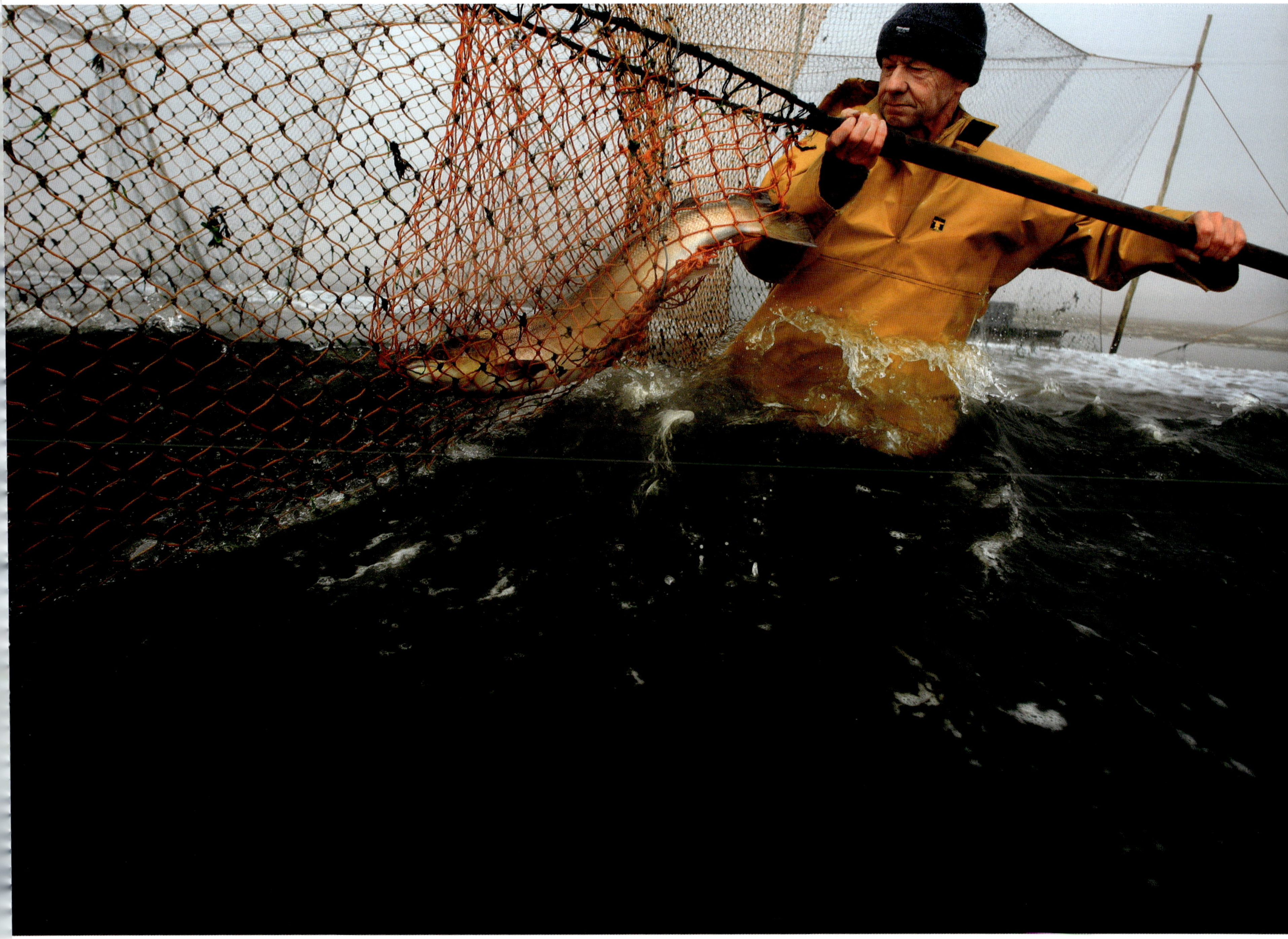

Stuart Ogston nets a kelt (a salmon that has already spawned). Kelts are rarely found in the sea nets since mortality rates are very high. They are distinguishable by their strange bronze phosphoric colour, their gills flaring to orange. Kelts are always released to return to their winter feeding grounds in Norway or Greenland.

WILDLIFE IN HD VIDEO

SPONSORED BY SKY+HD

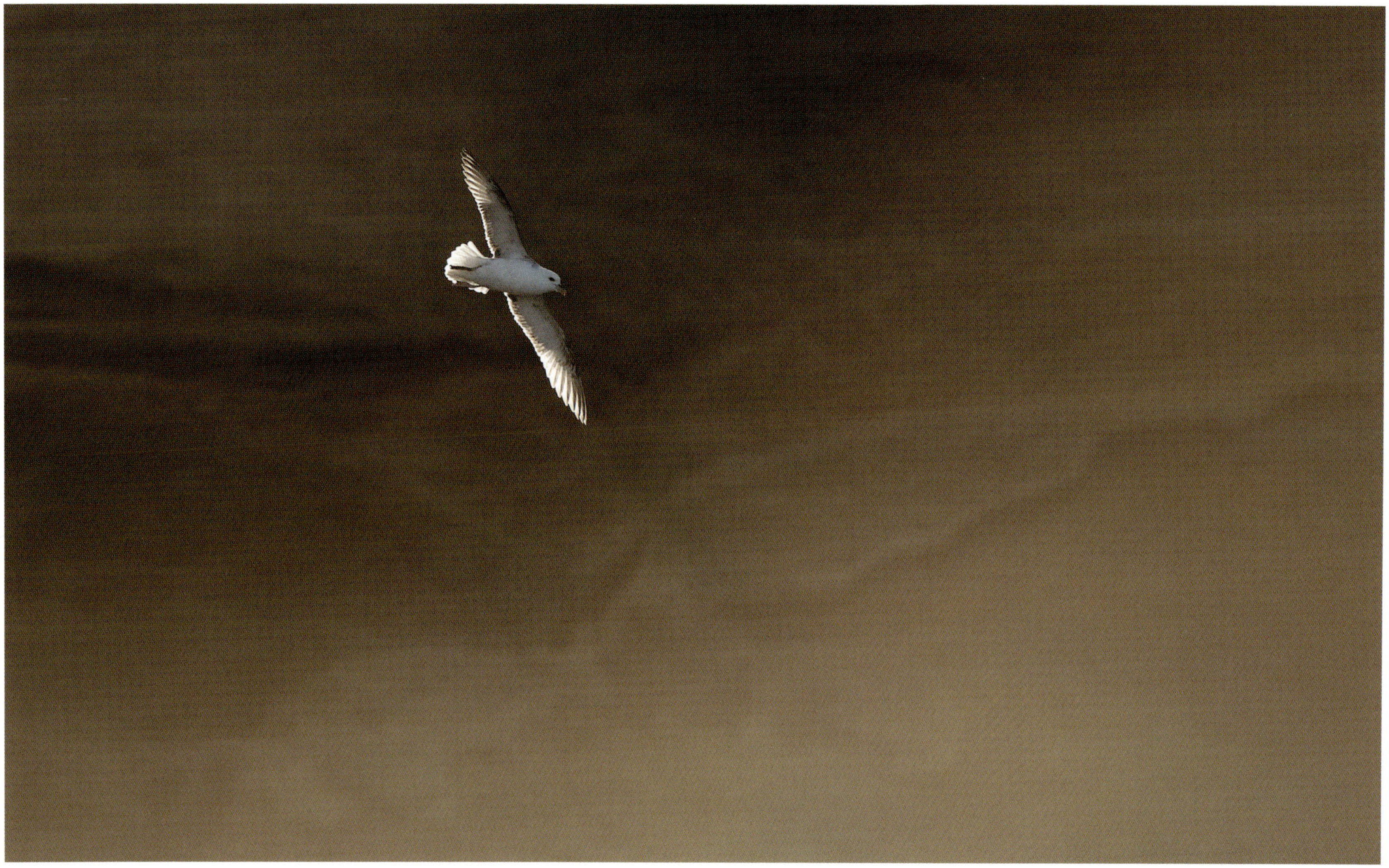

WILDLIFE IN HD VIDEO WINNER

RAYMOND BESANT

Fulmars
(Fulmar, *Fulmarus glacialis)*
Hoy, Orkney Islands, Scotland

The cliffs of Hoy are among the highest in the UK and they really come alive when fulmars return to nest on them. When the early morning sun sneaked over the Hoy hills it was captured in the spray from the crashing waves below. I knew I had to work quickly in order to film the special atmosphere. The flight of the fulmar seems so effortless, but on this morning they seemed especially serene, gliding through the beautiful mist. Ten minutes later the ethereal light had all but disappeared.

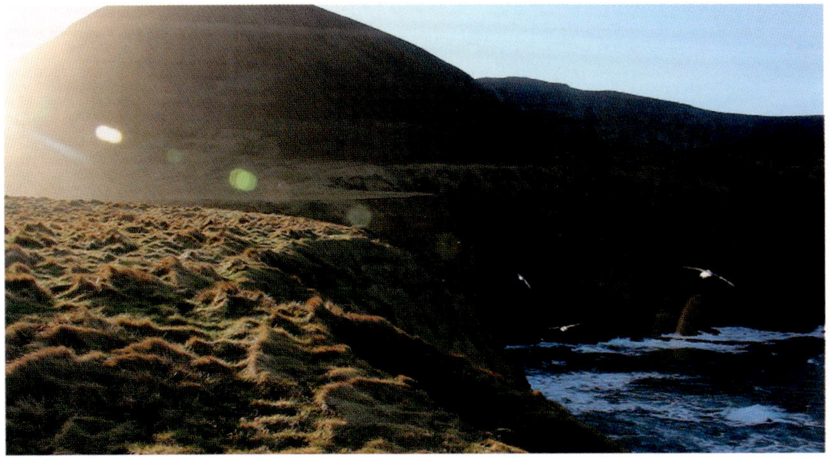
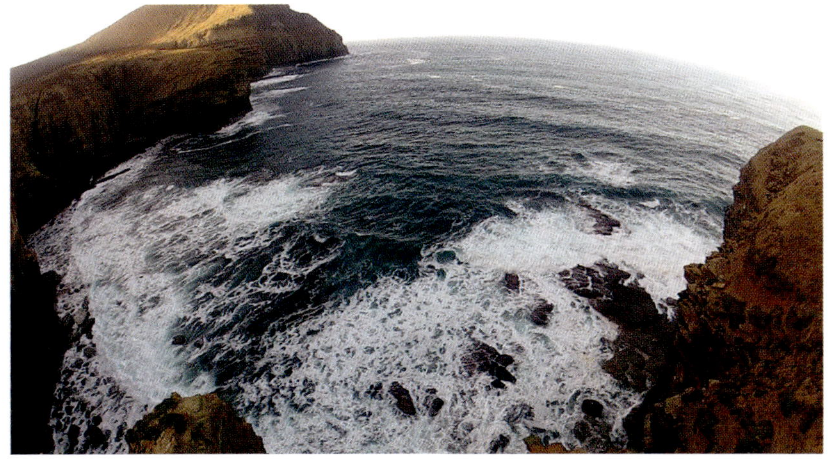
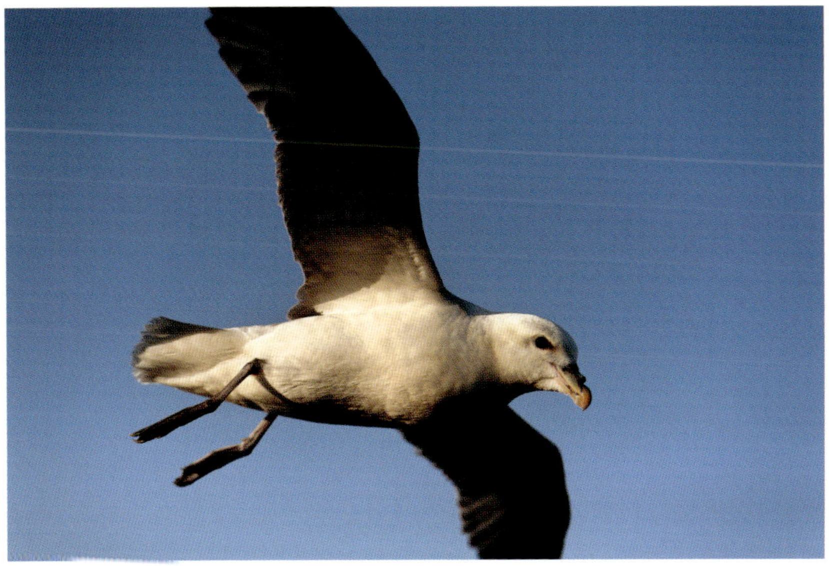
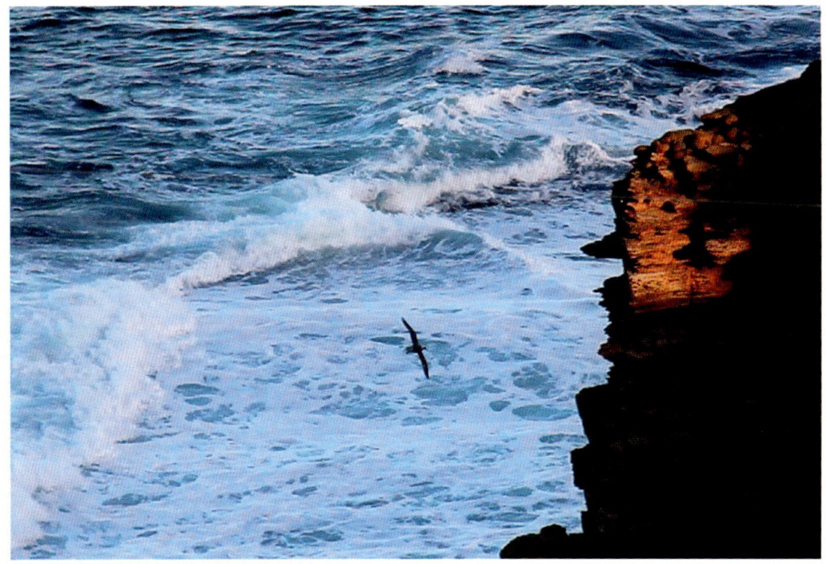
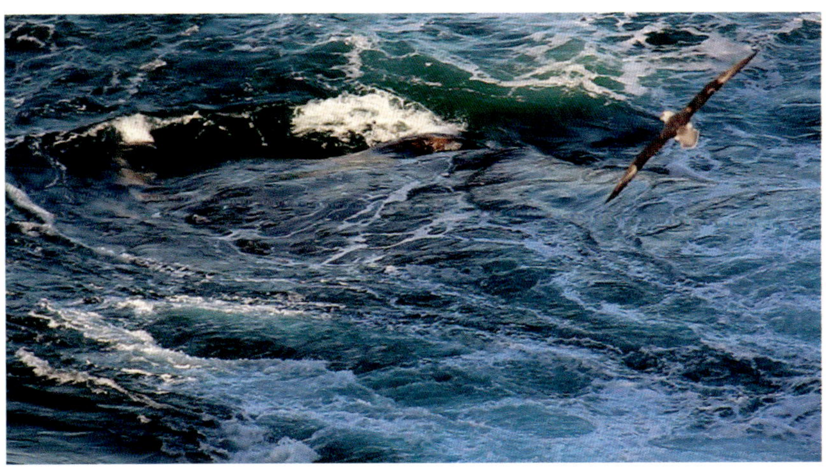

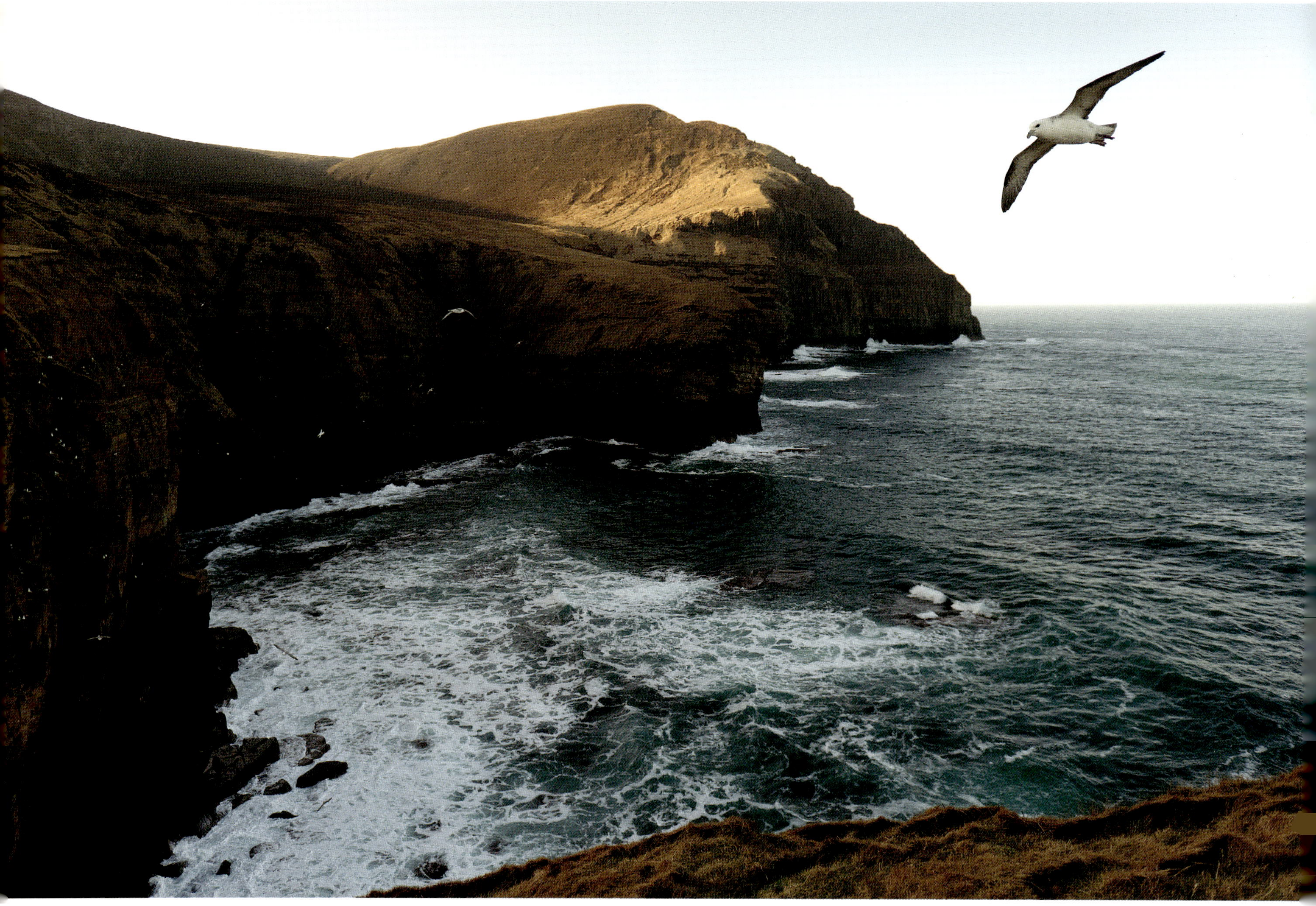

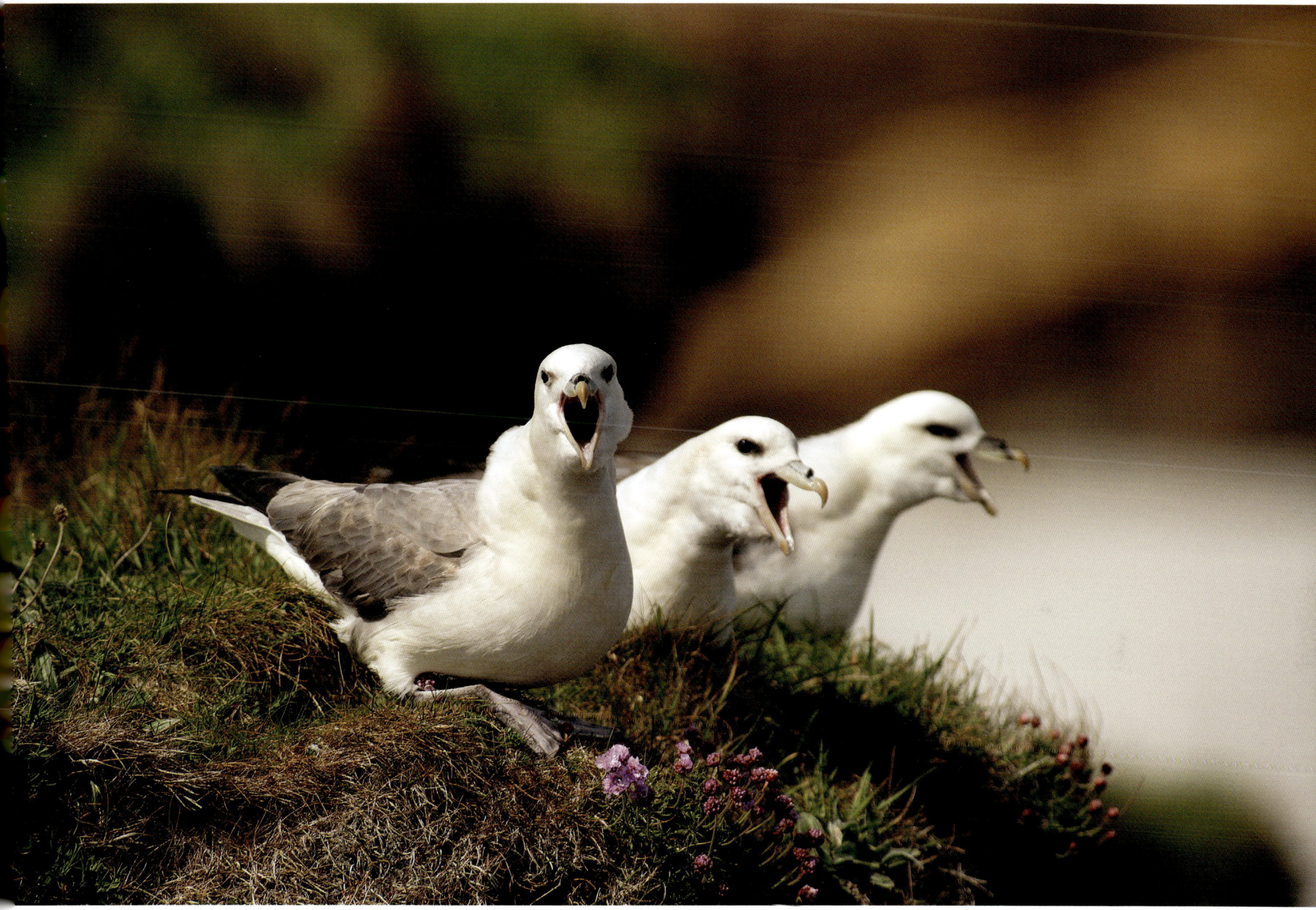

OUTDOOR PHOTOGRAPHY
EDITOR'S
CHOICE

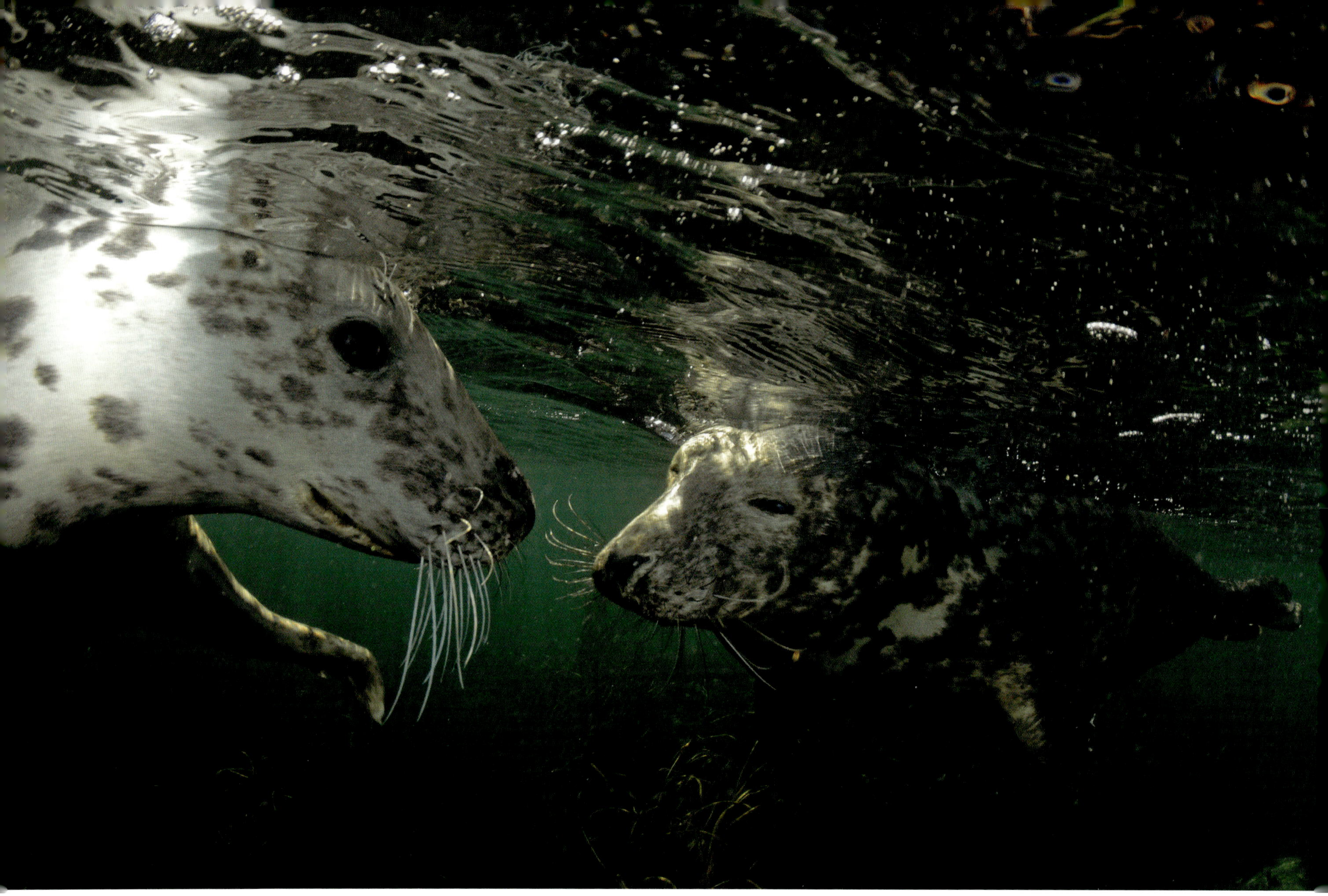

OUTDOOR PHOTOGRAPHY EDITOR'S CHOICE: FEBRUARY

ALEX TATTERSALL

Seals in Lundy
(Grey Seal, *Halichoerus grypus)*
Lundy Island, England

Friendly grey seals frolicking just below the surface of the water.

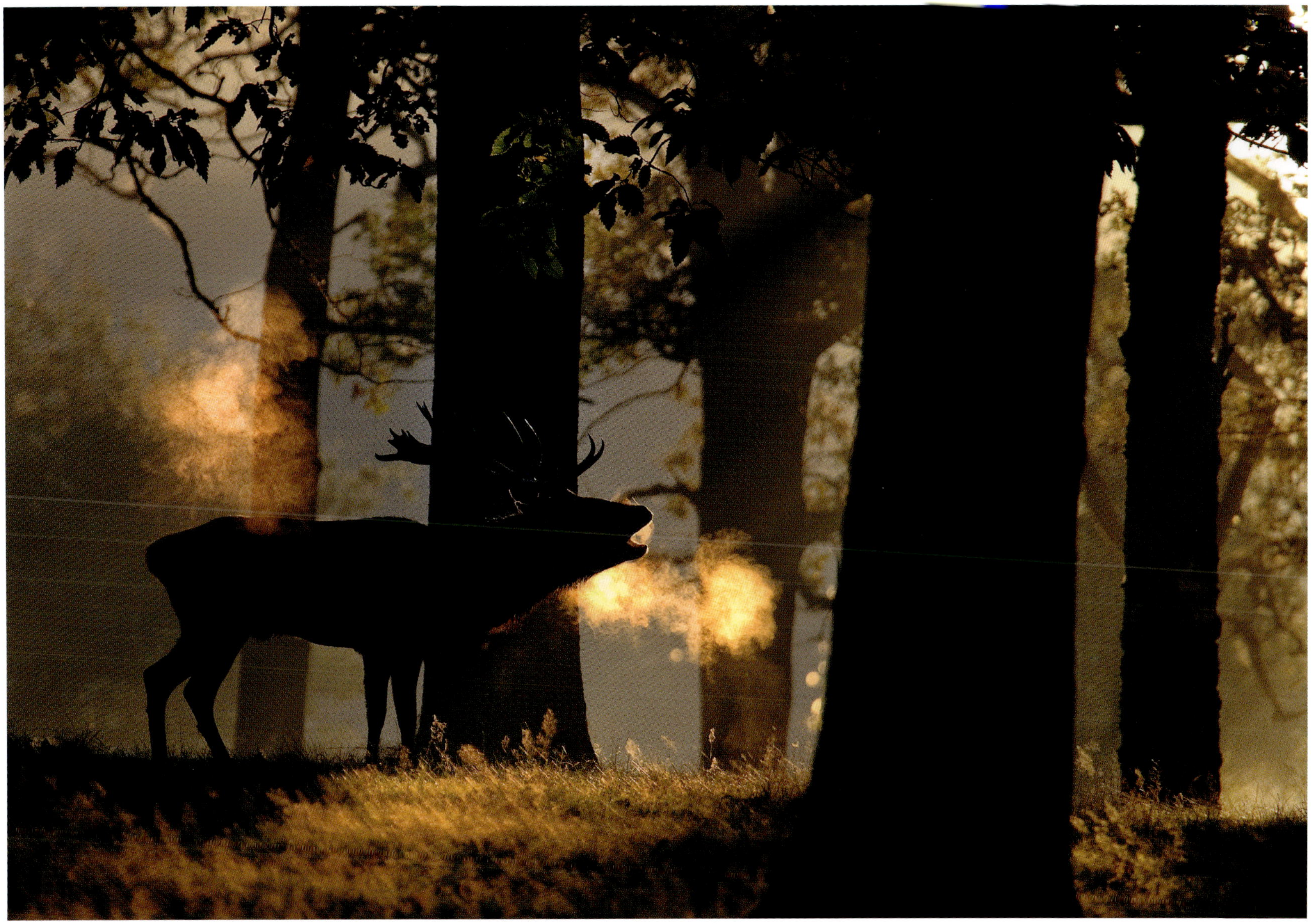

OUTDOOR PHOTOGRAPHY EDITOR'S CHOICE: MARCH

ALEX SABERI

Call of the Wild
(Red Deer, *Cervus elaphus)*
Richmond Park National Nature Reserve, Surrey, England

The autumn rutting season. A large male stag bellows in the early morning mist. I loved the light that morning as his breath was illuminated.

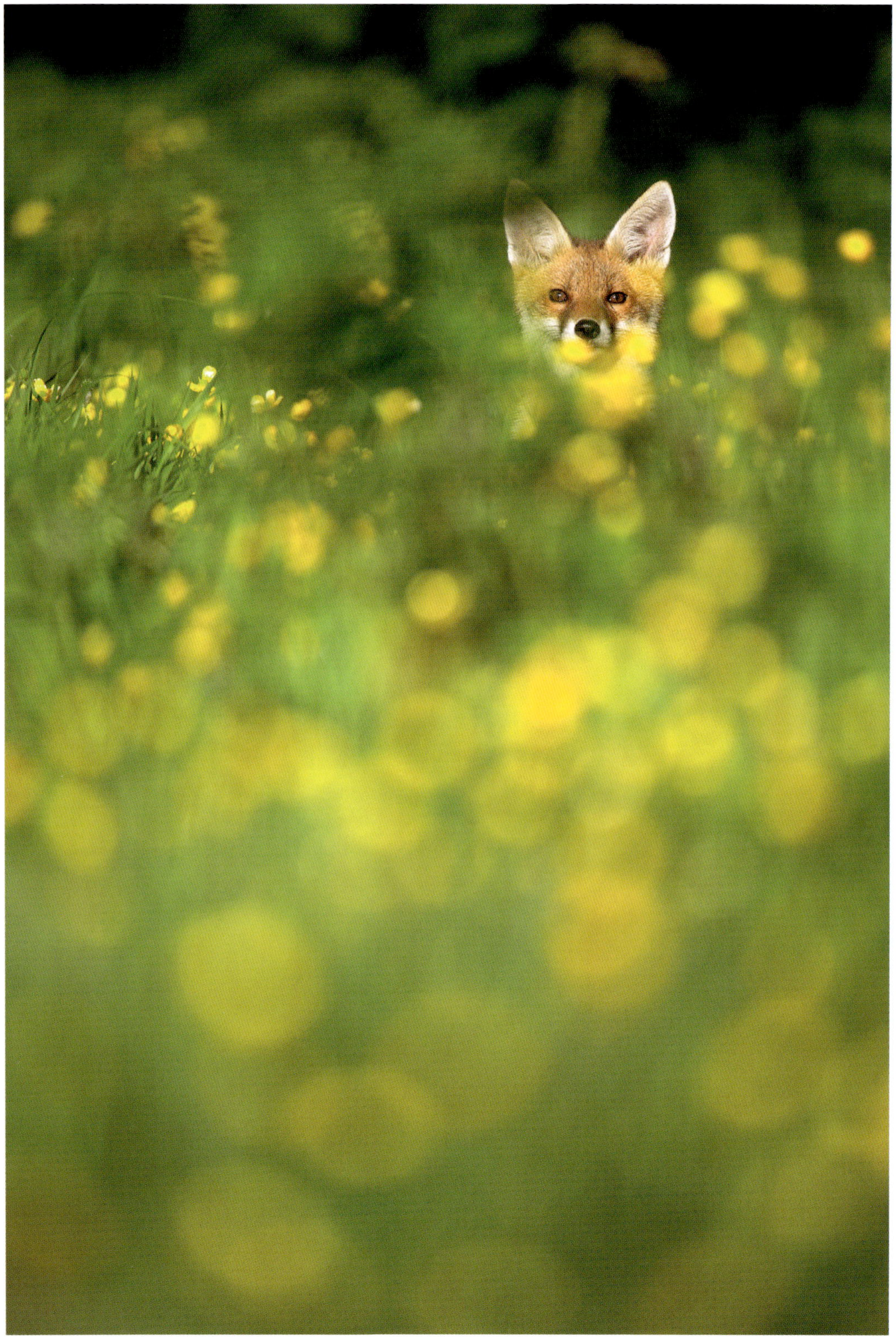

OUTDOOR PHOTOGRAPHY EDITOR'S CHOICE: APRIL

ANDREW PARKINSON

Inquisitive Red Fox Cub Among Summer Flowers
(Red Fox, *Vulpes vulpes)*
Derbyshire, England

I had spent several weeks photographing fox cubs from a hide in a quiet corner of a Derbyshire meadow and I was glad that they maintained their wary, watchful nature. I watched this little fellow creep through the undergrowth close to where I was hidden and was pleased that I was able to show him in his colourful and flower-rich habitat.

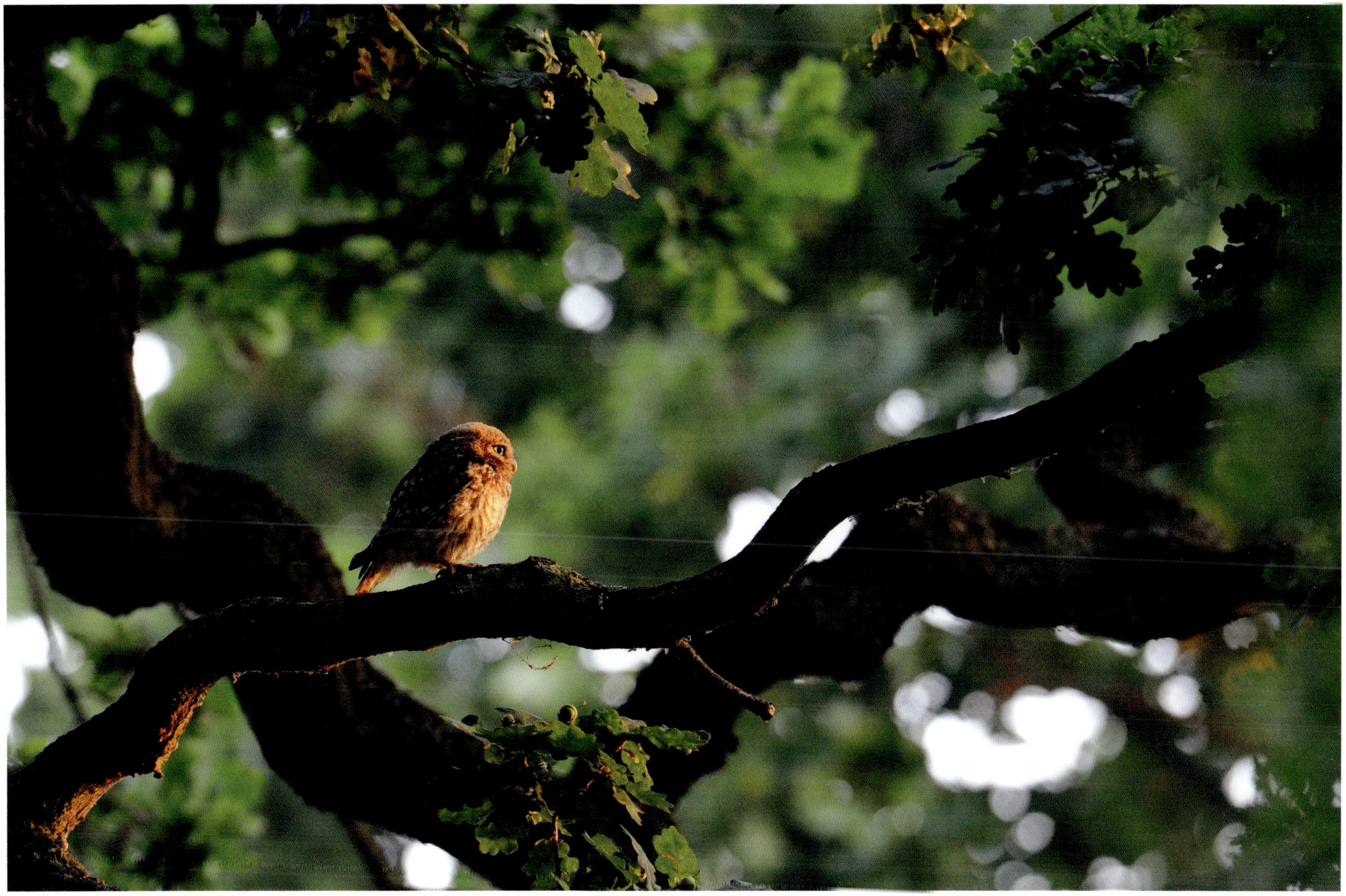

OUTDOOR PHOTOGRAPHY EDITOR'S CHOICE: MAY

RICHARD PETERS

Sunrise in the Old Oak Tree
(Little Owl, *Athene noctua)*
Richmond Park National Nature Reserve, Surrey, England

I arrived before sunrise to set up close to this oak tree where the owls were nesting. There were clouds in the sky and it was quite chilly. I knew there would be a small window of opportunity once it brightened up, before the first dog walkers and joggers arrived and the owls would be gone for the day. Thankfully, as the sun rose the clouds cleared. Right on cue this bird flew across from a nearby tree and perched in the first rays of sun as they penetrated the oak tree's foliage. Having this window into the owl's world, watching it bathe in the warm rays of sunrise, was a wonderful experience and made the early start well worth it.

YOUNG PEOPLE'S AWARDS

SPONSORED BY

RSPB WILDLIFE EXPLORERS

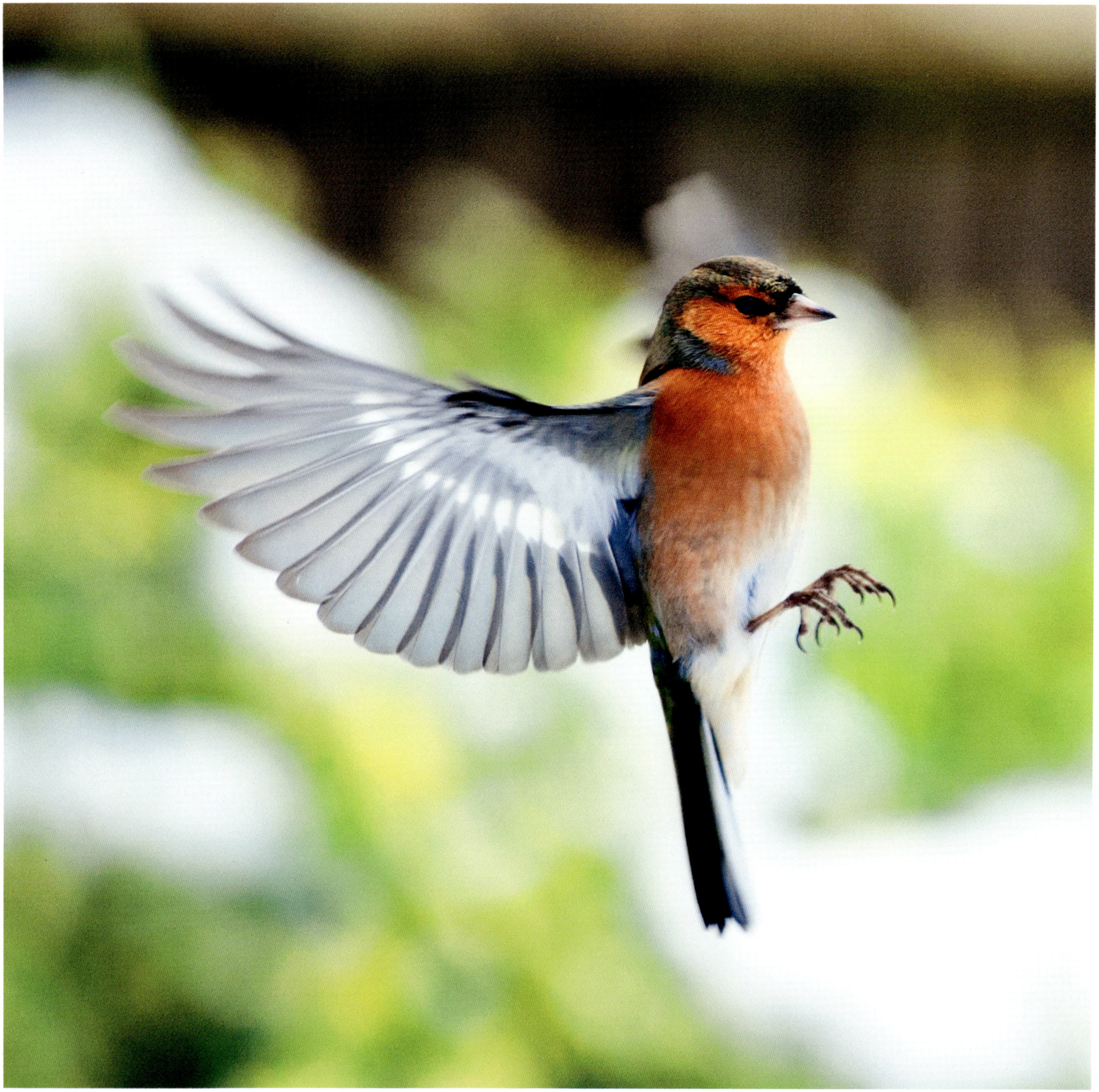

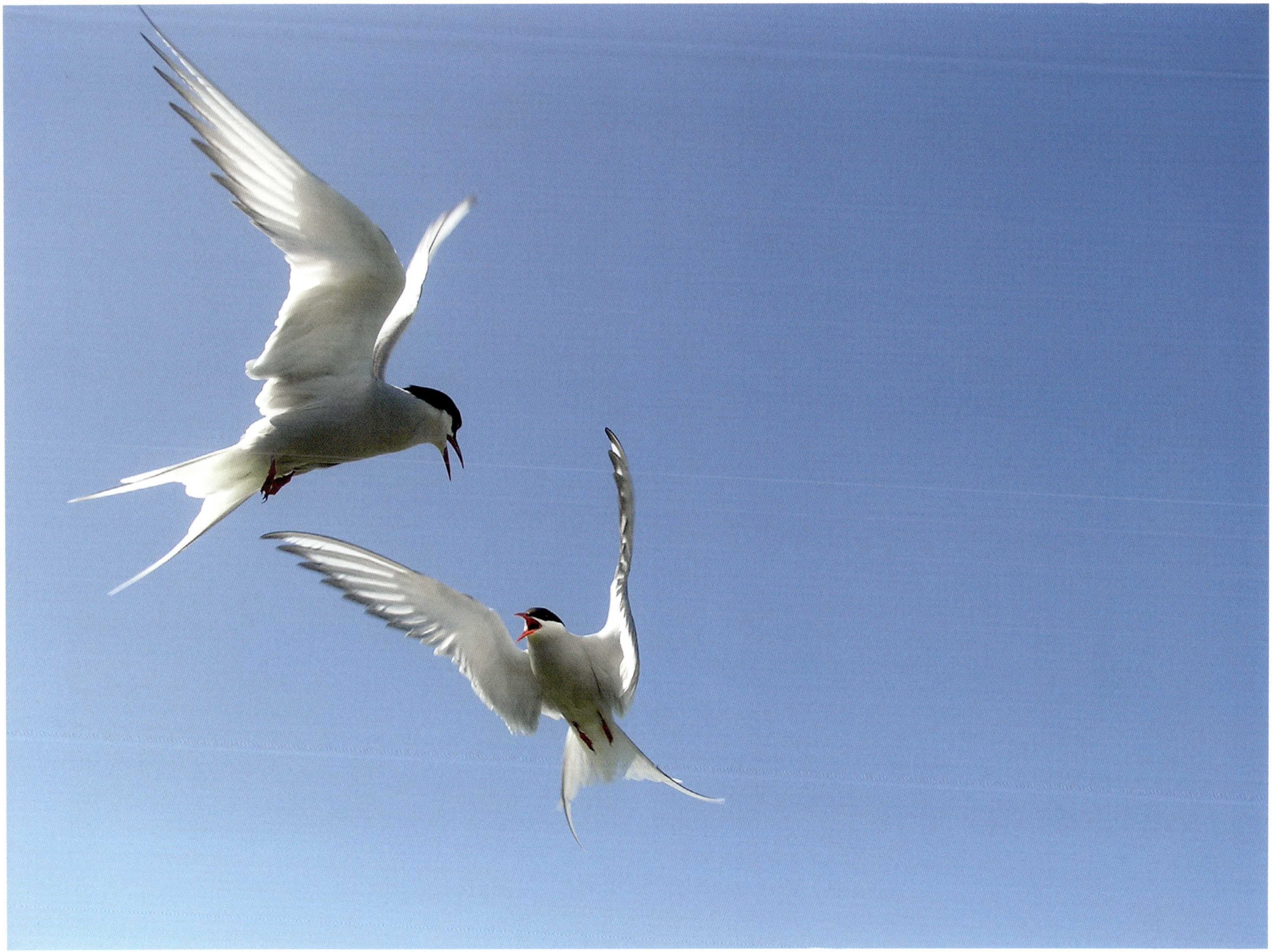

WINNER OF THE UNDER 12 CATEGORY

WILLIAM BRIGGS (AGE 8) <

Male Chaffinch Coming into Land
(Chaffinch, *Fringilla coelebs*)
Kent, England

WINNER OF THE 12 TO 18 YEARS CATEGORY

ASHLEY BUTTERWORTH (AGE 14) ∧

Arctic Terns on the Farne Islands
(Arctic Tern, *Sterna paradisaea*)
Northumberland, England

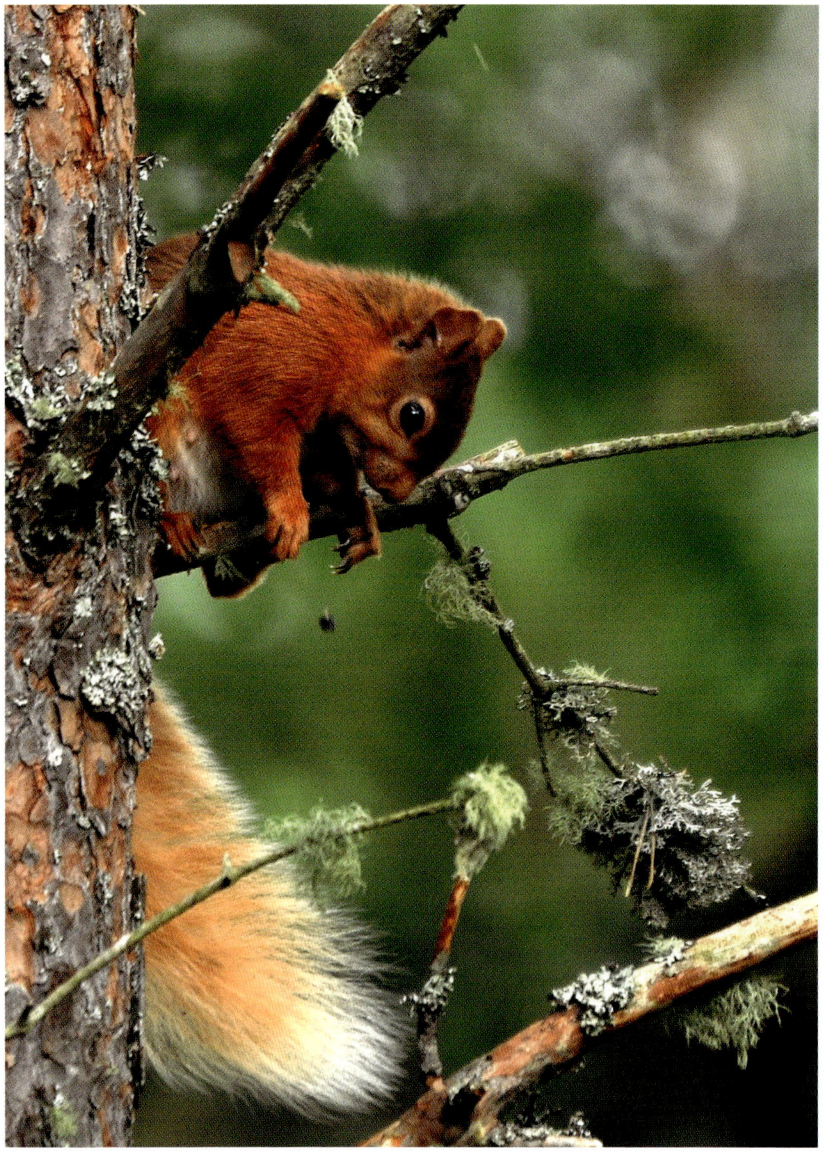

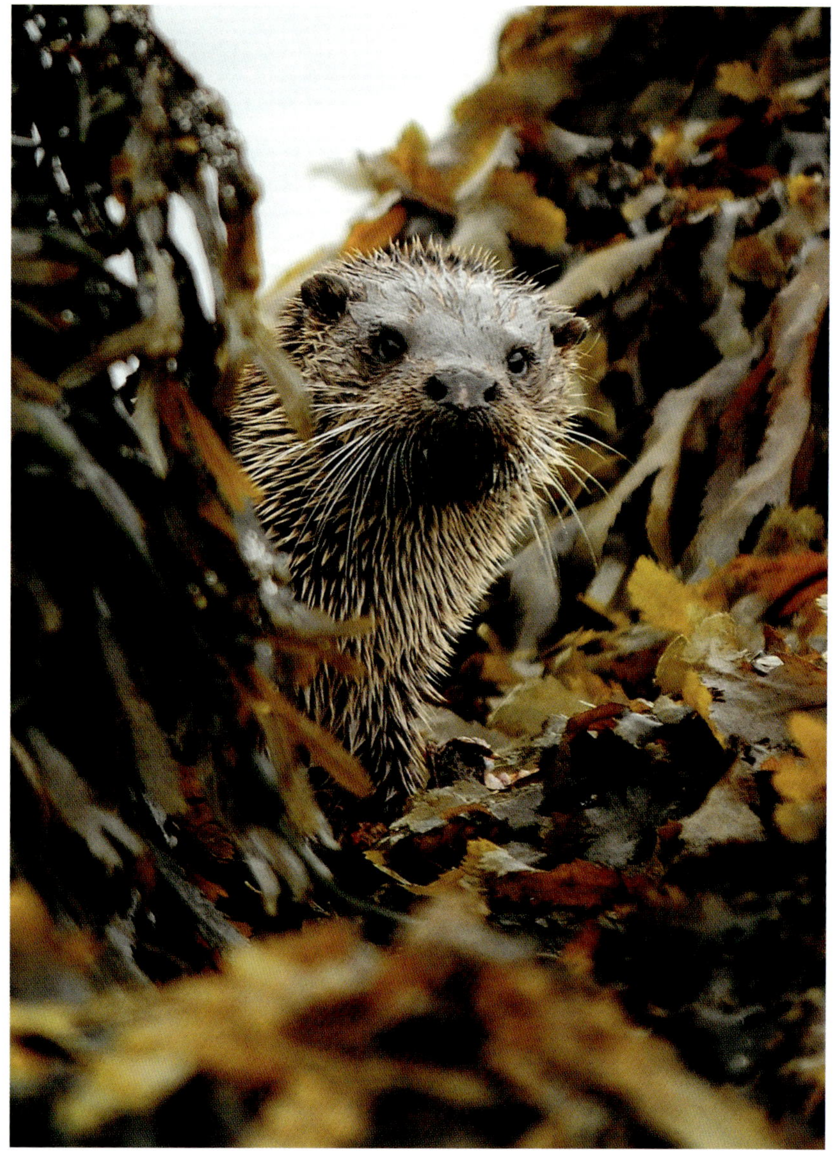

SAM BAYLIS (AGE 16) HIGHLY COMMENDED

Red Squirrel
(Red Squirrel, *Sciurus vulgaris)*
Aviemore, Scotland

MEGAN FRANCIS (AGE 16) HIGHLY COMMENDED

Otter
(European Otter, *Lutra lutra)*
Isle of Mull, Scotland

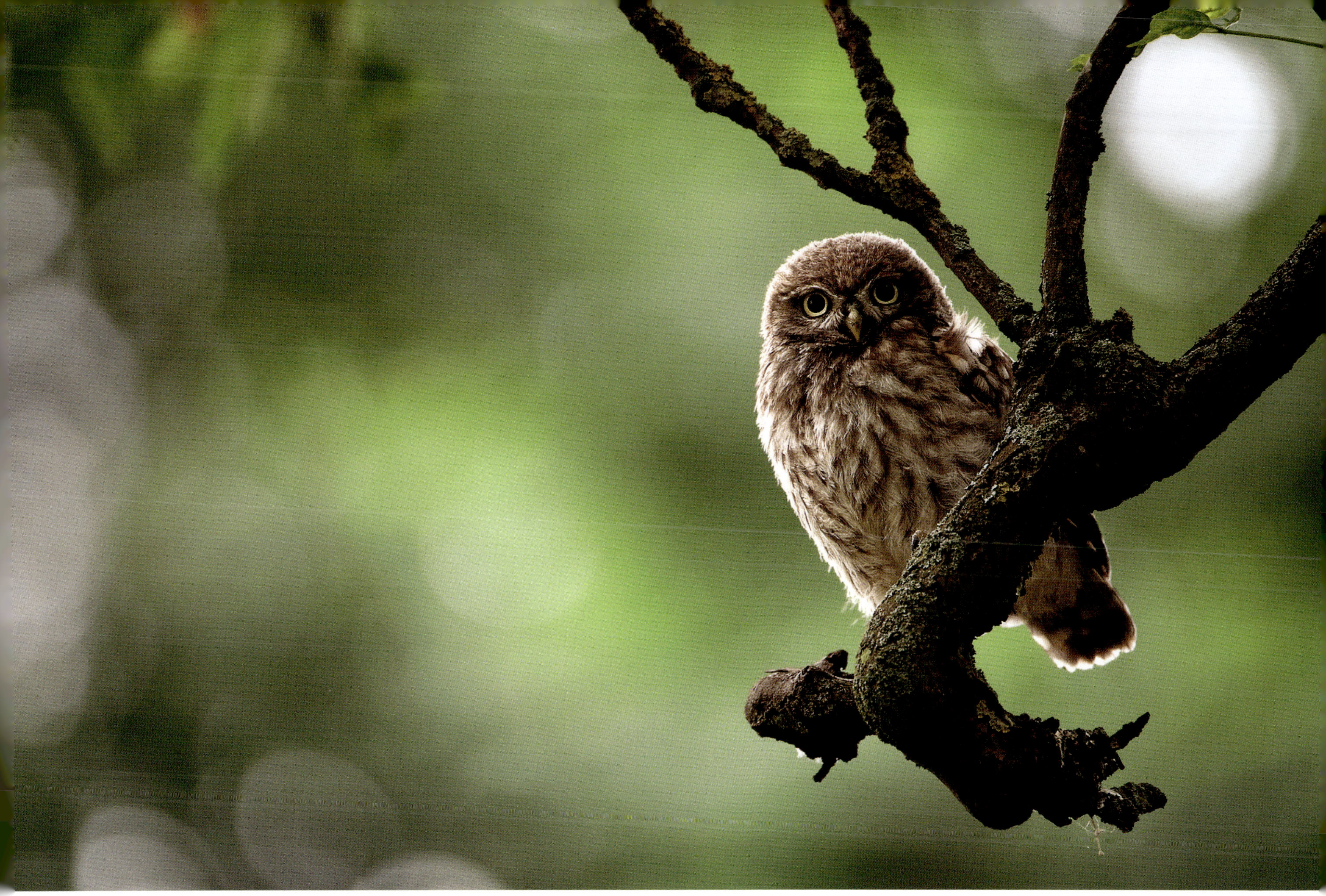

WILL NICHOLLS (AGE 17) HIGHLY COMMENDED

Immature Little Owl
(Little Owl, *Athene noctua)*
Northumberland, England

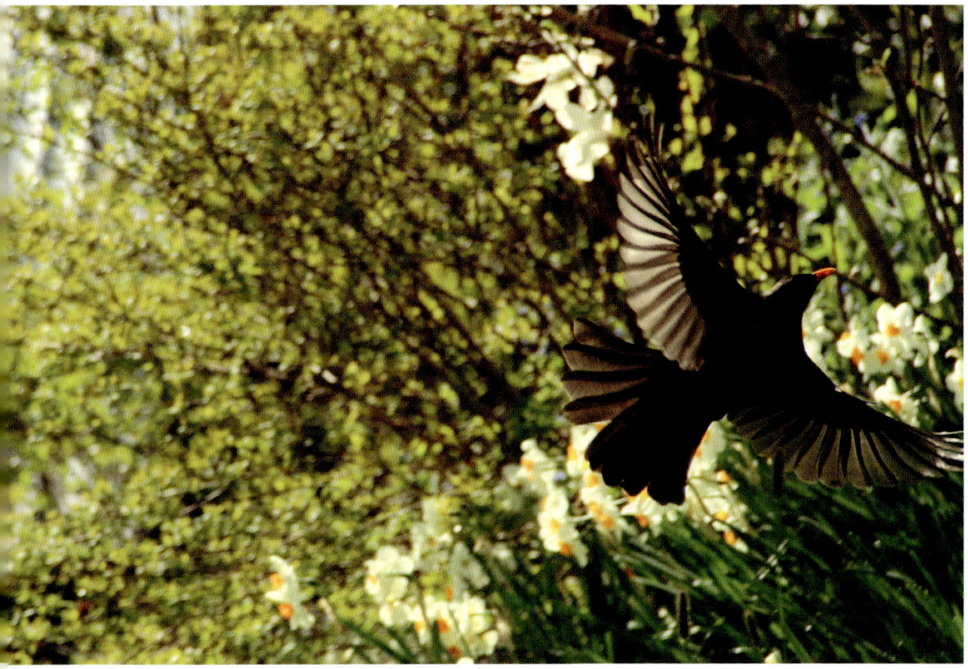

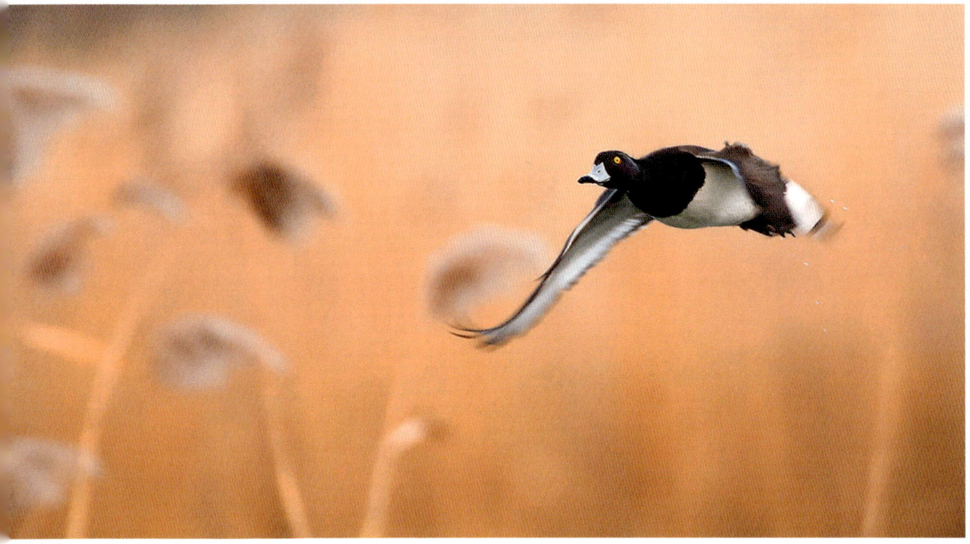

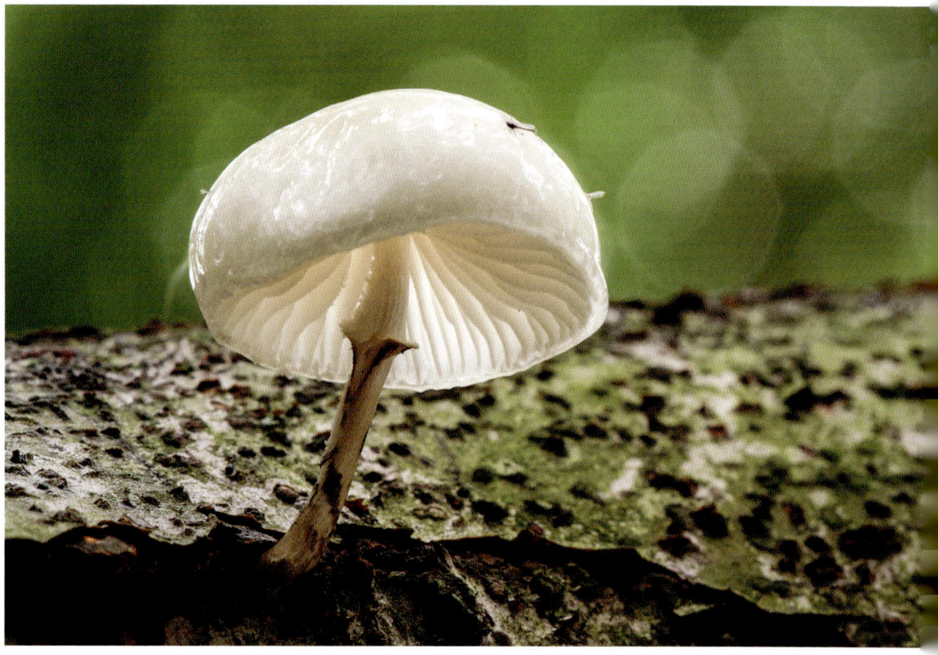

ADAM HAWTIN (AGE 17) *(above)* HIGHLY COMMENDED

Porcelain Fungus
(Porcelain Fungus, *Oudemansiella mucida)*
Clumber Park, Nottingham, England

MAIRI EYRES (AGE 13) *(top left)* HIGHLY COMMENDED

Blackbird in Flight
(Blackbird, *Turdus merula)*
Powys, Wales

OSCAR DEWHURST (AGE 17) *(bottom left)* HIGHLY COMMENDED

Tufted Duck
(Tufted Duck, *Aythya fuligula)*
London, England

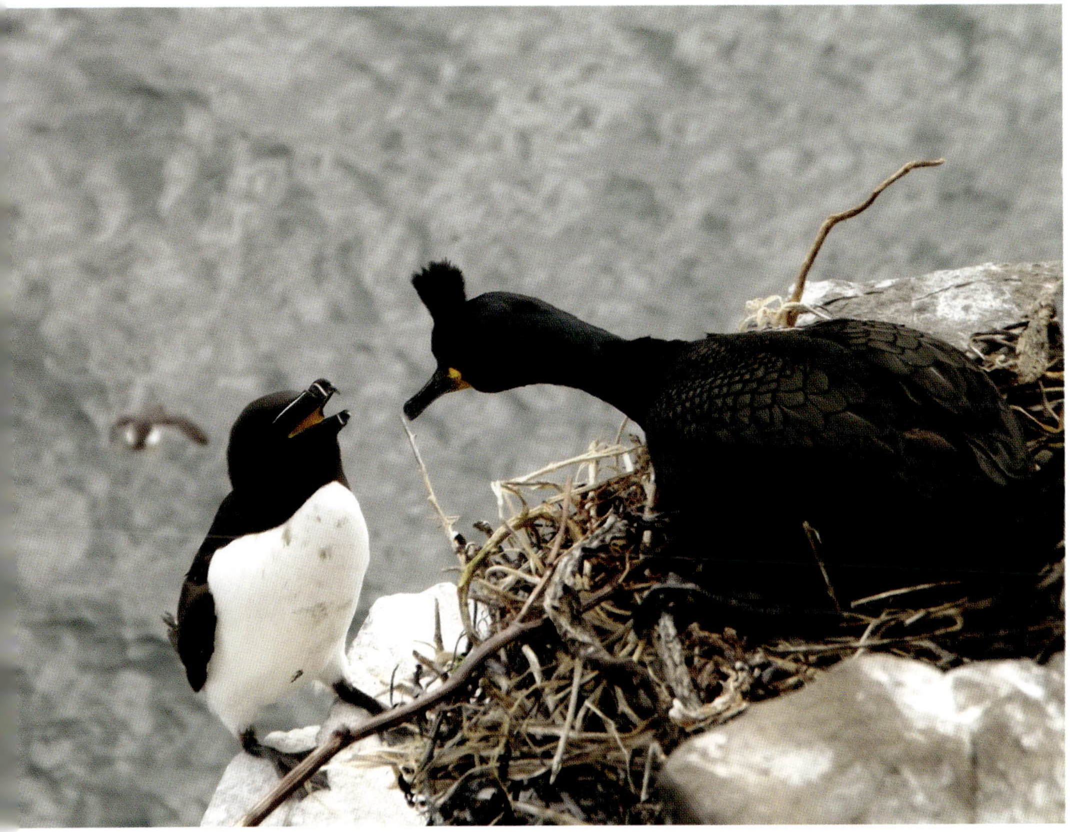

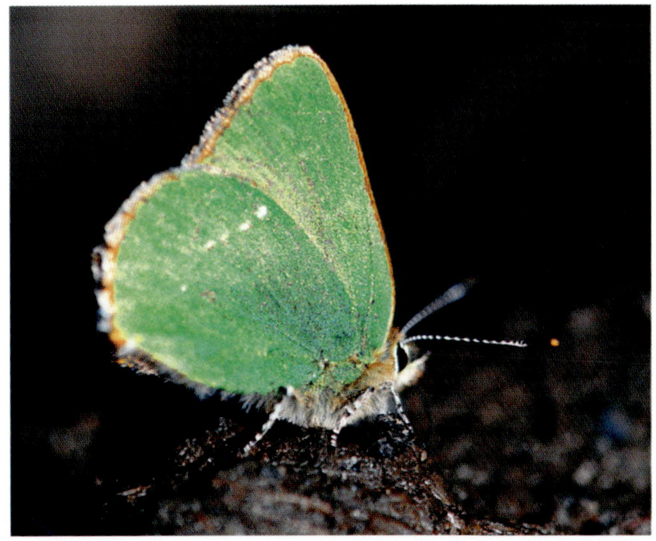

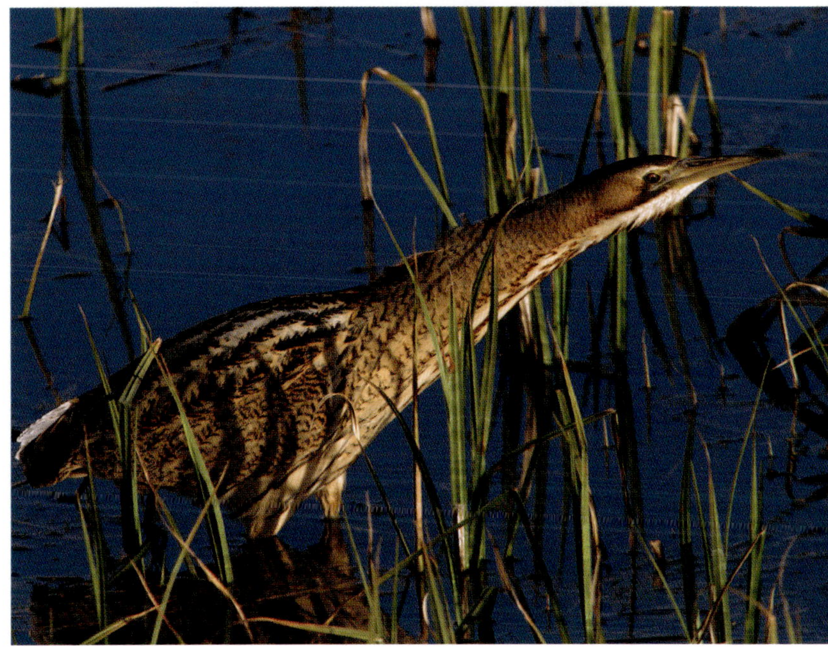

JOSEPH JENNER (AGE 11) *(above)* HIGHLY COMMENDED

Face-off Between Razorbill and Shag
(Razorbill, *Alca torda;* Shag, *Phalacrocorax aristotelis)*
Farne Islands, Northumberland, England

BECKY DUNLOP (AGE 10) *(top right)* HIGHLY COMMENDED

Green Hairstreak Butterfly by the Road
(Green Hairstreak Butterfly, *Callophrys rubi)*
Ardnamurchan, Scotland

NOAH SAMUEL WALKER (AGE 11) *(bottom right)* HIGHLY COMMENDED

Bittern at Minsmere
(Bittern, *Botaurus stellaris)*
Suffolk, England

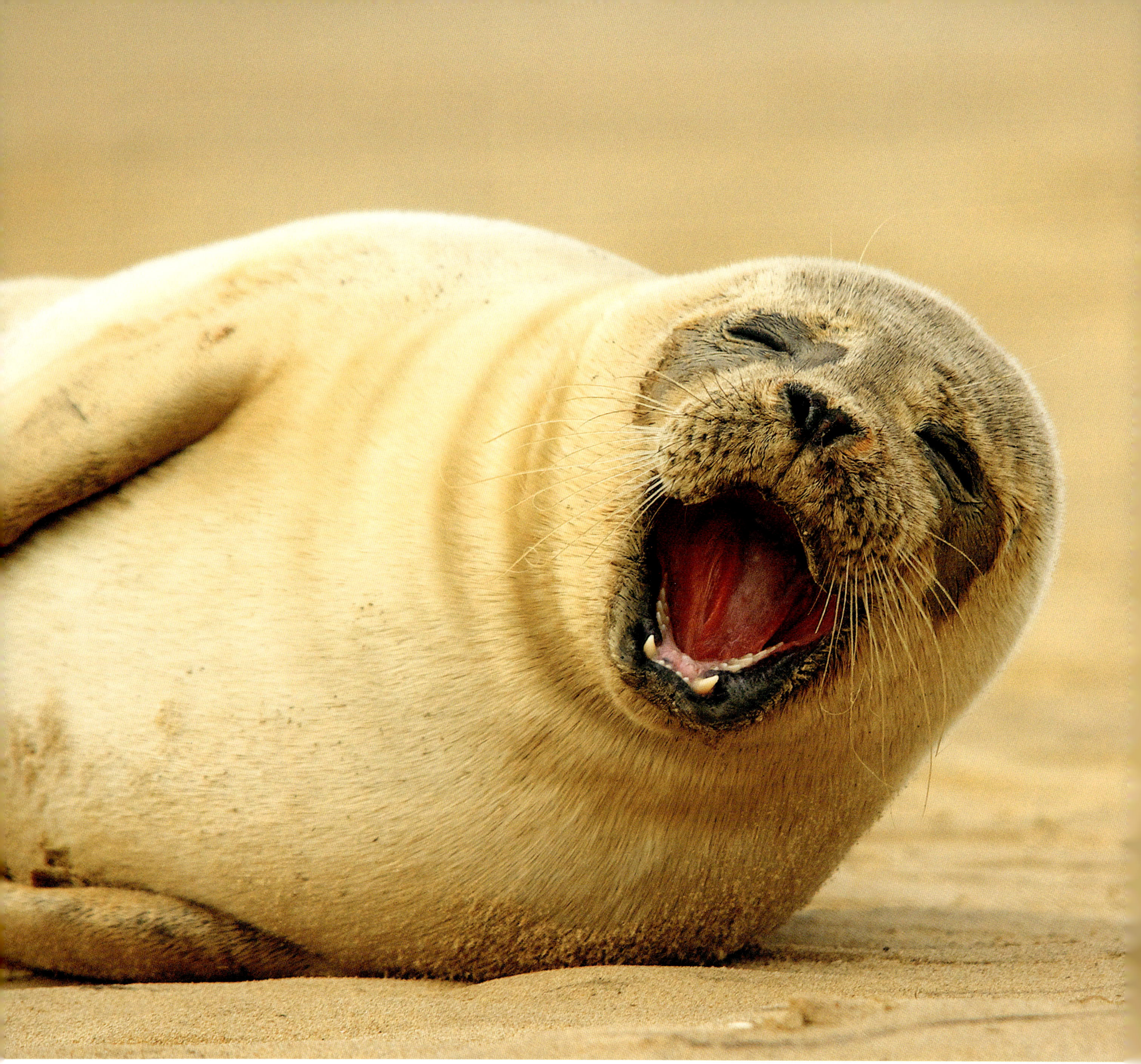

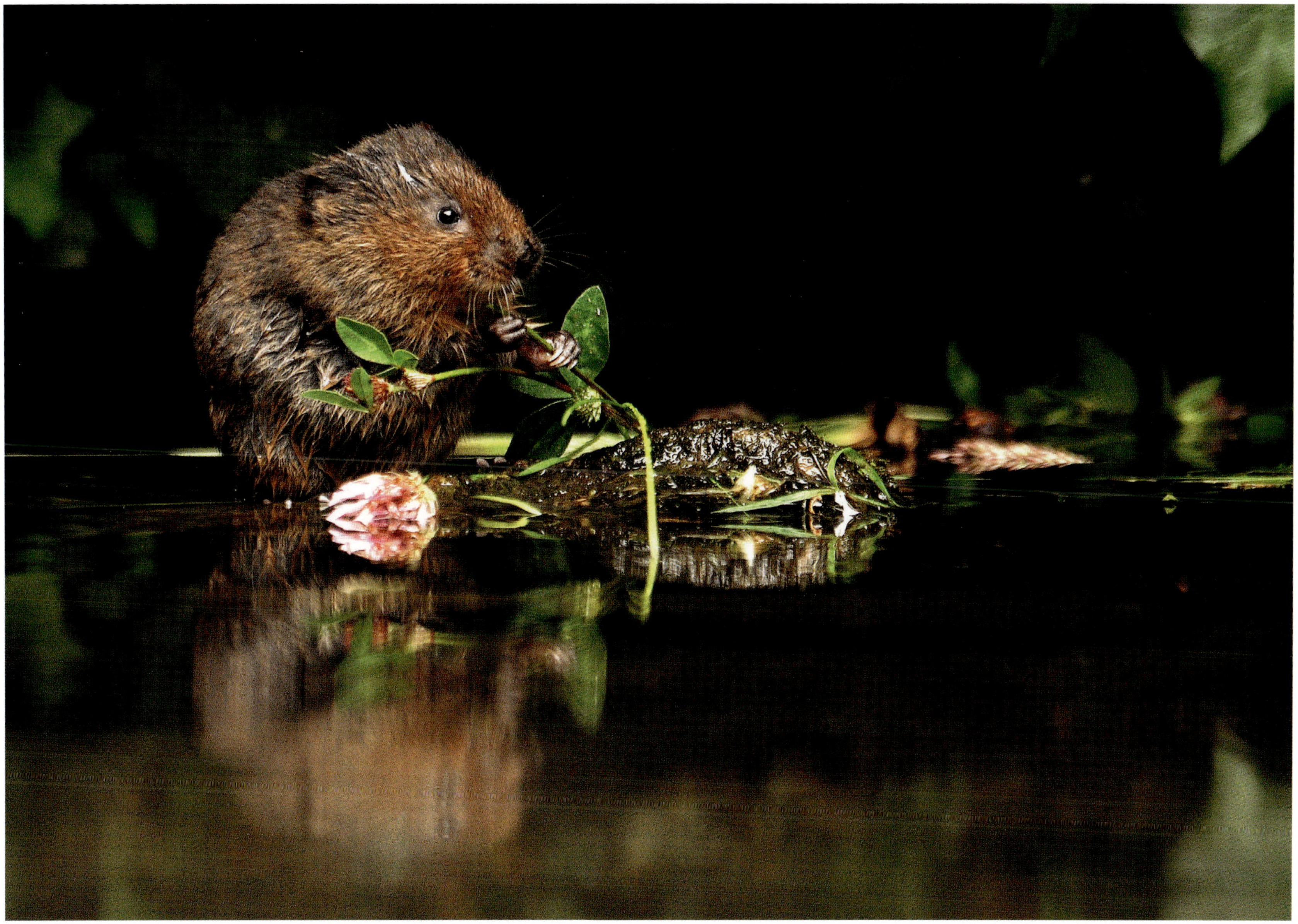

ALEX BERRYMAN (AGE 14) < HIGHLY COMMENDED

Seal on the Norfolk Coast
(Common Seal, *Phoca vitulina)*
Norfolk, England

SAM BAYLIS (AGE 16) Λ HIGHLY COMMENDED

Water Vole
(Water Vole, *Arvicola amphibius)*
East Malling, Kent, England

technical information

PRELIMS

SUE DALY (p.3, 166)
CAMERA: Canon EOS 5Dii
LENS: 17-40mmm at 17mm
SHUTTER SPEED: 1/80
APERTURE: F18
ISO: 320
OTHER: Sea & Sea underwater housing and 2 Sea and Sea strobes

LES GIBBON (p.5, 90)
CAMERA: Canon EOS-1D Mark IV
LENS: 65mm macro
SHUTTER SPEED: 1/500
APERTURE: F16
ISO: 200
OTHER: Twin macro flash with speedlite 580EX on wireless slave

ALEX SABERI (p.9, 139)
CAMERA: Canon 5d mk11
LENS: Canon 100-400
SHUTTER SPEED: 1/250
APERTURE: F5.6
ISO: 500

BWPA 2012 WINNER

DR MATT DOGGETT (p.20-21, 100-101)
CAMERA: Canon 5D Mark II
LENS: Canon EF 17-40mm L, at 20mm
SHUTTER SPEED: 1/250
APERTURE: F6.3
ISO: 200
OTHER: Ikelite u/w housing, 6" dome port

ANIMAL PORTRAITS

NEIL McINTYRE (p.24)
CAMERA: Canon EOD-5D Mark II
LENS: Canon 500mm f4 IS
SHUTTER SPEED: 1/100
APERTURE: F4
ISO: 640
OTHER: Gitzo tripod with Wimberley mark II head

MARK HAMBLIN (p.25)
CAMERA: Canon EOS1DMark IV
LENS: Canon 500mm f4 USM
SHUTTER SPEED: 1/125
APERTURE: F4
ISO: 800
OTHER: Beanbag from hide

CRAIG CHURCHILL (p.26)
CAMERA: nikon d3
LENS: Nikon 500 f4afs/2
SHUTTER SPEED: 1/200
APERTURE: F5.6
ISO: 640
OTHER: Beanbag/wws hide

GERALD ROBINSON (p.27)
CAMERA: Fuji X10
LENS: Fixed
SHUTTER SPEED: 1/850
APERTURE: F5.6
ISO: 100

ANDY SANDS (p.28)
CAMERA: Nikon D300
LENS: Nikon 500mm f4 VR
SHUTTER SPEED: 1/100
APERTURE: F8
ISO: 200
OTHER: Gitzo Systematic tripod with Wimberley head, standard fensman type hide

STEVE MAGENNIS (p.29)
CAMERA: Nikon D3
LENS: Nikon Nikkor VR 200-400 F4 G + 1.4 teleconvertor
SHUTTER SPEED: 1/400
APERTURE: F5.6
ISO: 500
OTHER: Tripod

JOHN MONCRIEFF (p.30)
CAMERA: Nikon D300
LENS: Nikon 300mm f4 afs
SHUTTER SPEED: 1/125
APERTURE: F5.6
ISO: 1250
OTHER: Nikon TC14 teleconverter attached. Shot handheld from prone position

ANDREW PARKINSON (p.31)
CAMERA: Nikon D300s
LENS: Nikon 200-400 F4 VR lens
SHUTTER SPEED: 1/1000
APERTURE: F5.6
ISO: 320
OTHER: Gitzo tripod, Wimberley head, various ropes, harnesses and other climbing equipment

MARK HAMBLIN (p.32)
CAMERA: Nikon D3
LENS: Nikon 200-400mm f4
SHUTTER SPEED: 1/250
APERTURE: F4
ISO: 1600

TOM LANGLANDS (p.33)
CAMERA: Canon EOS 7D
LENS: Canon 600mm F4 L IS USM
SHUTTER SPEED: 1/3200
APERTURE: F5.6
ISO: 400
OTHER: Manfrotto tripod with Wimberley Gimbal head

MATT BINSTEAD (p.34)
CAMERA: Nikon D300s
LENS: Nikon Fisheye 10.5mm
SHUTTER SPEED: 1/200
APERTURE: F8
ISO: 800

RICHARD PETERS (p.35)
CAMERA: Nikon D300
LENS: 600 f4 + 1.4x TC
SHUTTER SPEED: 1/640
APERTURE: F8
ISO: 720
OTHER: Beanbag

RICHARD COSTIN (p.36)
CAMERA: Nikon D3
LENS: Nikon 200-400 f4 and 1.4x teleconvertor (set to 550mm)
SHUTTER SPEED: 1/500
APERTURE: F7.1
ISO: 320
OTHER: Lens balanced on knees as I sat close to the ground

MATT BINSTEAD (p.37)
CAMERA: Nikon D300s
LENS: Nikon 70-200VRII f/2.8
SHUTTER SPEED: 1/640
APERTURE: F2.8
ISO: 200

BEN HALL (p.38)
CAMERA: Canon EOS 1D Mark II
LENS: Canon 500mm F4 L IS
SHUTTER SPEED: 1/200
APERTURE: F4.5
ISO: 250
OTHER: Beanbag and +1 exposure compensation

JULES COX (p.39)
CAMERA: Canon 1D mark IV
LENS: 500mm f/4L IS USM
SHUTTER SPEED: 1/1250
APERTURE: F4
ISO: 400
OTHER: Gitzo carbon fibre tripod and Wimberley mark II head

SIMON LITTEN (p.40)
CAMERA: Canon 1d Mk III
LENS: 500mm with a 1.4x extender
SHUTTER SPEED: 1/640
APERTURE: F7.1
ISO: 400
OTHER: Beanbag

JAN DAVIES (p.41)
CAMERA: Canon EOS 5 D Mark II
LENS: 70-200mm@188mm
SHUTTER SPEED: 1/160
APERTURE: F5.6
ISO: 100
OTHER: Handheld manual focus

PAUL HOBSON (p.42)
CAMERA: Canon 5Dmk11
LENS: 180 macro
SHUTTER SPEED: 1/400
APERTURE: F3.5
ISO: 800
OTHER: Handheld, stalked

JOAN O'HAGAN (p.43)
CAMERA: Canon Ixus 970 IS compact digital
OTHER: I just used the pre-set macro function on my camera. I really don't know anything about shutter speeds and apertures. I didn't use any other equipment, and my only 'technique' was to put the caterpillar on the ground because it was a very windy day and the willow tree kept moving

SIMON PHILLPOTTS (p.44)
CAMERA: Canon 1D Mark IV
LENS: Sigma 150mm f/2.8 Macro
SHUTTER SPEED: 1/10
APERTURE: F8
ISO: 800
OTHER: Tripod and ball head with remote release

LUKE MASSEY (p.45)
CAMERA: Canon 1dMk IV
LENS: Canon 500mm F4 IS
SHUTTER SPEED: 1/400
APERTURE: F4
ISO: 1250
OTHER: Gitzo Tripod, Wimberley head

MATT BINSTEAD (p.46)
CAMERA: Nikon D300s
LENS: Nikon 70-200VRII f/2.8
SHUTTER SPEED: 1/160
APERTURE: F5.6
ISO: 200
OTHER: Captive mouse set up on corn stalks planted in plant pot

PENNY DIXIE (p.47)
CAMERA: Canon EOS 7D
LENS: EF500mm f/4L IS USM +1.4x
SHUTTER SPEED: 1/640
APERTURE: F5.6
ISO: 400
OTHER: Gitzo tripod with swinghead

NEIL McINTYRE (p.48)
CAMERA: Canon EOS-1D Mark IV
LENS: Canon 500mm f4 IS
SHUTTER SPEED: 1/400
APERTURE: F8
ISO: 400
OTHER: Gitzo tripod with Wimberley mark II head

RICHARD SHUCKSMITH (p.49)
CAMERA: Nikon D300
LENS: Nikon 600mm
SHUTTER SPEED: 1/1250
APERTURE: F5.6
ISO: 800

JOHN TAYLOR (p.50)
CAMERA: D300
LENS: Nikon f4 +1.7 con
SHUTTER SPEED: 1/160
APERTURE: F8
ISO: 200
OTHER: Beanbag in friend's car

RICHARD PETERS (p.51)
CAMERA: Nikon D3s
LENS: 600 VR + 1.4x TC
SHUTTER SPEED: 1/1000
APERTURE: F7.1
ISO: 3200
OTHER: Gitzo tripod, Wimberley mark II

ANIMAL BEHAVIOUR

AMANDA HAYES (p.54)
CAMERA: Nikon D3s
LENS: Nikon 500mm f4 ED VR AF-S Nikkor
SHUTTER SPEED: 1/3200
APERTURE: F5.6
ISO: 1250
OTHER: Gitzo tripod, Wimberley mark II head. As I prefer to capture action and flight images, I had chosen an ISO and shutter speed that would freeze the action and avoid blur in the wings, and still produce a clean usable image

ROBERT LIND (p.55)
CAMERA: Canon EOS 5D mark II
LENS: Canon EF 600mm IS F4
SHUTTER SPEED: 1/8000
APERTURE: F4
ISO: 800
OTHER: Wimberley gimbal mark II head attached to a Gitzo 1348 tripod and a sense of good timing to press the shutter! A Canon 25mm extension tube II was used between lens and camera to allow a shorter focal distance

MARK SMITH (p.56)
CAMERA: Canon EOS 1D mk IV
LENS: Canon EF 500mm f4 IS L
SHUTTER SPEED: 1/2500
APERTURE: F4
ISO: 800
OTHER: Beanbag

MARK HAMBLIN (p.57)
CAMERA: Canon 1D Mark III
LENS: Canon 500mm f2.8 L
SHUTTER SPEED: 1/3200
APERTURE: F5.6
ISO: 400
OTHER: Stalked

SIMON PHILLPOTTS (p.58)
CAMERA: Canon 1D Mark IV
LENS: EF 500mm f/4.0L IS
SHUTTER SPEED: 1/1250
APERTURE: F4
ISO: 640
OTHER: Tripod and gimbal head

JAMES YAXLEY (p.59)
CAMERA: Canon EOS 5D Mark II
LENS: Canon 180mm Macro
SHUTTER SPEED: 1/40
APERTURE: F8
ISO: 400
OTHER: Tripod

TOM LANGLANDS (p.60)
CAMERA: Canon EOS 7D
LENS: Canon 600mm F4 L IS USM
SHUTTER SPEED: 1/3200
APERTURE: F5.6
ISO: 400
OTHER: Manfrotto tripod with Wimberley Gimbal head

ROBERT HOWARD (p.61)
CAMERA: Canon 1Ds MK II
LENS: 600mm f4
SHUTTER SPEED: 1/125
APERTURE: F5.6
ISO: 400
OTHER: 1.4x converter, beanbags on car window ledge

ALEX TATTERSALL (p.62)
CAMERA: Nikon D300s in Nauticam housing
LENS: Tokina 10-17mm Fisheye
SHUTTER SPEED: 1/320
APERTURE: F13
ISO: 320
OTHER: INON Z240 strobes x2

RICHARD WATERS (p.63)
CAMERA: Canon 40D
LENS: Canon 70-200 IS II + 1.4x Extender
SHUTTER SPEED: 1/1000
APERTURE: F4
ISO: 100
OTHER: Handheld

DR MATT DOGGETT (p.64)
CAMERA: Canon 5D Mark 11
LENS: Canon EF 17 -40 mm L, at 20mm
SHUTTER SPEED: 1/250
APERTURE: F5
ISO: 200
OTHER: Ikelite u/w Housing, 6" dome port

GARY COX (p.65)
CAMERA: Canon 1dmk2n
LENS: Canon 180mm macro
SHUTTER SPEED: 1/40
APERTURE: F8
ISO: 100
OTHER: Beanbag

ANDREW PARKINSON (p.66)
CAMERA: Nikon D3S
LENS: Nikon 600mm F4 VR lens
SHUTTER SPEED: 1/2500
APERTURE: F4
ISO: 200
OTHER: Kevin Keatley hide, Gitzo tripod, Wimberley head

NEIL McINTYRE (p.67)
CAMERA: Canon EOS-1D Mark IV
LENS: Canon 300mm f2.8 L IS
SHUTTER SPEED: 1/8000
APERTURE: F3.5
ISO: 800
OTHER: Gitzo tripod with Wimberley mark II head

ANDREW PARKINSON (p.68)
CAMERA: Nikon D3S
LENS: Nikon 600mm F4 lens
SHUTTER SPEED: 1/3200
APERTURE: F4
ISO: 400

MARTIN WASE (p.69)
CAMERA: Canon Eos 5D M11
LENS: Canon 100-400mm f4.5-f5.6 IS L
SHUTTER SPEED: 1/400
APERTURE: F6.3
ISO: 800

TOM LANGLANDS (p.70)
CAMERA: Canon EOS 7D
LENS: Canon 600mm F4 L IS USM
SHUTTER SPEED: 1/400
APERTURE: F5
ISO: 800
OTHER: Manfrotto tripod with Wimberley Gimbal head

JAMES YAXLEY (p.71)
CAMERA: Canon EOS7D
LENS: 100-400mm Canon Lens
SHUTTER SPEED: 1/3200
APERTURE: F9
ISO: 200
OTHER: Tripod

URBAN WILDLIFE

PHIL JONES (p.74)
CAMERA: Canon 7D
LENS: Sigma 300-800mm f5.6 (at 500mm)
SHUTTER SPEED: 1/13
APERTURE: F7.1
ISO: 100
OTHER: Manfrotto 055CXPRO4 tripod and 393 long lens bracket. Remote release. Raw file processed in Lightroom 3 and 4

JEREMY MOORE (p.75)
CAMERA: Canon 5d2
LENS: 70-200 f4 L zoom + x1.4 converter
SHUTTER SPEED: 1/160
APERTURE: F10
ISO: 400
OTHER: Tripod

JAMIE HALL (p.76-77)
CAMERA: Canon 1ds mkii
LENS: 17-40mm f/4
SHUTTER SPEED: 1/532
APERTURE: F22
ISO: 100
OTHER: I used a beanbag to support the camera and a cable release to activate the shutter

DAVID HERMON (p.78)
CAMERA: CANON 1D MK4
LENS: CANON 800mm
SHUTTER SPEED: 1/50
APERTURE: F11
ISO: 200
OTHER: Gitzo tripod with Wimberley head

LEWIS PHILLIPS (p.79)
CAMERA: mk11
LENS: 100 400
SHUTTER SPEED: 1/200
APERTURE: F5
ISO: 100
OTHER: Tripod

PAUL HOBSON (p.80)
CAMERA: Canon 1Dmk4
LENS: 70-200
SHUTTER SPEED: 1/320
APERTURE: F3.2 at -1/3
ISO: 2500
OTHER: Handheld

LIAM MARSH (p.81)
CAMERA: Canon 5D MKII
LENS: Canon 8-15mm fisheye (at 15mm)
SHUTTER SPEED: 1/0.4
APERTURE: F16 (for background definition and to produce the starburst effect)
ISO: 1600
OTHER: Canon Macro twin lite MT-24EX with

home-made diffusers and half blue gels. Manfrotto tripod with geared pan and tilt head. Torch. A white balance set to tungsten to balance the house lights and deepen the colour of the evening sky

HIDDEN BRITAIN

DALE SUTTON (p.84)
CAMERA: Canon 5d mk2
LENS: Canon 100mm
SHUTTER SPEED: 1/400
APERTURE: F18
ISO: 50
OTHER: High-speed flash, infra-red trigger, special designed leaf shutter, solenoid and control box, 4 tripods

ALEX SABERI (p.85)
CAMERA: Canon 5d
LENS: Canon 16-35
SHUTTER SPEED: 1/80
APERTURE: F4.5
ISO: 100

ROSS HODDINOTT (p.86)
CAMERA: Nikon D300
LENS: 150mm macro
SHUTTER SPEED: 1/2
APERTURE: F11
ISO: 200
OTHER: 36mm extension tube, remote cord, tripod

ALEX SABERI (p.87)
CAMERA: Canon 5d
LENS: Canon 70-200
SHUTTER SPEED: 1/800
APERTURE: F4.5
ISO: 250

SIMON WHITE (p.88-89)
CAMERA: Canon EOS 1D Mark IV
LENS: Canon EF 100mm f/2.8 Macro USM
SHUTTER SPEED: 1/30
APERTURE: F5
ISO: 800
OTHER: Tripod used

JOHN H BRACKENBURY (p.91)
CAMERA: Nikon 300s
LENS: 105 mm f 2.8 macro lens
SHUTTER SPEED: 1/160
APERTURE: F11
ISO: 500

ANDY SANDS (p.92)
CAMERA: Nikon D7000
LENS: Nikon 105mm f2.8 VR Micro
SHUTTER SPEED: 1/200
APERTURE: F29
ISO: 400
OTHER: Lighting from Bowens 500 studio flash single head with white umbrella

OLIVER CHARLES WRIGHT (p.93)
CAMERA: Canon 1d mark iv
LENS: Canon MP-E 65mm f2.8
SHUTTER SPEED: 1/300
APERTURE: F13
ISO: 400

LES GIBBON (p.94)
CAMERA: Canon EOS-1Ds Mark II
LENS: Canon 180mm macro
SHUTTER SPEED: 1/500
APERTURE: F16
ISO: 100
OTHER: The picture was taken handheld using two speedlite 580EX flash heads to create interesting lighting and maximum depth of field

DUNCAN COOKE (p.95)
CAMERA: Canon EOS 5D II
LENS: Canon MP-E65mm f2.8 1-5x macro
SHUTTER SPEED: 1/200
APERTURE: F11
ISO: 100
OTHER: Tripod, Canon Macro Twin Lite MT-24EX Flash

LES GIBBON (p.96-97)
CAMERA: Canon EOS-1D Mark IV
LENS: Canon 180mm Macro
SHUTTER SPEED: 1/250
APERTURE: F8
ISO: 200

COAST AND MARINE

TREVOR REES (p.102)
CAMERA: Nikon D90
LENS: Tokina 10-17 @ 11mm
SHUTTER SPEED: 1/250
APERTURE: F16
ISO: 400
OTHER: Single Sea & Sea YS110 strobe fitted with a diffuser

PAUL NAYLOR (p.103)
CAMERA: Nikon D200
LENS: 10.5 mm
SHUTTER SPEED: 1/15
APERTURE: F8
ISO: 200
OTHER: Camera was in Subal aluminium housing. Lighting from two Inon flashguns

GEORGE KARBUS (p.104)
CAMERA: Nikon D700
LENS: Nikkor 16.0–35.00 mm f/4.0
SHUTTER SPEED: 1/2500
APERTURE: F4
ISO: 800
OTHER: Subal underwater housing

ALEX TATTERSALL (p.105)
CAMERA: Nikon D7000 in Nauticam housing
LENS: Nikon 60mm macro and FIT Pro +5, +8 and +16 stacked wet diopters
SHUTTER SPEED: 1/125
APERTURE: F32
ISO: 640
OTHER: INON Z240 strobe. Fiber snoot

ANDREW PARKINSON (p.106)
CAMERA: Nikon D3S
LENS: Nikon 80-200mm F2.8 lens
SHUTTER SPEED: 1/640
APERTURE: F22
ISO: 800

MARK SMITH (p.107)
CAMERA: Canon EOS 1D mkIV
LENS: Canon EF 500mm f4 IS L
SHUTTER SPEED: 1/125
APERTURE: F5.6
ISO: 2500

GEORGE KARBUS (p.108)
CAMERA: Nikon D700
LENS: Nikkor Fisheye 16.0 mm f/2.8
SHUTTER SPEED: 1/1000
APERTURE: F2.8
ISO: 640
OTHER: Subal underwater housing

NICK UPTON (p.109)
CAMERA: Nikon D300
LENS: Nikon VR 18-200 F/3.5-5.6G at 95mm
SHUTTER SPEED: 1/250
APERTURE: F9
ISO: 250
OTHER: Handheld, trying hard not to get reflections off the thankfully incredibly clear Perspex windshield

technical information

PAUL NAYLOR (p.110)
CAMERA: Nikon D200
LENS: 17-70mm zoom@45mm
SHUTTER SPEED: 1/100
APERTURE: F13
ISO: 200
OTHER: Camera was in Subal aluminium housing. Lighting from two Inon flashguns

ALEXANDER MUSTARD (p.111)
CAMERA: Nikon D7000
LENS: Nikon 105mm VR
SHUTTER SPEED: 1/320
APERTURE: F8
ISO: 200
OTHER: Nauticam underwater housing. Inon Z240 strobes. +10 dioptre Subsee close-up lens

GRAHAM EATON (p.112-113)
CAMERA: Nikon D3
LENS: 60mm Micro Nikkor
SHUTTER SPEED: 1/180
APERTURE: F13
ISO: 1000
OTHER: Sea & Sea housing with 3 YS110 strobes (2 wired, 1 slave)

GRAHAM EATON (p.114)
CAMERA: Nikon D3
LENS: 60mm Micro Nikkor
SHUTTER SPEED: 1/1500
APERTURE: F13
ISO: 1250
OTHER: Sea & Sea underwater housing with 3 strobes

GEORGE STOYLE (p.115)
CAMERA: Nikon D700
LENS: 105mm Macro f2.8 VRII
SHUTTER SPEED: 1/160
APERTURE: F22
ISO: 400
OTHER: Nauticam underwater housing, 2x Inon Z240 strobes, +3 dioptre

RICHARD SHUCKSMITH (p.116)
CAMERA: Nikon D300
LENS: Nikkor 60mm micro
SHUTTER SPEED: 1/250
APERTURE: F8
ISO: 200

TREVOR REES (p.117)
CAMERA: Nikon D80
LENS: Nikkor 60mm
SHUTTER SPEED: 1/200
APERTURE: F16
ISO: 100
OTHER: Two Sea & Sea YS110 strobes used to provide flash lighting on a somewhat deep dark dive

DANY ROS (p.118)
CAMERA: Sony NEX-5
LENS: Sony SEL16F28+Fisheye converter VCL-ECF1
SHUTTER SPEED: 1/80
APERTURE: F7.1
ISO: 200
OTHER: Nauticam Housing + Dome Port

GEORGE STOYLE (p.119)
CAMERA: Nikon D70s
LENS: 60mm Macro f2.8
SHUTTER SPEED: 1/100
APERTURE: F18
ISO: 200
OTHER: Ikelite underwater housing, 1x Inon Z220 strobe

ALEX TATTERSALL (p.120)
CAMERA: Canon 7D in Nauticam housing
LENS: Tokina 10-17mm Fisheye
SHUTTER SPEED: 1/125
APERTURE: F22
ISO: 500
OTHER: INON Z240 strobes x2. Zen 100mm mindome

ROSS HODDINOTT (p.121)
CAMERA: Nikon D70
LENS: 105mm macro
SHUTTER SPEED: 1/6
APERTURE: F11
ISO: 200
OTHER: Tripod

WILDLIFE IN MY BACKYARD

DAVID THOMAS HANDLEY (p.124)
CAMERA: Canon eos 40d
LENS: Canon ef 100mm Macro
SHUTTER SPEED: 1/250
APERTURE: F8
ISO: 640
OTHER: To freeze the wasp as it was leaving the air brick I used a Canon Speedlite 550ex flashgun mounted on the camera

DENISE BISHOP (p.125)
CAMERA: Canon EOS 7D
LENS: Canon 100mm f2.8 macro lens
SHUTTER SPEED: 1/500
APERTURE: F3.5
ISO: 320
OTHER: Handheld

JESS F MASLEN (p.126)
CAMERA: Nikon D80
LENS: Nikon AF S Micro Nikkor 105mm
SHUTTER SPEED: 1/160
APERTURE: F3.3
ISO: 100
OTHER: Handheld in natural light; Skylight 1A filter

JAMES YAXLEY (p.127)
CAMERA: Canon EOS 5D mark II
LENS: Canon 180mm Macro
SHUTTER SPEED: 1/200
APERTURE: F8
ISO: 400

RACHEL PIPER (p.128-129)
CAMERA: Canon 7D
LENS: EFS-60
SHUTTER SPEED: 1/100
APERTURE: F4.5
ISO: 320

ANNE LATRÉMOLIÈRE (p.130)
CAMERA: Nikon D200
LENS: AF-S Nikkor 300mm f/2.8 VR with TC-14EII
SHUTTER SPEED: 1/250
APERTURE: F4
ISO: 320
OTHER: Tripod and makeshift hide

ANNE LATRÉMOLIÈRE (p.131)
CAMERA: Nikon D200
LENS: AF-S Nikkor 300mm f/2.8 VR with TC-14EII
SHUTTER SPEED: 1/400
APERTURE: F4
ISO: 250
OTHER: Tripod and makeshift hide

WILD WOODS

JEREMY WALKER (p.134-135)
CAMERA: Nikon D3X
LENS: Nikkor 24-70 f2.8
SHUTTER SPEED: 1/1.3
APERTURE: F14
ISO: 100
OTHER: Gitzo tripod. Arca Swiss head. Three images stitched in Adobe Photoshop CS5

IAN WADE (p.136)
CAMERA: Canon EOS 5D Mark 11
LENS: Canon 100-400 L Lens
SHUTTER SPEED: 1/100
APERTURE: F11
ISO: 640
OTHER: Handheld with the camera tight to the floor to achieve the feeling that viewing is down amongst the woodland flowers and grass

ANTHONY MURDEN (p.137)
CAMERA: Nikon D300
LENS: Nikon 16-85mm
SHUTTER SPEED: 1/4
APERTURE: F10
ISO: 200
OTHER: Beanbag on wall, circular polarising filter

CALUM DICKSON (p.138)
CAMERA: Canon EOS 1D-Mark IV
LENS: EF500mm f/4L IS USM
SHUTTER SPEED: 1/125
APERTURE: F4
ISO: 1000
OTHER: Manfrotto 190 XPRB tripod with MN393 long lens head

CALUM DICKSON (p.140)
CAMERA: Canon EOS 1D-Mark IV
LENS: EF500mm f/4L IS USM
SHUTTER SPEED: 1/200
APERTURE: F4
ISO: 1600
OTHER: Gitzo 5541LS tripod with WH-200 Wimberley head version II

DREW BUCKLEY (p.141)
CAMERA: Canon EOS 5D Mark II
LENS: EF 17-40 f/4L USM
SHUTTER SPEED: 1/4
APERTURE: F16
ISO: 100
OTHER: Circular polarising filter to remove reflection glare and tripod for stability during long exposure

JEREMY WALKER (p.142-143)
CAMERA: Nikon D3X
LENS: Nikkor 24-70 f2.8
SHUTTER SPEED: 1/8
APERTURE: F11
ISO: 100
OTHER: Gitzo tripod with Arca Swiss head. Three images stitched in Adobe Photoshop CS5

IAN PAUL HASKELL (p.144)
CAMERA: 5D Mark II
LENS: 17-40 at 17mm
SHUTTER SPEED: 1/0.6
APERTURE: F16
ISO: 50
OTHER: Tripod, cable release, gradient filter

ALEX SABERI (p.145)
CAMERA: Canon 5d mk11
LENS: Canon 100-400
SHUTTER SPEED: 1/250
APERTURE: F8
ISO: 250

MARTIN WASE (p.146)
CAMERA: Canon Eos 5D MK11
LENS: Canon 100-400mm f4.5-f5.6 IS L
SHUTTER SPEED: 1/60
APERTURE: F5.6
ISO: 400

ROSS HODDINOTT (p.147)
CAMERA: Nikon D70
LENS: 105mm macro
SHUTTER SPEED: 1/4
APERTURE: F16
ISO: 200
OTHER: Remote cord, tripod, reflector

JEREMY WALKER (p.148-149)
CAMERA: Nikon D3S
LENS: Zeiss 50mm f 1.4
SHUTTER SPEED: 1/10
APERTURE: F16
ISO: 200
OTHER: Gitzo tripod. Arca Swiss head. Three images stitched in Adobe Photoshop CS5

LEE BEEL (p.150)
CAMERA: Canon EOS 5D MKII
LENS: Canon EF L 100-400mm
SHUTTER SPEED: 1/4
APERTURE: F16
ISO: 100
OTHER: Manfrotto 190CXPRO3 tripod plus geared head. I used a cable release so I could stand in front of my camera to shade the front lens element from direct sunlight to prevent flare

NEIL McINTYRE (p.151)
CAMERA: Canon EOS-1D Mark IV
LENS: Canon 500mm f4 IS
SHUTTER SPEED: 1/320
APERTURE: F4.5
ISO: 640
OTHER: Gitzo tripod with Wimberley mark II head

JEREMY WALKER (p.152-153)
CAMERA: Nikon D3X
LENS: Nikkor 24-70 f2.8
SHUTTER SPEED: 1/13
APERTURE: F11
ISO: 100
OTHER: Gitzo tripod. Arca Swiss head. Three images stitched in Adobe Photoshop CS5

HABITAT

ALEXANDER MUSTARD (p.156-157)
CAMERA: Nikon D700
LENS: Sigma 15mm
SHUTTER SPEED: 1/60
APERTURE: F11
ISO: 800
OTHER: Subal underwater housing. Two Inon Z240 underwater strobes

MARK HAMBLIN (p.158)
CAMERA: Canon EOS 1D Mark IV
LENS: Canon 500mm f4L IS USM + 1.4x teleconvertor
SHUTTER SPEED: 1/1000
APERTURE: F5.6
ISO: 100
OTHER: Beanbag from car window

BEN HALL (p.159)
CAMERA: Canon EOS 7D
LENS: Canon 500mm F4 L IS
SHUTTER SPEED: 1/4000
APERTURE: F6.3
ISO: 320
OTHER: Gitzo tripod

technical information

ANDREW PARKINSON (p.160)
CAMERA: Nikon D300s
LENS: Nikon 200-400mm F4 VR lens
SHUTTER SPEED: 1/500
APERTURE: F10
ISO: 320
OTHER: Gitzo tripod, Wimberley head, ropes, harnesses and various bits of climbing equipment

ANDREW PARKINSON (p.161)
CAMERA: Nikon D300s
LENS: Nikon 80-200mm F2.8 lens
SHUTTER SPEED: 1/500
APERTURE: F7.1
ISO: 400
OTHER: Gitzo tripod, Wimberley head, cable release

MARK HAMBLIN (p.162)
CAMERA: Canon EOS 1D Mark IV
LENS: Canon 500mm f4L IS USM
SHUTTER SPEED: 1/6400
APERTURE: F5.6
ISO: 400

MARCIN ZAGORSKI (p.163)
CAMERA: Canon 40D
LENS: 17-40/f4 L USM
SHUTTER SPEED: 1/50
APERTURE: F13
ISO: 320

BEN HALL (p.164)
CAMERA: Canon EOS 1D Mark II
LENS: Canon 100-400mm L IS at 115mm
SHUTTER SPEED: 1/500
APERTURE: F5.6
ISO: 160

DALE SUTTON (p.165)
CAMERA: Canon 5d mk2
LENS: Canon 100mm
SHUTTER SPEED: 1/400
APERTURE: F16
ISO: 100
OTHER: High-speed flash, infra-red trigger, solenoid and control unit, special designed leaf shutter

BEN HALL (p.167)
CAMERA: Canon EOS 1D Mark IV
LENS: Canon 100-400mm L IS
SHUTTER SPEED: 1/8
APERTURE: F16
ISO: 50
OTHER: Gitzo tripod and cable release

PETER CAIRNS (p.168)
CAMERA: Canon 1dIV
LENS: 300mm
SHUTTER SPEED: 1/250
APERTURE: F4
ISO: 400

MATT COLE (p.169)
CAMERA: Canon 7D
LENS: Tokina 10-17mm fisheye (at 17mm)
SHUTTER SPEED: 1/640
APERTURE: F11
ISO: 400
OTHER: Tripod

JAMIE HALL (p.170-171)
CAMERA: Canon 50d
LENS: Canon 100mm Macro
SHUTTER SPEED: 1/320
APERTURE: F4
ISO: 400
OTHER: Cable release

ALEX WINSER (p.172)
CAMERA: Nikon D300
LENS: Sigma 180mm f/3.5 EX APO Macro IF HSM
SHUTTER SPEED: 1/160
APERTURE: F7.1
ISO: 200
OTHER: Tripod-mounted

ANDREW FOX (p.173)
CAMERA: Canon 5D Mk II
LENS: Canon 70-200mm
SHUTTER SPEED: 1/250
APERTURE: F6.3
ISO: 320

LESLIE HOLBURN (p.174)
CAMERA: Canon 1d mk4
LENS: Canon 500mm f4
SHUTTER SPEED: 1/250
APERTURE: F8
ISO: 800
OTHER: Beanbag

JAMES YAXLEY (p.175)
CAMERA: Canon EOS7D
LENS: 100-400mm Canon Lens
SHUTTER SPEED: 1/400
APERTURE: F5.6
ISO: 500

BRITISH SEASONS
JULES COX (p.178 – 1)
CAMERA: Canon 1D mark IV
LENS: 500mm f/4 IS USM + x 1.4 extender
SHUTTER SPEED: 1/2000
APERTURE: F5.6
ISO: 640
OTHER: Gitzo carbon fibre tripod and Wimberley mark II head

JULES COX (p.178–2)
CAMERA: Canon 1D mark IV
LENS: 500mm f/4 IS USM + x 1.4 extender
SHUTTER SPEED: 1/1250
APERTURE: F5.6
ISO: 640
OTHER: Gitzo carbon fibre tripod and Wimberley mark II head

JULES COX (p.178–3)
CAMERA: Canon 1D mark IV
LENS: 500mm f/4 IS USM
SHUTTER SPEED: 1/1250
APERTURE: F4
ISO: 400
OTHER: Gitzo carbon fibre tripod and Wimberley mark II head

JULES COX (p.179–4)
CAMERA: Canon 1D mark IV
LENS: 500mm f/4 IS USM + x 1.4 extender
SHUTTER SPEED: 1/500
APERTURE: F5.6
ISO: 640
OTHER: Gitzo carbon fibre tripod and Wimberley mark II head

LIVING LANDSCAPE: CONNECTIVITY
STEPHEN WILLIAM POWLES (p.182-183)
CAMERA: Nikon D200
LENS: Nikkor 80-400
SHUTTER SPEED: 1/30
APERTURE: F20
ISO: 200
OTHER: Tripod

PETER CAIRNS (p.184)
CAMERA: Canon 5dII
LENS: 17-40
SHUTTER SPEED: 1/500
APERTURE: F4
ISO: 200

RACHEL PIPER (p.185)
CAMERA: Canon 7D
LENS: EFS-60
SHUTTER SPEED: 1/60
APERTURE: F9
ISO: 400

DOCUMENTARY SERIES
MARK HAMBLIN (p.188-191)
CAMERA: Canon EOS 1D Mark IV & 5D Mark II
LENS: Canon 17-40mm, 70-200mm & 500mm
SHUTTER SPEED: Various
APERTURE: Various
ISO: Various

FIONA HALLIDAY (p.192-195)
CAMERA: Canon 5D
LENS: 20-24mm
SHUTTER SPEED: 1/3200 – 1/8000
APERTURE: F2.8 – F5.0
ISO: 400
OTHER: I always underexpose by at least one third and then darken further in Photoshop. (duplicate layer, multiply, then adjust opacity.) I hate a well-exposed photo! Some of them are cropped

WILDLIFE IN HD VIDEO
RAYMOND BESANT (p.198-201)
CAMERA: Canon EOS 600D and NIKON D700

***OUTDOOR PHOTOGRAPHY* EDITOR'S CHOICE**
ALEX TATTERSALL (p.204)
CAMERA: Nikon D7000 in Nauticam housing
LENS: Tokina 10-17
SHUTTER SPEED: 1/200
APERTURE: F18
ISO: 320
OTHER: INON Z240 strobes x2. Zen 100mm mindome

ALEX SABERI (p.205)
CAMERA: Canon 5d mark II
LENS: Canon 100-400
SHUTTER SPEED: 1/400
APERTURE: F5.6
ISO: 500

ANDREW PARKINSON (p.206)
CAMERA: Nikon D3S
LENS: Nikon 600mm F4 VR lens
SHUTTER SPEED: 1/1600
APERTURE: F4
ISO: 200
OTHER: Kevin Keatley hide, Gitzo tripod, Wimberley head

RICHARD PETERS (p.207)
CAMERA: Nikon D3s
LENS: 600 VR
SHUTTER SPEED: 1/200
APERTURE: F5
ISO: 2800
OTHER: Gitzo tripod and Wimberley mark II gimbal head

Index

Photographers' Websites

LEE BEEL www.lee-beel-photography.co.uk

MATT BINSTEAD www.mattbinstead.blogspot.co.uk

JOHN H BRACKENBURY www.seeingsmall.com

DREW BUCKLEY www.drewbuckleyphotography.com

PETER CAIRNS www.northshots.com

CRAIG CHURCHILL www.craigchurchill.co.uk

MATT COLE www.mattcolephotography.co.uk

RICHARD COSTIN www.richardcostin.com

JULES COX www.julescoxphotography.co.uk

JAN DAVIES www.jandaviesphotography.co.uk

SUE DALY www.suedalyproductions.com

CALUM DICKSON www.calumdickson.co.uk

PENNY DIXIE www.pennydixie.co.uk

DR MATT DOGGETT www.earthinfocus.com

ANDREW FOX www.foxphotographer.com

FIONA HALLIDAY www.lightstalkers.com/fiona-halliday

MARK HAMBLIN www.markhamblin.com

BEN HALL www.benhallphotography.com

JAMIE HALL www.jamiehallphotography.co.uk

AMANDA HAYES www.amandahayes.co.uk

DAVID HERMON www.bawdseyhall.co.uk

PAUL HOBSON www.paulhobson.co.uk

ROSS HODDINOTT www.rosshoddinott.co.uk

PHIL JONES www.philjonesphotography.com

GEORGE KARBUS www.emerald-vision.com

TOM LANGLANDS www.tomlanglandsphotography.com

ANNE LATRÉMOLIÈRE www.latremoliere.co.uk/anne

ROBERT LIND www.roblind.com

SIMON LITTEN www.simonlitten.photoshelter.com

STEVE MAGENNIS www.stevemagennis.co.uk

LIAM MARSH www.liammarsh.co.uk

JESS F MASLEN www.jessfmaslenphotography.com

LUKE MASSEY www.lmasseyimages.com

NEIL McINTYRE www.neilmcintyre.com
www.scottishwildimages.com

JEREMY MOORE www.jerry@wild-wales.com

JOHN MONCRIEFF www.flickr.com/crieffy

ANTHONY MURDEN www.anthonymurdenphotography.co.uk

ALEXANDER MUSTARD www.amustard.com

PAUL NAYLOR www.marinephoto.co.uk

ANDREW PARKINSON www.andrewparkinson.com

RICHARD PETERS www.richardpeters.co.uk

LEWIS PHILLIPS www.lewisphillipsimages.com

SIMON PHILLPOTTS www.wilddales.co.uk

RACHEL PIPER www.rachelpiper.me.uk

TREVOR REES www.trevorreesphotography.co.uk

DANY ROS www.danyros.co.uk

ALEX SABERI www.alexsaberi.com

RICHARD SHUCKSMITH www.rshucksmith.co.uk
www.earthinfocus.com

MARK SMITH www.marksmithphotography.net

GEORGE STOYLE www.georgestoyle.com

DALE SUTTON www.hsfnature.co.uk

ALEX TATTERSALL www.uwvisions.com

NICK UPTON www.naturepl.com

IAN WADE www.ianwadephotography.co.uk

JEREMY WALKER www.jeremywalker.co.uk

MARTIN WASE www.mjwasephotography.co.uk

RICHARD WATERS www.500px.com/RichardWaters

SIMON WHITE www.simonwhitephotography.com

ALEX WINSER www.alexwinserphotography.co.uk

OLIVER CHARLES WRIGHT www.oliverwrightphotography.com